ASSET MANAGEMENT OF BRIDGES

Asset Management of Bridges

Editor

Khaled M. Mahmoud
Bridge Technology Consulting (BTC), New York City, USA

CRC Press
Taylor & Francis Group
Boca Raton London New York

CRC Press is an imprint of the
Taylor & Francis Group, an **informa** business

A BALKEMA BOOK

Front Cover:
Yavuz Sultan Bridge, Istanbul, Turkey
Photo courtesy of İçtaş İnşaat, Turkey

Back Cover:
Yavuz Sultan Bridge, Istanbul, Turkey
Photo courtesy of İçtaş İnşaat, Turkey

Cover Design:
Khaled M. Mahmoud
Bridge Technology Consulting (BTC)
New York City, USA

CRC Press
Taylor & Francis Group
6000 Broken Sound Parkway NW, Suite 300
Boca Raton, FL 33487-2742

First issued in paperback 2020

© 2017 by Taylor & Francis Group, LLC
CRC Press is an imprint of Taylor & Francis Group, an Informa business

No claim to original U.S. Government works

ISBN-13: 978-1-138-56903-4 (hbk)
ISBN-13: 978-0-367-73592-0 (pbk)

Typeset by V Publishing Solutions Pvt Ltd., Chennai, India

**Visit the Taylor & Francis Web site at
http://www.taylorandfrancis.com**

**and the CRC Press Web site at
http://www.crcpress.com**

Table of contents

Preface

Maintaining bridges in good condition extends service life and is proven to be more cost effective than allowing degradation to advance, necessitating costlier bridge rehabilitation or replacement projects. Preventive maintenance is therefore an important tool to retard deterioration and sustain the safe operation of bridges. This includes a continuous effort of periodic inspections, condition assessment and prioritizing repairs accordingly. The above measures define the framework for asset management of bridges. Effective bridge management and preservation strategies employ long-term practices, which are aimed to preserve the safe operation of bridges and extend their useful life, and ensure that the necessary repair measures are applied at the appropriate time. All this is contingent on the availability of sustained and adequate funding sources.

On August 21–22, 2017, bridge engineering experts from around the world convened at the 9th New York City Bridge Conference to discuss issues of construction, design, inspection, monitoring, preservation and rehabilitation of bridge structures. This volume documents their contributions to the safe operation of bridge assets.

The main cables are the most critical and vulnerable elements of a suspension bridge. Operating in harsh moist environments, cables are susceptible to wire breaks, cracking, and embrittlement. All these phenomena culminate in a reduction of the cable load carrying capacity. It is therefore imperative to the long-term asset management and preservation strategy for the main cables to ensure that the condition of cables is known. Several cable inspection and strength assessment strategies have been used over the years in an attempt to preserve cables against deterioration. Some of these strategies are focused on the visual assessment of corrosion along the wire surface, whereas other strategies depend on reliability-based techniques and measured mechanical properties. These proceedings lead off with a paper by Mahmoud et al. on "Management strategies for suspension bridge main cables." The paper presents a comparative analysis of the evaluation of remaining strength and residual life of bridge cables utilizing Report 534 Guidelines and the BTC method. The Federal Highway Administration (FHWA) "Primer on the Inspection and Strength Evaluation of Suspension Bridge Cables" recognizes these two methods for the management of suspension bridge cable assets. Dehumidification is being utilized to reduce moisture inside main cables of suspension bridges. Many suspension bridge main cables, in Japan, Europe and very recently in the United States, are fitted with dehumidification systems to decrease the relative humidity inside the cable. Dehumidification will not reverse the damage due to cracked and broken wires, and thus will not allow the cable to regain strength loss. With that in mind, dehumidification must be regarded as a measure of slowing down the cable deterioration, but certainly not a measure of ceasing deterioration and loss of cable strength. Future inspections and strength evaluations of main cables that have had dehumidification systems will provide more knowledge regarding the efficiency of dehumidification on slowing down loss of cable strength. The flow length has a strong impact on the efficiency of the dehumidification system. In "Main cable dehumidification—flow testing and other innovations", Bloomstine and Melén present flow testing as a design parameter for an optimal system. The Delaware Memorial Bridge comprises two parallel suspension bridges crossing the Delaware River. Each bridge has a main span of 655 m and side spans of 229 m. The First Bridge was opened in 1951 and the Second Bridge in 1968. The bridges are owned and operated by the Delaware River & Bay Authority (DRPA). Elnahal et al. present the effort undertaken by DRPA for "The dehumidification of the main cables of the Delaware Memorial Bridge". In 2005 an internal cable inspection was carried out at the Severn Bridge in England, which revealed broken wires and strength loss. This prompted the implementation of a series of measures, starting with acoustic monitoring and followed by a cable dehumidification system. Two further internal cable inspections took place in 2010 and 2016. In "M48 Severn Bridge—managing the main cables since 2005", Cocksedge et al. describe investigation and rehabilitation work carried out since 2005. It includes some special investigations into low stress/high cycle fatigue testing of cracked wires and friction tests to refine the evaluation of broken wire redevelopment length. The Angus L. Macdonald Bridge, completed in 1955, connects Dartmouth and Halifax, NS, Canada. The suspension bridge is 762 m long, with a 441 m long main span. The bridge deck reached the end of its functional life and was replaced segment-by-segment during bridge weekend closures, with traffic using the bridge during the weekdays. New deck segments were fully prefabricated, including an initial layer of wearing surface, and erected in a way that allowed traffic to use the bridge immediately following replacement. Fabrication began in early 2015 and erection in late 2015. The last segment was replaced in February 2017. In "Superstructure replacement works for the Macdonald Suspension Bridge, Canada", Radojevic and Kirkwood describe some aspects of the superstructure fabrication, segment erection, and challenges encountered during erection. Deck segments for the Macdonald Suspension Bridge were either 10 m or 20 m long and approximately 18 m wide. The new segments had to be matched to adjacent replacement segments as well as matched to temporary deck connectors to provide continuity with the existing, soon to be replaced, sections. A high level of dimensional control was required, in addition to the fatigue sensitive nature of orthotropic decks. In his paper, Ross presents details for "Fabricating orthotropic deck panels for the Macdonald Bridge, Halifax, Canada." The deteriorating condition and associated high maintenance costs of the existing Champlain Bridge, which opened to traffic in 1962, prompted the accelerated need to design, build, operate, maintain, and finance the New Champlain Bridge Corridor Project. Part of the largest transportation infrastructure currently underway in North America, the 3.4 km New Champlain Bridge is comprised of a signature cable-stayed bridge, an east approach, and a west approach. Performance and design criteria of this lifeline structure must meet the design-life requirement of 125 years. In "Design and construction of the New Champlain Bridge, Montreal, Canada", Nader et al. provide an overview of the design and construction of the new bridge.

The Kosciuszko Bridge is a vital transportation link carrying the Brooklyn-Queens Expressway (I-278) over Newtown Creek. With a construction cost of $554,700,000, the Kosciuszko Bridge Phase 1 bridge replacement project is NYSDOT's largest Design-Build Project to date. Choosing the Design-Build delivery method allowed NYSDOT to reduce the number of bid packages and the duration of construction by several years. With Notice to Proceed on May 23, 2014 for design and construction, the new bridge is scheduled to open in the spring of 2017. The existing bridge is being replaced with two new parallel bridges, and the work will be performed in two phases. Phase 1 represents the design and construction of the eastbound structure and westbound Brooklyn connector, with the remainder of the westbound structure in Phase 2. In "Design-Build replacement of the I-278 Kosciuszko Bridge Phase 1—approaches and connectors", D'Ambrosio et al. provide details of the replacement project. In 2009, the New York City Department of Transportation (NYCDOT) embarked on an ambitious, 12 year, $750 million reconstruction program of the historic Belt Parkway between Exit 9 (Knapp Street) and Exit 14 (Pennsylvania Avenue) in Brooklyn, New York. Included within the five mile construction limits are six bridges of varying span lengths, configurations and design features. In their paper, Hom et al. provide details of "A new Belt for Brooklyn—The five mile Belt Parkway reconstruction project". The use of Accelerated Bridge Construction (ABC) is becoming more critical, as many of the nation's bridges approach the end of their service life and require replacement while maintaining traffic flow. Most states have built at least one bridge with some aspect of ABC. Many states, however, remain unsure of the best ways to utilize these streamlined technologies and would like to know how other states have achieved successful ABC implementation. To assist the states, the ABC University Transportation Center (ABC-UTC) at Florida International University has imported the Federal Highway Administration's ABC project database and enhanced it to provide additional functionality and capacity including the incorporation of ABC research projects. In "Accelerated bridge construction project and research databases", Garber & Ralls describe the ABC-UTC's project and research databases. The South Road Bridge is part of the $620 million Darlington Upgrade design/build project, which provides for the improvement of approximately 3.2 kilometers of one of the most important transit corridors in Adelaide, Australia. The bridge is 180 meters long and consists of three-span continuous twin curved steel tub-girders carrying a multi-use path as well as vehicular traffic over a major expressway. In "Analysis and design of the South Road double composite steel tub girder bridge", Loureiro et al. present details of the design of the bridge. Hurricane Sandy hit New York City with a devastating blow in October 2012. Every portion of the City's infrastructure was damaged and needed repair. This included New York City Department of Transportation (NYCDOT) roads and bridges. Particularly hit hard were NYCDOT's movable bridges, which are located in low-lying areas. In "Repairs to 13 movable bridges in New York City after Hurricane Sandy", Kelly and Gusani discuss the impact that Sandy had on the bridges, the measures and procedures that NYCDOT had to follow to ensure permitting and funding was obtained as well as the hurdles that needed to be crossed to develop biddable repair contracts in a short time and in compliance with Federal guidelines.

Cable-supported bridges that carry Light Rail Transit are vulnerable to fatigue stresses due to the cyclic loading from passage of trains. This raises the risk of fatigue-induced damage to stay cables. In "Fatigue damage assessment of stay cables for the light rail transit bridges", Jiang and Coughlin present a recent study completed on assessing fatigue induced damage to stay cables for an in-service 616 m long light rail transit bridge with a main span of 340 m. The Kemaliye Bridge in Turkey is a 3-span with 140 m central span and two side spans each of 75 m length, and width of 17 m. As measured from the top level of the foundations, the pier heights are 60 m. The bridge deck has been designed as prestressed post-tensioned concrete by cantilever method. In "The design and construction of Kemaliye Bridge, Turkey", Caculi and Namlı summarize the details of the segment design and construction. In rural Kenya rivers create barriers that restrict access to education, health care, and commerce. Bridging the Gap Africa (BtGA) is a non-profit that enables communities to build footbridges across these dangerous rivers. To date, the organization has facilitated the construction of 60 footbridges and it is estimated that hundreds of bridges are still needed throughout Kenya. BtGA believes that their approach of enabling local communities to build these bridges will result in a sustainable bridge program that is capable of reaching hundreds of communities throughout Kenya. In "Design of a short span suspension footbridge: detailing for success in rural Kenya", Smith et al. provides details of the design for the Peace Bridge, a 45 m (148 ft) span footbridge. Cables produced with high strength fibers have gained acceptance in many applications due to their enhanced mechanical properties and resistance to the elements. In "Innovative cable system designs", Klein describes the design requirements and advancements of high performance fiber cable systems with a focus on structural applications. The bending stresses induced in parallel wire suspension bridge cables at cable bands are correlated with the forces in the suspenders. Investigators have showed, 20 years ago, the ratio between the bending stress and the total stress in the cable wires, as a function of the ratio between the forces in the suspenders and the tension in the main cables. In "Bending stresses in parallel wire cables of suspension bridges", Gjelsvik and Yanev reproduce that relationship.

Structural health monitoring of bridges offers an important tool in the asset management of bridges. The monitoring feedback provides the bridge owners with important input regarding the condition of the structure and assists in the prioritization of funds and required repairs. Owned and managed by the West Virginia Department of Transportation, the Market Street Bridge is a suspension bridge over the Ohio River connecting Market Street in Steubenville, Ohio with West Virginia Route 2 in Follansbee, West Virginia. The bridge was constructed in 1905 with an overall length of 547 m (1,794 ft) and a roadway width of approximately 6.7 m (22 ft), with a main span length of 210 m (700 ft). in "Structural health monitoring of a historical suspension bridge" William et al. describe the development of a monitoring system for real-time condition assessment of the Market Street Bridge after the completion of its rehabilitation in November 2011. The sensory system deployed, monitors the global structural displacements of the bridge towers under operating and environmental conditions. A common challenge faced by foundation engineers while working in urban settings is the maintenance and protection of adjacent structures. The standard practice to mitigate damage caused by adjacent foundation construction work is to limit the existing structures' vibrations within acceptable safe constraints. In spite of these efforts, numerous existing structures have suffered damage from adjacent foundation construction activities even in full compliance to the standard practice. In recent years, monitoring techniques have been recognized as an effective approach for the protection of existing structures from ground excitation caused by adjoining foundation construction work. In "Protection of existing structures using health monitoring", Ramakrishna and Mankbadi present case histories on protecting existing structures from adjacent foundation construction activities using structure heath monitoring. Rail Network is an important segment of the United States' transportation portfolio and has revolutionized transportation and catalyzed economic development. In "Data-to-decision framework for monitoring railroad bridges", Alampalli et al. present the data-to-decision framework where data from commercial remote sensing and spatial information technologies are used in a decision-making tool software with built-in predictive analytical models for making cost-effective predictive maintenance and management decisions. The framework is applied on a selected sub-network of the Union Pacific railroad.

The p-y method is widely used in bridge design to model the pile behavior under lateral loads. Appropriate p-multipliers and y-multipliers are usually applied to the "backbone" p-y curves in order to account for group effect or any other possible influence on the pile lateral behavior. In "A numerical model of lateral response of a drilled shaft adjacent to a caisson foundation", Wei et al. provide details of the design of deep foundations with groups of drilled shafts, proposed to replace the existing caisson foundations of the Pulaski Skyway in New Jersey. The Mississippi DOT's plans to widen northbound I-55 at the interchange with I-220 near Jackson, MS were impeded by various geometric constraints, the most challenging being the limited vertical clearance under an existing CIP multi-cell box structure. Shallow depth steel plate girders were found to be the only viable solution to accommodate the maximum span length within the limited structure depth. An in-depth analysis using 3D FEM methods was conducted to assess the interaction between the existing and new structures. The relatively flexible steel structure was found to deflect and twist away from the existing resulting in large bending stresses within the slab. In "Widening an existing multi-cell box structure with shallow depth steel girders", Rolwes provides the details of the design that allowed the DOT to overcome the impasse and proceed with their plans to widen this important transportation corridor. During inspections of Gowanus Expressway Viaduct in New York City, cracks in stringer-end copes, stringer webs, and stringer to floor-beam connection angles have been found on a regular basis. In order to understand the behavior of the structural system and determine remaining fatigue life, analytical and experimental studies of the stringer to steel bent connections on the viaduct were performed. 3D finite element models, including a global viaduct model and a detailed connection model, were developed. In "3D structural modeling of stringer-bent connections on Gowanus Expressway Viaduct, New York", Wei et al. provide details of the analytical studies.

The evolution of the modern bridge, from around 1850, both coincided with, and influenced the growth of modernist art movements. Many European avant-gardists used the new bridges as subject material for their revolutionary art. That trend continued in New York, as modern art permeated the city, particularly after the celebrated Armory Exhibition of 1913. The bridges of New York were a rich source of inspiration for a plethora of twentieth century artists. In "Modern Art and New York City Bridges", Rothwell discusses some bridges of New York City, both the iconic and unsung, the people who built them, the artists who were inspired by them, and the paintings they produced. The Hell Gate Arch Bridge of the New York Connecting Railroad was dedicated 100 years ago on March 9, 1917. When constructed, it was the longest arch bridge in the world with a span of 304 m (997.5 ft) between centers of bearings and 310 m (1017 ft) between the faces of abutments. The Chief Engineer of this project was Gustav Lindenthal and working under him were Othmar H. Ammann and David B. Steinman, two future giants of long span bridge engineering in the United States. In "The Hell Gate Arch Bridge in New York City", Gandhi describes the development of this project and the design and construction of this monumental bridge. At the time of its construction, the 549 m (1800 ft) span cantilever bridge across the St. Lawrence River in Quebec, Canada was going to be the longest cantilever bridge in the world. However, on August 29, 1907, during its erection, the bridge collapsed killing 75 workers. A commission of prominent international engineers was formed by the Canadian Government to investigate the collapse of the Quebec Bridge. It was decided to build a new, but much heavier and stronger, cantilever bridge adjacent to the old failed bridge. On September 11, 1916, after the center span was raised successfully 3.7 m to 4.6 m (12 to 15 feet), it suddenly fell into the St. Lawrence River killing eleven workers and injuring six. The St. Lawrence Bridge Company, which was erecting it, took full responsibility for the collapse of the second bridge and placed orders for the new steel. The new center span was successfully hoisted for the third time and put in place on September 18, 1917 using the same lifting procedure that was used in 1916. The new bridge was opened to traffic 100 years ago, on December 3, 1917. This is featured in Gandhi's "The Failure and Reconstruction of the Quebec Bridge".

The editor is thankful to all the authors and reviewers for their efforts in producing this volume.

Khaled M. Mahmoud, PhD, PE
Chairman of Bridge Engineering Association
www.bridgeengineer.org
Chief Bridge Engineer
BTC
www.kmbtc.com
New York City, USA

New York City, August 2017

Cable-supported bridges

Chapter 1

Management strategies for suspension bridge main cables

K. Mahmoud
BTC, New York City, USA

W. Hindshaw
Transport Scotland, Glasgow, Scotland, UK

R. McCulloch
Amey, Edinburgh, Scotland, UK

ABSTRACT: There are two recognized methods for the evaluation of the remaining strength of cables based on internal inspection findings and testing results, NCHRP Report 534 Guidelines (Report 534) and the BTC method. Report 534 provides visual-based assessments of the corrsion on wire surface. The Guidelines assign a proportion of the cable cross-section to each of the stage of corrosion observed in the wedge openings. The evaluation of Report 534 then proceeds with the evaluation of remaining strength based on wire sample test results and the hypothized proportions of the stages of corrosion in the cable cross-section. The BTC method provides a reliability-based evaluation of the remaining strength and residual service life of cables. The method includes random sampling without regard to wire appearance, mechanical testing of wire samples, determining the probability of broken and cracked wires, and fracture-based analysis of cracked wires. The probabilistic-based method forecasts the residual life of the cable by predicting the increasing rate of detrioration and strength degradation. The BTC method is published in the latest FHWA Primer for the Inspection and Strength Evaluation of Suspension Bridge Cables. The BTC method is currently being applied alongside Report 534 Guidelines at the Forth Road Bridge in Scotland. The dual application of the two methods provides the bridge owners with high level of confidence in the estimated cable strength.

1 INTRODUCTION

The main cables are the most critical and vulnerable elements of a suspension bridge. Operating in harsh moist environments, cables are susceptible to wire breaks, cracking, and embrittlement. All these phenomena culminate in a reduction of the cable load carrying capacity. It is therefore imperative to the long-term asset management and preservation strategy for the main cables to ensure that the condition of cables in known. Several cable inspection and strength assessment strategies have been used over the years in an attempt to preserve the cable against deterioration. Some of these strategies are focused on the visual assessment of corrosion along the wire surface, whereas other strategies depend on reliability-based techniques and measured mechanical properties.

There are two recognized methods for to assess degradation and estimate the remaining strength of bridge cables Report 534 Guidelines and the BTC method. Report 534 Guidelines rely on the visual appearance of corrosion observed on individual wire surface. The wire sampling employed in the visual-based method assigns a number of wires to be extracted from each corrosion stage. The strength assessment then follows by a hypothesized proportioning of the cable cross-section to the observed corrosion stages along the wedge openings

(NCHRP Report 534 2004). This produces cable strengths that are derived by the visual assessment of surface corrosion on wire surface. The BTC method is a patented methodology that employs a reliability-based analysis to estimate the remaining strength and service life of both, parallel wire and helical wire bridge cables (NYSDOT Report C-07-11 2011). It is included in the Federal Highway Administration (FHWA) Primer for the Inspection and Strength Evaluation of Suspension Bridge Cables (FHWA 2012). The BTC method applies to both zinc-coated and bright bridge (non-galvanized) wire. In the BTC method, wires are collected from the wedge openings, utilizing random sampling of individual wires, in each investigated panel. The randomly selected sample is tested to obtain the mechanical properties, including ultimate strength, ultimate elongation, yield strength, Young's modulus and fracture toughness. The probability of broken wires is estimated based on inspection observation of broken wires, and probability of cracked wires is estimated based on the cracks detected from fractographic examination of wire fracture surfaces. The ultimate strength of cracked wires is determined using fracture toughness criteria. All these data is utilized to assess the remaining strength of the cable in each of the investigated panels. The BTC method employs a probabilistic-based approach to assess the remaining service life of the cable by determining the rate of change of broken and cracked wires detected over a time frame, and available data from previous cable investigations, and measuring the rate of change in effective fracture toughness over the same time frame. The words reliability and probabilistic are used to describe the method of assessing the remaining strength and service life of the cable.

This paper presents a comparative analysis of the evaluation of remaining strength and residual life of bridge cables utilizing Report 534 Guidelines and the BTC method.

2   METHODS FOR THE ASSESSMENT OF CABLE STRENGTH

There are two recognized methods for the evaluation of remaining strength of bridges cables:

- Report 534 Guidelines; depends on the visual assessment of wire surface corrosion, and
- BTC method; employs reliability-based analysis of inspection findings and mechanical properties of sampled wires.

The two methods present two different techniques to the modeling of wire degradation and the cable strength evaluation. This in fact is advantageous to bridge owners, as it would allow owners to make well-informed decisions about cable degradation and future maintenance strategies. The two separate methods give owners confidence as although different, the results provide upper and lower bound envelopes for the cable strength.

The BTC method has been applied alongside Report 534 at the Bronx-Whitestone Bridge, Mid-Hudson Bridge, in New York, USA. Currently, the two methods are being applied at the Forth Road Bridge, in Edinburgh, Scotland. It is noted that in the dual application of the two methods, there is no duplication in the inspection, sampling or testing effort.

2.1   *Visual evaluation of wire degradation*

The first major cable investigation was conducted by Hopwood and Havens of the numerous breaks of wires in the helical strands of the main suspension cables of General U.S. Grant Bridge over the Ohio River in 1984. In their study, (Hopwood and Havens 1984) first classified the visual corrosion for the galvanized wires in the helical strands into four stages (Hopwood and Havens 1984):

Stage 1: wire surfaces have a shiny metallic appearance, though some signs of white zinc corrosion product may be visible in spots.
Stage 2: wire surfaces dull as the zinc corrodes. The wires eventually are covered with the white corrosion product. However, there is no ferrous corrosion under the white corrosion product.

Stage 3: signs of ferrous rust are visible on wire surfaces. The zinc coating is almost completely consumed.

Stage 4: ferrous rust stains displace most of the white corrosion product on wires. Wire surfaces become very rough and pitted.

The Williamsburg Bridge cables were investigated almost 30 years ago (1988) also utilizing a visual-based method. It was the first bridge with parallel wires whose cables conditions were inspected in-depth. The Williamsburg Bridge main cable is composed of parallel bright wires. Therefore the bridge investigation adopted a modified six (6) degrees of corrosion classification system for the metal surface of wires, ranging from Grade 0; no corrosion, almost new condition, to Grade 5; worst corrosion (Steinman 1988).

When published in 2004, Report 534 generalized the definition of Hopwood and Havens, which was originally developed for helical wire strands, to classify visual corrosion on surface of parallel wire cables (NCHRP Report 534 2004).

2.2 Reliability-based evaluation of wire degradation

A typical suspension bridge cable is composed of thousands of wires, and the assessment for cable strength is based on a small sample size. Therefore it is essential to employ reliability-based method, which infers the strength of the cable from a small sample of wires.

With this understanding of the limited sample from a large population of wires, the BTC method utilizes modern assessment techniques that employ reliability criteria (very similar to the Load and Resistance Factor Design "LRFD" criteria), in which the wire mechanical properties obtained from a random sample, including strength and ductility (strain) are known as probabilistic entities, from which a "probability of failure" could be estimated. If an evaluation is conducted using these criteria, it can help establish, with high level of confidence, the tempo of inspection and further evaluations in the future.

The BTC method employs random sampling to eliminate visual bias, and fracture mechanics principles to assess the strength of cracked wires. Designing the sampling plan in each investigated panel, which marks the wires to be sampled from each wedge opening, prior to the field inspection ensures that the sampling is random. The inspectors and the Contractor are strictly instructed to sample only the wires marked on the sampling plan. With the use of a random sample, the analysis evaluates the sampling error in the estimated cable strength. The method forecasts the remaining life of the bridge cable based on strength degradation and rates of growth in broken and cracked wires proportions detected over a time frame, and the rate of change in fracture toughness over same time frame (NYSDOT Report C-07-11 2011) and (FHWA 2012).

The BTC method was first utilized to evaluate the strength and residual life of the main suspension cables at the Bronx-Whitestone Bridge in New York City. The BTC method was afterwards applied at the Mid-Hudson Bridge, in New York, and to check and validate the previously applied NCHRP Report 534 Guidelines, at the Forth Road Bridge, in Scotland.

The following section presents a brief comparative analysis of the BTC method and Report 534 based on the dominant metrics of degradation, which affect the estimated strength and residual life of the cable.

3   METRICS FOR WIRE DEGRADATION

The accuracy and reliability of the estimated cable strength depend on the influential metrics of degradation listed below:

- Hydrogen embrittlement
- Broken wires
- Cracked wire
- Forecast of cable life

3.1   *Hydrogen embrittlement*

The main inputs for the assessment of remaining strength of bridge cables are the test results of wire specimens and data collected during inspection. Following Report 534 Guidelines, during inspection, a number of wires are sampled, based on the visual assessment of the corrosion stages, and tested in the laboratory. The strength of each corrosion stage is assigned to the fraction of the cable cross-section identified during inspection. Test results have shown tenuous correlation between the stages of corrosion and measured ultimate strength and ultimate elongation. The wire shown in Figure 1(a) displays half of the elongation displayed by the wire shown in Figure 1(b), therefore the wire in Figure 1(a) is more embrittled. It is noted that the two specimens demonstrate the same yield plateau and almost the same ultimate strength. The major and significant difference displayed by the two stress-strain curves is the value of the ultimate elongation. In other words, the wire could be embrittled but its ultimate strength would not be significantly affected. This phenomenon is not measurable by the four stages of corrosion approach. In the evaluation of the strength, Report 534 considers only the ultimate strength, without including the strain (elongation) in the analysis. This has a very important ramification on the estimated strength as explained below.

The area under the stress-strain curve of a wire specimen represents the energy that it takes to break the wire. Therefore disregarding the elongation data in the analysis does not reflect the full extent of wire degradation. The BTC method includes in the analysis the ultimate strength, ultimate elongation, Young's modulus and yield strength. Thus the BTC method accounts for the embrittlement of wires in its estimation of the cable strength.

The hydrogen embrittlement causes significant reduction in the elongation capacity of the wire. The critical hydrogen concentration to cause embrittlement is 0.7 ppm (Nakamura and Suzumura 2009). With a concentration higher than 0.7 ppm, hydrogen can degrade the fracture resistance of the high strength wire material. The mechanics of this process pivots around the generation of atomic hydrogen, which causes embrittlement and eventual reduction in the fracture toughness of the wire material (Mahmoud 2003). In the presence of moisture, the released atomic hydrogen diffuses into the interior of the wire, at location of surface imperfection, weakens the inter-atomic bond and causes the embrittlement, Equation (1).

$$xFe + H_2S \underset{H2O}{\Rightarrow} Fe_xS + 2[H]_{diffusible} \tag{1}$$

The hydrogen embrittlement mechanism is confirmed by evidence of hydrogen concentration testing. According to testing of wire specimens from the cables of several suspension bridges, the average hydrogen concentration measured values that are higher than the 0.7 ppm critical threshold to cause embrittlement.

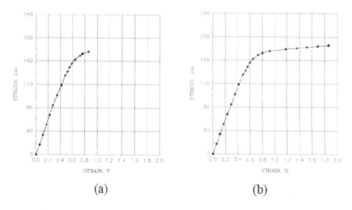

(a)   (b)

Figure 1.   Stress-strain curves for tested wire specimens.

The following factors, which all exist in the bridge cable environment, are required to cause hydrogen embrittlement:

- Source of hydrogen
- Susceptible material; i.e., high strength steel
- High tensile load

The sources of hydrogen are defined as follows:

- Trapped Hydrogen: During manufacturing process, freshly drawn wires are dipped in a bath of molten zinc to provide zinc coating. Hydrogen gets trapped, with random levels of concentration in the wire.
- Diffusible Hydrogen: During the bridge service, free hydrogen generates, in the presence of moisture, as byproduct of the corrosion reaction.

The corrosion reaction releases the atomic hydrogen, which causes the embrittlement and the subsequent reduction in both; the elongation and load carrying capacity of the wire. Wires that are found broken during inspection display embrittled patterns, where the hydrogen assists the growth of a preexisting crack to critical depth, at which the wire breaks, Figure 2.

It is clear that the broken wire shown in Figure 2, which was observed in an over 80 years old suspension cable, has no section loss, or necking, and that the wire displays unmistakable embrittled pattern. The break is driven by the growth of a crack on the lower plane of the fracture surface. In the BTC method, estimation of cable strength, wire embrittlement is quantified through the use of measurements of ultimate elongation, and evaluation of fracture toughness.

3.2 Broken wires

The BTC method evaluates the proportion of broken wires in each of the investigated panels as a probability. The outer wires at the surface of the cables are fully accessible for inspection, and the number of broken wires in the outer ring is observed and identified. The probability of broken wires, in the interior rings of the cable, is assessed based on the observed broken wires, as a fraction of the total observed interior wires. This approach is consistent with the random pattern of wire breaks observed during internal inspection of main suspension cables.

In a recent investigation, it was noted that there is an inverse correlation between the number of broken wires and percentages of Stages 2, 3 and 4 in different investigated panels. The inverse correlation is defined as a contrary relationship between two variables such that they move in opposite directions. According to Report 534, it is expected to have less broken wires in the presence of higher percentage of Stage 2 wires and more broken wires in the presence of higher percentage of Stage 3 and 4 wires. However, the number of broken wires,

Figure 2.   Embrittled wire break.

observed during inspection of a major suspension bridge, demonstrated a contrary correlation to the above expectation. For instance, Figure 3 shows a correlation coefficient of 0.513 between number of broken wires and percentage of stage 2 wires. This implies that a higher proportion of stage 2 corresponds to a larger number of broken wires, which is contrary to the definition in Report 534.

Figures 4 and 5 show correlation coefficients of –0.495 and –0.373, between broken wires and proportions of Stage 3 and Stage 4 respectively. These correlation coefficients suggest that the presence of higher proportions of either Stage 3 or Stage 4 wires correspond to lower number of broken wires. This inverse correlation is inconsistent with the definition of the stages of corrosion, which assign the least damage to Stage 1 wires and the worst damage to Stage 4 wires. This observation echoes similar inverse correlation between the stages of corrosion and broken wires on other suspension bridge main cable investigations.

This suggests that a supplemental evaluation, which employs a reliability-based approach to assess broken wires, will provide more accurate assessment and better picture of the distribution of wire breaks.

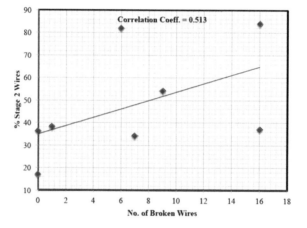

Figure 3.   Correlation between number of broken wires and % stage 2 wires.

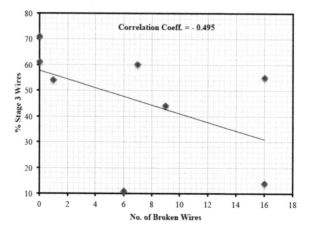

Figure 4.   Correlation between number of broken wires and % stage 3 wires.

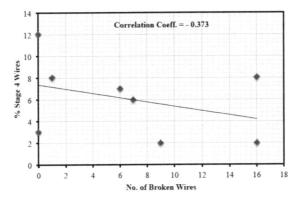

Figure 5.   Correlation between number of broken wires and % stage 4 wires.

3.3   *Cracked wires*

Cracking is the major driver of wire degradation. Therefore accurate assessment of both; the proportion and ultimate strength of cracked wires is of great impact on the reliability and confidence in the estimated cable strength. There are two important factors that affect the contribution of the cracked wires in the estimated cable strength, as follows.

3.3.1   *Cracked wire proportion*

In the BTC method, the cracked wire proportion is determined, based on fractographic examination of all of the fracture surfaces of tested samples, as the ratio of number of wire samples, which contain preexisting cracks, to the total number of tested wire samples. The preexisting crack depth, shown in Figure 6, is used to assess the strength of the cracked wire, as presented in subsection 3.2.2. This cracking proportion, in each investigated panel, is treated as a probabilistic quantity.

In the assessment of the proportion of cracked wires in the investigated panel and the effect of wires that are cracked in adjacent panels, Report 534 apply a reduction factor of 0.33 to cracked wires in Stage 3. This reduction factor is derived from fitting cracked wire data of 64% cracking in Stage 4 wires and 13% cracking in Stage 3 wires. There is no evidence to support the above proportions of cracking.

It is evident that the use of the reduction factor of 0.33 can lead to minimization of cracking damage and overestimation of the remaining strength in the bridge cable. Conversely, having a clear picture of the cracking extent will provide more accurate assessment of the cable strength.

The two methods; Report 534 and the BTC method, may produce different results, but this dual application of the two methods would provide the bridge owner with the benefits of having a truly independent check, rather than relying only on one method.

3.3.2   *Strength of cracked wires*

The BTC method determines the strength of cracked wires utilizing fracture-based analysis. The cracking typically develops on the inside of the cast of the wire, which is subject to tensile stress when the wire cast is straightened out. The average value for the ultimate capacity of a cracked wire, $\sigma_{cracked}$, is determined from the following relationship (Mahmoud 2007):

$$\sigma_{cracked} = \frac{K_c}{Y\left(\dfrac{a}{D}\right)\sqrt{\pi a_c}} \tag{2}$$

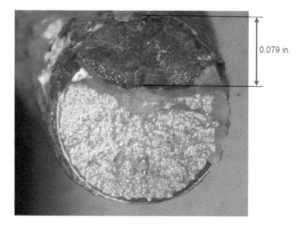

Figure 6.   Preexisting crack in a tested wire specimen.

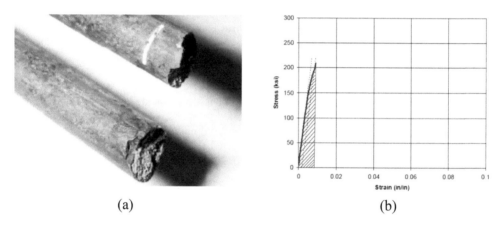

(a)   (b)

Figure 7.   Cracked wire fracture and stress-strain curve.

where, K_c is the fracture toughness of the wire determined from test data, a_c is the critical crack depth measured in specimens with preexisting cracks, see Figure 6, and $Y\left(\frac{a}{D}\right)$ is a crack geometry factor.

Figure 7 helps the understanding of the strong effect of cracking on the estimated cable strength. The wire specimen shown in Figure 7(a) contains a preexisting crack in the fracture surface, which was detected following the tensile test. It is noted that the wire cross section did not suffer any loss, and that the fracture profile shows no signs of necking. In other words, it is clear that the wire is embrittled. This is mirrored in the stress-strain curve of the same wire specimen shown in Figure 7(b), which shows an ultimate elongation of only about 1%. Consequently, the area under the stress-strain curve is only the shaded sliver, in Figure 7(b), which represents the very small amount of energy that it took to break the cracked wire.

If the wire, shown in Figure 7, was not sampled out for testing and remained in the cable, the crack would have grown under load application, until it reaches a critical depth, at which point the wire breaks, following the same mechanism of wire break shown in Figure 2. This demonstrates the strong influence of cracks on reducing the load carrying capacity of cracked wires and the overall remaining strength of the cable.

From the above explanation, it is evident that the accurate assessment of both; the cracked wire proportion and the strength of cracked wires are essential to arrive at an accurate estimate of the remaining strength of the cable.

The BTC method considers the effect of cracking, through fracture-based analysis and thus it provides an accurate assessment for the estimated cable strength.

However, Report 534 does not apply fracture mechanics to estimate the ultimate strength of cracked wires.

3.4 *Forecast of cable life*

In the BTC method, wire degradation is quantified as a function of the wire measured mechanical properties. The BTC method correlates the decline of wire properties with degradation kinetics. To forecast the degraded strength of the cable, at anytime in the future, t_2, the BTC method assesses the following quantities:

- Degraded strength of intact wires, $(\sigma_u)_{t2}$
- Effective fracture toughness, $(K_c)_{t2}$
- Degraded strength of cracked wires, $(\sigma_{cracked})_{t2}$
- Proportions of broken and cracked wires at time t_2, including effect of degradation in adjacent panels.

The forecast of cable life, in the BTC method, includes the effect of proportions for broken and cracked wires in adjacent panels.

Report 534 does not provide a forecast for cable life. However, the accompanying document, (NCHRP Project 10-57 2004), proposes a linear model and a nonlinear model for the future prediction of cable strength. The linear model calculates a linear rate of strength loss, starting at the first appearance of Stage 3 corrosion. However, this linear model does not account for an increasing rate of deterioration (NCHRP Project 10-57 2004), and may lead to an underestimation of future strength loss. The nonlinear model requires experimental data to develop a graph showing the rate of deterioration of a wire in an environment similar to that inside a cable. As the experimental data required for the nonlinear model is unlikely to exist, a hypothetical graph with correlation factors is assumed in which the onset of each corrosion stage is plotted against the start time of the test. It is recognized and suggested in Project 10-57 that due to this lack of data a linear relationship for loss of strength be used. Project 10-57 limits the extent of the use of the assumed linear relationship, stating *"...provided that the period over which the cable strength deterioration is estimated does not exceed 10% of the age at the time of inspection."* This is recognized in (Colford 2014), which states *"...if inspection and strength evaluation is carried out when a bridge is 30 years old then the future strength would only be projected for a period of just over three years. This makes long term planning quite difficult."*

The forecast of cable strength degradation and prediction of service life are of critical importance to bridge owners. It is acknowledged that this task is carried out based on limited data to predict uncertainties in the future. Therefore it is necessary to account for the measurable metrics of wire degradation and the increasing rate of deterioration. The BTC method employs a prediction model, which is based on strength degradation for both intact and cracked wires and includes for an increasing rate of deterioration. On the other hand Report 534 has to rely on a linear model of future deterioration, which does not allow for an increasing rate of deterioration. Report 534 also assumes a strength loss only from the onset of Stage 3 corrosion and in many cases galvanized bridge wires, which exhibit significant corrosion when subject to tensile tests have not suffered any loss of tensile strength (Colford 2014). Indeed in some cases, the measured average strength for Stage 3 wires has been higher than the average strength for Stage 2 wires.

4 CONCLUSIONS

Suspension bridges are strategic structures with a significant value to the economy and environment enabling social and business connectivity. Therefore the assessment of the residual strength and service life of their main cables is of paramount importance. Only conducting internal inspection, sampling, and testing of wire specimens from different panels along the cables can provide the inputs needed for the strength evaluation. Random sampling of wires, from each of the investigated panels, eliminates visual bias and allows for the evaluation of the sampling error in the estimated cable strength. This paper presents a comparative analysis of the evaluation of the strength and residual life of suspension cables based on Report 534 Guidelines and the BTC method. The paper shows that cracking is the main driver of deterioration of the cable strength. Thus the fracture-based analysis of cracked wires is required to achieve accurate assessment of residual strength. The damage resulting from the corrosion reaction is assessed more accurately by the byproduct of the corrosion reaction that releases atomic hydrogen. At locations of micro deficiencies on wire surface, atomic hydrogen penetrates the interior of the wire, breaks the interatomic bond and causes wire embrittlement. Over time, and with the application of load, the surface crack grows, leading to strength reduction and ultimately to the wire break. It is this understanding of the corrosion process, and deterioration, that is based on measureable metrics of the effect of the corrosion reaction byproducts that is significant to the strength evaluation. The forecast of residual life in the BTC method is based on strength degradation model for intact and cracked wires, and increasing rate of deterioration.

The main cables are the most critical and vulnerable elements of a suspension bridge. Therefore it would be sensible and logical to use the two recognized methods to provide confidence and correlation by using a separate and independent checking mechanism/method rather than just relying on one method. This would provide owners with greater assurances when dealing with perhaps the most challenging decisions they may face when mapping out maintenance and preservation strategies, based on the results of an internal inspection and strength evaluation of their main cables.

Supplementing Report 534 Guidelines by the BTC method will provide bridge owners with an enhanced understanding that considers another approach to bridge cable degradation. The FHWA Primer recognizes the BTC method as a valid methodology for the evaluation of remaining strength and residual life of bridge cables. The application of the BTC method alongside Report 534 Guidelines will provide bridge owners a high level of confidence in the estimated cable condition.

REFERENCES

Colford, B.R., 2014. Suspension bridges – design, maintenance, and inspection of main cables. *15th European Bridge Conference, Imperial College of London, London, United Kingdom, July 8–9.*

Hopwood, T. and J.H. Havens, J.H., 1984. Corrosion of cable suspension bridges, Kentucky Transportation Research Program, University of Kentucky Lexington, Kentucky: Kentucky.

Mahmoud, K.M., 2003. Hydrogen embrittlement of suspension bridge cable wires, *System-based Vision for Strategic and Creative Design,* Bontempi (ed.), Swets & Zeitlinger, Lisse, ISBN 90 5809 599 1.

Mahmoud, K.M., 2007. Fracture strength for a high strength steel bridge cable wire with a surface crack, *Theoretical and Applied Fracture Mechanics, 48.*

Mahmoud, K.M., *NYSDOT Report C-07-11,* 2011. New York State Department of Transportation (NYSDOT), BTC method for evaluation of remaining strength and service life of bridge cables, *NYSDOT, cosponsored by FHWA and New York State Bridge Authority.*

Mayrbaurl, R.; Camo, S., 2004. NCHRP Report 534 - Guidelines for inspection and strength evaluation of suspension bridge parallel wire cables, Transportation Research Board, Washington DC, USA.

Nakamura, S., and Suzumura, K., 2003. Hydrogen embrittlement and corrosion fatigue of corroded bridge wires, *Journal of Constructional Steel Research,* Volume 65, Issue 2, 2009.

NCHRP Project 10-57, Structural safety evaluation of suspension bridge parallel wire cables, Final Report, 2004.

Steinman, 1988. Williamsburg bridge cable investigation program: Final Report. New York State Department of Transportation & New York City Department of Transportation.

U.S. Federal Highway Administration, 2012. Primer for the inspection and strength evaluation of suspension bridge cables, FHWA, Washington, DC, FHWA Report No. FHWA-IF-11-045.

Chapter 2

Main cable dehumidification—flow testing and other innovations

M.L. Bloomstine & J.F. Melén
COWI A/S, Denmark

ABSTRACT: Corrosion of main cables on suspension bridges is a well-known and very serious international problem. Over the last 20 years dehumidification systems for main cables have been developed and applied worldwide. Dehumidification is generally accepted as the worldwide best practice for corrosion protection of main cables. Bridge authorities around the world are requiring dehumidification of main cables on new suspension bridges and retrofits on existing bridges. Over the last 20 years the understanding and design of main cable dehumidification systems has been constantly improved, as illustrated by this paper and the papers including in the references. This paper presents: Flow testing as a design parameter for an optimal system—experience and recommendations, Various other innovations and optimizations that ensure the effectiveness and minimize life cycle costs.

1 INTRODUCTION

Many examples of serious suspension bridge main cable corrosion have been discovered during inspections in various countries worldwide. The most critical cases are generally found on older bridges, and in a number of cases the safety factor for the main cables has been significantly reduced due to wire cracking, which is promoted by moisture (Mahmoud et al, 2016). There are however, numerous reports of serious corrosion of main cables within the age range of 5–30 years in Europe and Japan, see Figure 1. These cables are potentially even more critical than cables on older bridges, as the design factor of safety on main cables has been reduced over the years to roughly half of the original level. It is therefore strongly recommended that all suspension bridge owners/operators, no matter the age of the their bridge, instigate measures to determine the condition of the main cables and protect them from corrosion by the best means possible, which has been proven to be dehumidification.

Dehumidification is generally accepted as the worldwide best practice for corrosion protection of main cables—including both cables made up of parallel wires and of helical strands. Bridge authorities around the world are requiring dehumidification on new suspension

Figure 1. Corrosion on main cables, roughly 30 (left) and 15 (right) years old.

bridges, such that these bridges are born with a dehumidification system integrated in the design of the bridge. Bridge authorities in numerous countries are also requiring dehumidification systems on their existing bridges and many bridges have been or are currently being retrofitted to protect the main cables from corrosion. Dehumidification ensures a long lifetime for the bridge, as all other superstructure bridge elements except the main cable are replaceable without major structural modifications of the bridge.

COWI has continuously developed the field of corrosion protection of steel bridges by dehumidification, starting back in the 1960s, when dehumidification systems were designed for the bridge box girder and the splay chambers of the Little Belt Suspension Bridge (Bloomstine et al, 1999). Since then, dehumidification systems have been integrated in all COWI major bridge designs—including new bridges and retrofits. This experience has been utilized to develop, design and optimize dehumidification systems for main cables since the late 1990s. In connection with this work, flow testing on existing bridges as a design parameter that ensures a viable and cost-effective system has been developed.

2   FLOW TESTING

2.1   *Background and purpose of flow testing*

The current best practice for viable flow lengths on main cables of parallel wires is a maximum length of approximately 200 m. For main cables of helical strands there is currently no best practice for flow lengths. The voids are fewer and larger, which enables much longer flow lengths, for example approximately 380 m on the Älvsborg Bridge in Sweden. In order to design an optimal system for an existing bridge it is recommended that a pre-design flow test be carried out. Experience has shown that a flow length of 200 m for main cables can be either too conservative or even worse, not feasible. The actual optimal flow length depends among other things on the condition of the main cables, e.g. the corrosion state and if they have been oiled or not. A further benefit of a flow test is determining the points of leakage that will need to be sealed in connection with a dehumidification system. Hence, the test also provides design basis for the sealing system and ensures an effective system.

The purpose of the flow test is twofold:

1. Establish the maximum viable flow length in order to develop the optimal layout of the system.
2. Detect where the cable bands leak air in order to design necessary further sealing. Typical leaks are at the caulking of cable band joints, around the cable band bolt heads, nuts and washers, at drain holes and at other cable band details.

To date we have performed pre-design flow tests on 6 suspension bridge, 4 with main cables of parallel wires and 2 with main cables of helical strands. The results from these tests have been utilized to design optimal systems and have also provided valuable knowledge to the understanding of air flow conditions in main cables. Further, we have optimized the flow test itself over the last 15 years, providing an efficient method to establish the design basis for an optimal main cable dehumidification system for each individual bridge. The optimal system is defined as the system with the lowest possible life cycle cost that provides reliable corrosion protection.

2.2   *General description of a flow test*

The test is generally carried out in the main span with an injection point near mid-span. This allows the easiest possible access and long lengths of main cables for flow testing. At the injection point the wrapping wire is removed over a length of approximately 1–2 m, depending on the diameter of the main cable. Before cutting and removing the wrapping wire, it is secured at each end with banding. The exposed surface of the main cable is then thoroughly cleaned, removing all the paste on the surface to allow the injection air to enter the main cable.

A temporary injection sleeve is then erected over the open area. There are two options for the temporary injection sleeve:

1. A flexible sleeve, see Figure 2. This type of sleeve is inexpensive to fabricate and erect, it can be used on both the main cables and it is not sensitive to the exact diameter or shape of the main cable. This type of sleeve is recommended if the flow test will be performed in connection with an in-depth inspection of the respective panels, where the panels will be re-wrapped after the inspection.
2. A lightweight stainless steel sleeve, as shown in Figure 2. The advantage of this type of sleeve is that it can be used for temporary corrosion protection of the main cable in the open area (instead of the paste, wrapping wire and paint) and it can remain in service for 5–10 years until dehumidification is applied or the main cable panel is rewrapped. A nozzle for connection of the hose on the bottom side will act as a drain opening after the test.

The remaining equipment for the flow test is shown in Figure 3 and is composed of a fan, a hose connecting the sleeve and the fan, instruments for measuring pressure and flow and a generator for power supply.

The flow testing is generally carried out in the normal operating range of pressure, dependent on the type of main cable, but generally between approximately 1,000 and 3,000 Pa. The upper limit has been established as the maximum allowable pressure on the basis of pressure tests on main cable models and bridges and the ensures that the sealing system is not

Figure 2.   Temporary injection sleeves, flexible (left) and stainless steel (right).

Figure 3.   Equipment for flow test.

Figure 4.   Leakage at cable band bolt (left) and at drain hole (right).

overloaded as well as minimizing leakage. Greater pressure will overload the sealing and result in unnecessary leakage and a less effective dehumidification system.

When the equipment is installed and tested, testing with overpressure commences. The initial activity is to discover where there are leaks at the cable bands, so that appropriate sealing can be designed in connection with a dehumidification project. Soapy water is sprayed on all possible areas of leakage and bubbles occur where there is leakage (see Figure 4). Typical points of leakage at cable bands include caulking, cable band bolts including washers, nuts and bolt heads and drain holes. It is important to note, that there can be other points of leakage, depending on the design of the cable bands. On most suspension bridges without main cable dehumidification there is a drain hole at the lower end of the cable band, see Figure 4. This is generally the most consistent indicator of leakage/airflow for the testing, see Figure 4. On some suspension bridges the cable bands are vertically split, which gives a long gap on the bottom side, which also functions as a drain.

Upon completion of the testing, a comprehensive report should be prepared, which includes all the test results, an analysis of the results and a viable layout for an optimal main cable dehumidification system. As the flow testing is performed without sealing the main cable panels and cable bands, the test results will be somewhat conservative, as these will be effectively sealed in connection with installation a main cable dehumidification system. Therefore, utilizing the maximum flow length from the test is acceptable, as it will provide an appropriate margin of safety and ensure a well-functioning system. In practice, the final flow lengths are generally shorter than the maximum test length, as the flow lengths are also typically determined by dividing the bridge's span lengths into equally long flow lengths.

2.3   *First flow test—Little Belt Bridge, main cables of helical strands*

We carried out the first flow test in 2001 in connection with research for developing the retrofit design on the Little Belt Bridge in Denmark (Bloomstine & Thomsen, 2004). This is believed to be the first system for main cables of helical strands. At this time, the dehumidification technique for main cables was still quite new and not well developed. Even though there were main cable corrosion problems that needed to be solved, the Authority was hesitant to invest in developing a dehumidification system. In order to document the viability of a dehumidification system for this bridge, a simple flow test was developed that could be performed at a minimal cost and without opening the main cables, see Figure 5. This was the very first pre-design flow test.

Air was injected into a main cable inside the splay chamber at a pressure of approximately 1,500 Pa. Although most of the air flowed out in the splay chamber, a sufficient amount of air for the test did flow into the main cable. A small amount of a non-toxic trace gas was

Figure 5.   Flow test in 2001 on Little Belt Bridge, Denmark.

intermittently added to the injection air. Further up the main cable, at a distance of approximately 170 m, a sensor to detect the trace gas was held at an opening adjacent to at a cable band. After approximately eight minutes the trace gas was detected and the procedure was repeated several times with the same result.

2.4   *Flow test—Högakusten Bridge, main cables of parallel wires*

The first flow test we carried out on main cables of parallel wires was on the Högakusten Bridge in Sweden in 2005. Due to very serious water intrusion and relatively serious corrosion of the main cables of this bridge that was only 8 years old at the time, the authority fast tracked the installation of a dehumidification system. Therefore, there was not time for a pre-design flow test. A flow test was therefore first carried out during the start of the construction of the dehumidification system, with the main purpose of confirming the design flow lengths.

The injection point near the center of the main span was prepared by removing the wrapping wire and cleaning off the zinc paste. A simple temporary injection sleeve was prepared and installed over the cable opening and air was injected. By using soapy water to detect leakage sufficient flow lengths were confirmed. The test is illustrated in Figure 6.

2.5   *Flow test—Älvsborg Bridge, main cables of helical strands*

Based on the successful project on Little Belt Bridge, where the main cables also are made up of helical strands, it was believed that it would be possible with just one injection point on each main cable on the Älvsborg Bridge in Sweden. The flow test was based on this assumption and performed to confirm it. As it was only necessary to confirm that air could flow from the center of the main span to both splay chambers, it was decided to apply non-toxic trace gas, as this could be easily monitored in the splay chambers. The flow testing was successfully carried out in 2009 and confirmed the viability of a flow length of app. 380 m, from the center of the main span and to each splay chamber. The main cables were also inspected for leakage, which was fortunately quite limited and did not have a significant effect on the results. The flow test is illustrated in Figures 7 and 8.

The wrapping wire was first removed over a length of approximately 0.5 m at the center of the main span. As the main cables are made up of helical strands, there are filler strips to form a circular cross section. Holes were bored in the filler strips to ensure an easy access for the injection air. A simple injection sleeve was erected and sealed, the injection equipment

Figure 6.   Flow test on Högakusten Bridge, air injection and soapy water test.

Figure 7.   Älvsborg Bridge, opening of main cable and injection sleeve.

Figure 8.   Älvsborg Bridge, injection equipment (left) and trace gas monitoring in splay chamber (right).

was connected and the sensors were installed in the splay chambers. Air was injected at approximately 2,000 Pa and the trace gas was detected in both splay chambers after about 15 minutes, giving a flow speed of approximately 0.4 m/s.

2.6   *Flow test—Storebælt Bridge, main cables of parallel wires*

During the conceptual design of the main cable dehumidification system for the Storebælt Bridge in Denmark, a flow test was recommended to ensure a viable and optimal design. The testing was successfully carried out in 2012 and resulted in by far the longest flow lengths on any bridge with main cables of parallel wires, up to 285 m, as compared up to approximately 200 m on other bridges (Nielsen et al, 2016).

The initial testing for leakage showed leakage at the cable band drain holes, at drain holes in cable band bolt washers and in the caulk of the longitudinal joints in the cable bands (see Figure 9). Specifications for sealing of these leaks were developed and successfully carried out during the dehumidification system installation in 2014–15. The equipment for the flow test (injection sleeve, hose, fan and sensors) is illustrated in Figure 9.

The flow test results for both directions from the injection point were identical as follows:

- Pressure 2,000 Pa – flow length 320 m (horizontal)
- Pressure 2,500 Pa – flow length 345 m (horizontal)
- Pressure 3,000 Pa – flow length 390 m (horizontal)

The test results, which are conservative due to the leakage, resulted in optimized design, as illustrated in Figure 10. The flow lengths were increased from 190 to 285 m in the side

Figure 9.   Storebælt Bridge, equipment setup for flow test and leakage at longitudinal joint.

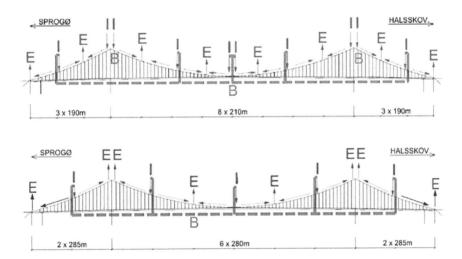

Figure 10.   Storebælt Bridge, Conceptual Design (above) and Final Design (below).

spans and from 210 to 280 m in the main span, still well below the proven possible lengths, allowing a comfortable margin of safety. The overpressure at the injection points is relatively low, approximately 2,200 Pa, well below the established best practice of maximum 3,000 Pa (confer section 2.2).

The modifications from the conceptual to the final design gave the following advantages:

- The design flow lengths were verified before starting construction
- The low design pressure minimizes electrical consumption and leakage
- The number of injection/exhaust sleeves with corresponding sensors, wiring and air pipes was reduced from 24 to 14
- Only one plant instead of three and this plant is actually a modification of an existing plant

All these changes give lower construction, operation and maintenance costs, i.e. a substantially lower life cycle cost. The flow testing was an inexpensive investment and resulted in an optimal system with a high level of safety and major life cycle savings.

2.7   *Flow test—Storda Bridge, main cables of parallel wires*

Storda and Bømla Bridges in Norway are two quite similar suspension bridges, which were built at the same time and opened for traffic in 2001. In connection with the design of main cable dehumidification systems for these bridges we proposed to carry out a flow test and this was accepted by the Authority. The testing was carried out on Storda Bridge, which is about 15% longer than Bømla Bridge, in 2016. The main cables have a relatively small diameter, 326 mm on Storda Bridge and 305 mm on Bømla Bridge (both exclusive wrapping wire). These are the smallest main cables that we have performed a flow test on and the experience from the test is very relevant for future projects with main cables of similar diameter. The exposed section of the main cable (after the zinc paste has been cleaned off) with a length of approximately 1 m is shown in Figure 11.

The initial testing for leakage at the cable bands showed leakage at the drain holes (as expected) and at the washers on the lower end of the cable bands. The lower washers are placed in recesses on the cable bands and the small gap between the recess and the washer is not in all cases filled with paint, which results in leakage, see Figure 11. Further, these washers also have a drain channel that also allows air leakage.

As the main cable has a much smaller diameter than earlier tested cables, the air flow volume for a given overpressure was substantially lower. This meant that the air leakage had a significantly greater effect on the testing. Initial testing showed very short flow lengths, approximately 72 m at approximately 1,500 Pa. It was then decided to seal the drain holes on 2 cables bands on each side of the injection point. This resulted in flow lengths of approximately 160 m at 1,500 Pa and 190 m at 3,000 Pa, with the same results in both directions. Another 6 drain holes were sealed in one direction and the results were only marginally better, approximately 205 m at 3,000 Pa.

Figure 11.   Storda Bridge, cleaned cable at injection point and leakage at washer on cable band.

The final conclusion was that it was not possible to further optimize the layouts for the two bridges, which were based on the best practice of an approximate maximum flow length of 200 m. (This best practice is based is based on experience from earlier systems, including information from colleagues in Japan.) In fact, it was deemed necessary to make a small modification of the layout for the Storda Bridge, where there is a long back stay on one end of the bridge. One injection point was moved as shown in Figure 12, so the maximum flow length is 183 m instead of 234 m. The maximum design flow length is therefore shorter than the maximum test flow length, ensuring that the layout is viable. It was not necessary to modify the flow lengths for Bømla Bridge, as all the flow lengths were well below the test flow length, with a maximum length of 166 m.

Based on the flow test results it was evaluated that the dehumidification systems (including sealing of main cables) on both bridges would function well with an injection overpressure well within the best practice maximum of approximately 3,000 Pa. Taking into consideration the much better sealing of the main cables that will be included in the work, the evaluated necessary pressure is in the range of 1,500–2,000 Pa for Bømla Bridge and 2,000–2,500 Pa for Storda Bridge.

An interesting and valuable observation, which was not directly related to the test, was made during the second day of the testing. It had been dry weather during the night and the morning and it was quite windy during the morning. Despite these weather conditions, there were two adjacent areas on the bottom of the cable with standing water, see Figure 13. These areas were dried off and inspected, see Figure 13. There were defects in the wrapping wire (gaps) and defective paint, which could not cover the gaps. After drying off, it was evident that water was seeping out of the main cable, as water soon accumulated again. It was concluded that the overpressure from the flow test was forcing the water out, which would normally be trapped in the main cable. This further confirmed the need for dehumidification, and is a valuable general warning—cables that appear to be in good condition on the outside are most likely concealing trapped water and conditions that cause corrosion.

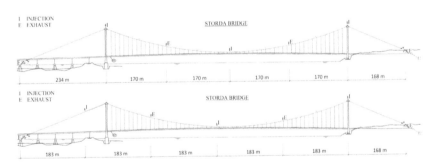

Figure 12.   Storda Bridge, Conceptual Design (upper) and Final Design (lower).

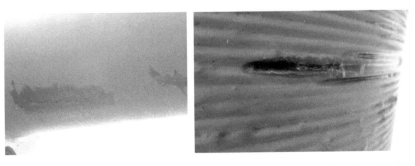

Figure 13.   Water on bottom of main cable and close-up off defective wrapping wire and paint.

3   INNOVATIONS

3.1   *Buffer chamber solution*

In 1994–95 we assisted the Humber Bridge Authority with 2 dehumidification projects. The first project was dehumidification of the stiffening box girder, which accumulated condensation water, causing corrosion of the inner surfaces. When this was successfully concluded, work with the water intrusion problem in the tower saddles commenced.

The authority had earlier attempted to seal the saddles, but none of the attempts succeeded in keeping water out. Our solution was to create an overpressure in the saddles by a constant flow of dry air. To achieve this a small dehumidification plant was installed in one leg of each tower with ducting to the two saddles. The dry air produced by a dehumidification unit has a relative humidity of just over 0%, which is much drier than necessary for corrosion protection. In order to conserve energy, the volume of the tower leg was utilized as a buffer chamber, where ambient air is mixed with air from the dehumidification unit to achieve a relative humidity of approximately 40%. This reduces the running time of the dehumidification unit by approximately 70–80% and saves a substantial amount of electrical energy, as well as significant reduction of wear on the dehumidification unit. This was the original buffer chamber solution, which thereafter has been successfully utilized as described below.

Although this system was limited to a small portion of the main cables, it was the first case of main cable corrosion protection by dehumidification and installed in 1995 (Bloomstine, 2013). When the Authority later decided to expand to a complete dehumidification system for the main cables in 2010, this system was no longer necessary, as the new system has an air flow through the saddles.

The calculated ideal volume for a buffer chamber is approximately 10 times the volume of the hourly flow to the main cables, which accommodates the variation in ambient relative humidity that occur during a day-night 24 hour period. This theoretical value has been proven by practice on bridges where there has not been enough room for an ideal buffer chamber. For example on the Högakusten Bridge in Sweden, the Authority could not allow the volume of a tower leg to be utilized as a buffer chamber, hence only the volume of the upper cross beam of the tower was available. The volume of the cross beam is roughly 5% of the ideal volume. On the same bridge there is also a buffer chamber in the stiffening box girder, where the ideal volume was available. The energy savings for the plant in the box girder are approximately 80%, whereas the energy savings for the plants in the tower cross beams are only approximately 10%. Substantial energy savings are possible for a buffer chamber with a volume slightly smaller than the ideal volume, but if the volume is substantially smaller, the savings are minimal. Placing the plant in a small room does not achieve the buffer chamber effect and does not provide any savings in electrical consumption or reduction of wear on the dehumidification unit.

Ideally, the buffer chamber should utilize the volume of an existing structural element, as in the case above, in order to minimize construction costs. In addition to the above mentioned structural elements, part or all of the anchorage structure could be utilized as a buffer chamber, a solution that has been applied on several projects.

3.2   *One large plant contra multiple plants*

In general, we strive to design the dehumidification system for main cables with just one plant, which can provide sufficient dry air for the entire length of both main cables, as this is by far the most economical solution. In order to provide redundancy, the plant is generally designed with two dehumidification units and two fans, as these are the essential mechanical elements that are necessary for production and injection of dry air in the main cables at the target pressure. The plants are generally designed such that these units run alternately, so they are running regularly, which is essential to keep them functional. If a unit should break down, the other unit will automatically take over full time until repair work has been completed.

This solution is maintenance friendly, as all the maintenance is carried out in one location and there is much less to maintain. In our projects the plant is always placed in the buffer chamber (refer 3.1), which provides a protected atmosphere for the plant and an easily accessible location. If there are multiple plants, at least some of these will be located at areas, which are relatively difficult and/or time consuming to access. This solution can require somewhat longer air pipes, but this is a minor disadvantages when compared to the many benefits.

The one plant solution ensures a much lower life cycle cost than a multiple plant solution for the following reasons:

- One plant cost much less to construct
- One plant requires much less maintenance
- Ease of access and only one area to access saves time
- One large plant consumes much less electrical energy

This solution has been successfully applied on two retrofit systems that are in service – Storebælt Bridge (see Figure 10 and 14) and Älvsborg Bridge. The dehumidification project for Macdonald Bridge in Canada includes this solution and is currently under construction. The project for George Washington Bridge also has this solution and is expected to be constructed within the next 2 years.

3.3 *Improved sealing detail at cable bands*

We have cooperated with the D. S. Brown company concerning the development of their Cableguard™ wrap system for sealing of main cable panels in connection with dehumidification. This started with some initial modifications of their system, which were applied to the design of the dehumidification system for Little Belt Bridge in Denmark. The latest improvement we have developed and applied concerns the detailing at the interface to the cable band. The original design included a neoprene wedge that was installed over the wrap at the band. The groove in the cable bands was first filled with caulk and the wedge was then pressed up against the band. Supplemental caulking was also applied in order to fill out the area between the neoprene wedge and the cable band. After caulking, a finish strip of Cableguard™ wrap was installed on top of the neoprene wedge.

Experience from several bridges, especially the Högakusten Bridge in Sweden, have shown that this detail is problematic and has the following disadvantages:

- The caulk is in a closed area and does not fully cure, as it needs air to cure.
- It is not possible to see the caulk and how well the area is filled out with caulk.
- There are often leaks and it is not possible to locate and repair these.

Figure 14.   Dehumidification plant for main cables, Storebælt Bridge.

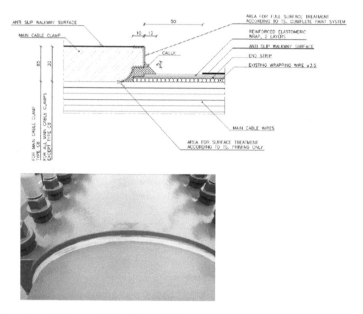

Figure 15.   Drawing and picture of detail at cable band.

- It is not practical to maintain or replace the caulk at the end of its lifetime. (The finishing strip is bonded to the normal wrap and it would be destructive to remove it, as would removal of caulk under the wedge. Maintenance or replacement would be very difficult and costly.)

Therefore, a detail that solves all the above mentioned problems and eliminates the use of the neoprene wedge was developed. Instead of using a finish strip of Cableguard™ wrap, a start/end strip is applied, which extends into the groove of the cable band. Normal Cableguard™ wrap is installed over the start strip, starting at the edge of the cable band. After heat bonding all three layers of the wrap, the caulk is installed in the groove with an overlap on the Cableguard™ wrap outside the groove. The detail is illustrated in Figure 15.

This detail has been applied in connection with main cable dehumidification on two bridges – The Hardanger Bridge in Norway, in service since 2013 and the Storebælt Bridge in Denmark, in service since 2015. On both of these bridges the sealing result is excellent, with a maximum loss of 10% of the air per flow length. Taking the long flow stretches into account, the results are exceptionally good. The longest flow lengths on Hardanger Bridge are 177 m and on Storebælt Bridge to the best of our knowledge the longest flow lengths in the world with a length of 285 m.

3.4   *Improved monitoring for new bridges and new understanding*

On the Hardanger Bridge in Norway, we cooperated closely with the Authority who was responsible for the overall design of this bridge. Hence, it was possible to incorporate the dehumidification system design in the original design of the bridge, which allowed innovations that would not be possible on a retrofit. This included nipples on all the cable bands (see Figure 16), where a pressure gauge can be easily attached and pressure measured. Hence, the overpressure at each cable band can be easily measured and any leakage that may develop over time or due to accidental damage can be readily located, as an unusual pressure drop between two cable bands would indicates a leak in the cable panels between these 2 cable bands.

Flow testing carried out in connection with a control of the sealing system has provided valuable understanding of main cable dehumidification technology. During this testing air was injected at approximately 2,500 Pa and the pressure was measured at each cable band. The main cables have a diameter of 617 mm. An evaluation of the results at this pressure level provide the following valuable information, which is consistent for all the flow lengths:

Figure 16.   Nipple for pressure gauge and installation of pressure gauge during flow testing.

- The singular pressure loss at the injection point is nearly negligible, approximately 100 Pa.
- The pressure loss due to the resistance in the main cables is approximately 14–15 Pa per m.

During commissioning and adjustment of the system the final injection overpressure was set at approximately 1,300 Pa, giving the following results:

- The singular pressure loss at the injection point is approximately 50 Pa.
- The average pressure loss due to the resistance in the main cables is approximately 7–8 Pa per m.

On the Storebælt Bridge the main cables have an outer diameter of 834 mm and the injection overpressure is approximately 2,200 Pa. Here the average pressure loss due to the resistance in the main cables is also approximately 7–8 Pa per m.

These are both excellent examples of optimized systems, where the following features are especially notable:

- The flow lengths are very long, especially on Storebælt Bridge, where the flow test results allowed this.
- The flow rate is optimized giving a very low average pressure loss of approximately 7–8 Pa/m. This is approximately half of what has earlier been reported for main cables of parallel wires.
- The leakage rate for both bridges is max. 10% per flow stretch.

3.5   *Pressure curve adjustment—dampers at exhaust sleeves*

Monitoring results of the exhaust air have shown that the humidity conditions inside the cable vary somewhat during operation. Under perfectly ideal conditions, with a 100% airtight cable and completely constant absolute water content (AWC) in the injection air, the exhaust air should have the same AWC as the injection air once the drying out process is completed. In the real world this is not the case, as the cable can never be 100% air-tight and the AWC varies somewhat as the systems are controlled by relative humidity, which gives variations according to variations in the temperature.

The overpressure inside the main cables along a given flow stretch varies more or less linearly from a maximum at the injection point to zero at the exhaust point. The low overpressure in the area near the exhaust points makes these areas susceptible to intrusion of moisture during extreme weather conditions such as high atmospheric pressure, hard rain and high speed turbulent wind. Due to the height, exposure and locations of many suspension bridges, these types of weather conditions occur relatively often. In order to counteract this effect a pressure curve adjustment solution was developed. By applying a damper in the exhaust pipe on the exhaust sleeve a small overpressure can be maintained instead of falling to zero, and thereby better protect these sensitive areas from moisture intrusion.

This solution is incorporated in all of our recent projects and two of these are in service – The Hardanger Bridge in Norway, since 2013 and the Storebælt Belt Bridge in Denmark, since 2015. During the commissioning of these systems the exhaust dampers were adjusted and locked at the correct position (determined during commissioning/adjustment of the system) and ensure an overpressure of approximately 100–150 Pa, thereby ensuring better protection of the end stretches of each flow length. An exhaust sleeve is shown in Figure 17. The damper is placed in the pipe in the lower part of the instrument box.

3.6   *Monitoring sleeves near anchor chambers*

Thorough monitoring of the dehumidification system is essential in order to document its effectiveness. This is quite straightforward for all points except at the exhaust flow in the anchor chambers. At this point it is possible to measure the relative humidity and the temperature, but it is not generally possible to measure the flow, as the exhausting air continues to flow between the wires or strands and out into the anchor chamber. It is in some cases possible to catch some of the exhausting air by building an enclosure around the strands and this has been done e.g. on the Älvsborg Bridge and Högakusten Bridges in Sweden, see Figure 18.

In order to improve the quality of the exhaust monitoring we have developed the concept of a monitoring sleeve, which is placed on the main cable a short distance from the anchor chamber. The monitoring sleeve includes sensors for relative humidity, temperature and pressure. During the commissioning of the dehumidification system the relation between flow and pressure can

Figure 17.   Exhaust sleeve with damper, bolt for adjustment of damper – see arrow, Hardanger Bridge.

Figure 18.   Enclosure of strands, Älvsborg Bridge and monitoring sleeve, Storebælt Bridge.

be accurately established and the monitoring system will then be able to indirectly measure the flow close to the anchor chambers. In order to still monitor the short stretch between the monitoring sleeve and the anchor chamber the sensors for relative humidity and temperature are still applied inside the anchor chamber. This solution provides better monitoring of the flow and saves the difficult and somewhat ineffective enclosures. It has been successfully applied on the Storebælt Bridge (see Figure 18) and will soon be installed on several other bridges.

4   WORLDWIDE STATUS

To the best of the authors' knowledge, the following is an overview of the worldwide status for dehumidification of main cables. There are most likely more bridges with dehumidification of main cables installed or planned than listed below.

USA:
- 2 bridges with dehumidification of main cables installed (retrofits)
- 2 bridges with dehumidification of main cables under construction (retrofits)
- 2 bridges with dehumidification of main cables tendered (retrofits)
- 1 bridge where design of dehumidification of main cables has commenced (retrofit)

Canada:
- 1 bridge with dehumidification of main cables under construction (retrofit)

Denmark:
- 2 bridges with dehumidification of main cables installed (retrofits)

Norway:
- 1 bridge with dehumidification of main cables installed (new bridge)
- 1 bridge with dehumidification of main cables under construction (new bridge)
- 2 bridges with dehumidification of main cables, construction 2018–19 (retrofits)

Sweden:
- 2 bridges with dehumidification of main cables installed (retrofits)

UK:
- 3 bridges with dehumidification of main cables installed (retrofits)

Turkey:
- 2 bridges with dehumidification of main cables installed (new bridges)
- 2 bridges with dehumidification of main cables installed (retrofits)
- 1 bridge with dehumidification of main cables planned (new bridge)

France:
- 2 bridges with dehumidification of main cables installed (retrofit)

Japan:
- 8 bridges with dehumidification of main cables installed (retrofits)
- 7 bridges with dehumidification of main cables installed (new bridges)
- 1 bridges with dehumidification of main cables under construction (retrofits)

South Korea:
- 1 bridge with dehumidification of main cables installed (retrofit)
- 1 bridge with dehumidification of main cables installed (new bridge)

China:
- 1 bridge with dehumidification of main cables installed (new bridge)

Qatar:
- 2 bridges with dehumidification of main cables (new bridges)

There are currently at least 36 suspension bridges with main cable dehumidification installed and in service, including 22 retrofits and 14 systems installed on new bridges. At least another 11 suspension bridges have main cable dehumidification systems under construction or other planning stages. This gives a grand total of at least 47 suspension bridges with installed or planned main cable dehumidification systems in a total of 12 different countries. This corresponds to roughly 25% of the world's major suspension bridges.

Further, six countries have installed main cable dehumidification systems on all their major suspension bridges or are in the process of doing so. These include Denmark, Japan, Norway, Sweden, Turkey and the UK (exclusive Tamar Bridge).

5   CONCLUSIONS

As described in this paper, the state of the art for main cable dehumidification has greatly advanced in recent years. COWI has led the way with continuous innovation and optimization, as well as international involvement and knowledge sharing. Based on the statistics above, it can be seen that the international suspension bridge community has truly embraced main cable dehumidification as the best practice method for protecting main cables from corrosion and ensuring a long life for their suspension bridges.

A pre-design flow test is strongly recommended in connection with designing a main cable dehumidification system on an existing bridge, as this allows the design of an optimal system and ensures the viability of the flow lengths. It is not sufficient to use the earlier established best practice of a maximum flow length of approximately 200 m, as this can lead to an ineffective system or even worse, as system that does not function properly. A flow test is particularly important if the main cables are corroded and/or oiled or the diameter is relatively small. The flow test also reveals where there is leakage at the cable bands, such that appropriate sealing can be designed in connection with the dehumidification system.

Flow tests have led to new understanding of airflow conditions in main cables. By optimizing the flow in the main cables, it has been possible to reduce the pressure loss for main cables of parallel wires by approximately half, from approximately 14–15 Pa/m to approximately 7–8 Pa/m. This results in a more effective system with less leakage. Flow test results prove that flow lengths for new parallel wire main cables (or main cables in good condition) with a large diameter can have flow lengths up to approximately 400 m, corresponding to twice the earlier best practice of maximum approximately 200 m. This allows systems that are even more efficient in the future. For main cables of helical strands it is possible to have even longer flow lengths. As suspension bridges with helical strand main cables are generally shorter, it is generally viable to just have one injection point per main cable at mid-span with dry air flow to both anchor chambers. In short – flow tests give great value for a very small investment!

It is strongly recommended that the other innovations described in this paper—buffer chamber solution, one large plant, improved sealing, improved monitoring, pressure curve adjustment and monitoring sleeves – all be implemented in future projects. Together with flow testing, these innovations provide an optimal, durable and maintenance friendly system with the lowest possible life cycle cost. Further, areas of the main cables close to exhaust points are better protected from corrosion and documentation of corrosion protection from the monitoring system is improved.

REFERENCES

Bloomstine, M.L., 2013, Latest Developments in Main Cable Dehumidification, *7th New York City Bridge Conference*, New York, USA.

Bloomstine, M.L., Thomsen, J.V., 2004, Little Belt Suspension Bridge—Corrosion Protection of Main Cables and Maintenance of Major Components, *4th International Cable Supported Bridge Operators' Conference*, Copenhagen, Denmark.

Bloomstine, M.L., Rubin, F., Veje, E., 1999, Corrosion Protection by Means of Dehumidification, *IABSE Symposium, Structures of the Future—The Search for Quality*, Rio de Janeiro, Brazil.

Mahmoud, K., Hindshaw, W., Mc Culloch, R. 2016. Inspection and Evaluation of Remaining Strength of Suspension Bridge Cables, *International Cable-Supported Bridge Operator's Conference (ICSBOC)*, Halifax, Canada, June 19–22.

Nielsen, K.A., Hansen, M.D., Bloomstine, M.L., 2016, The Storebælt East Bridge Main Cable Dehumidification—World's Largest Retrofit and Latest Design Optimization, *8th International Cable Supported Bridge Operators' Conference*, Halifax, Canada, June 19–22.

Chapter 3

The dehumidification of the main cables of the Delaware Memorial Bridge

S. Elnahal & S. Scindia
The Delaware River and Bay Authority, New Castle, Delaware, US

B. Colford & S. Beabes
AECOM, Philadelphia, Pennsylvania, US

ABSTRACT: The Delaware Memorial Bridge comprises two parallel suspension bridges crossing the Delaware River. Each bridge has a main span of 2150 ft. and side spans of 750 ft. The First Bridge was opened in 1951 and the Second Bridge in 1968. The bridges are owned and operated by the Delaware River & Bay Authority (DRBA). AECOM are the DRBA's Engineer for this modified Design Build Contract to install a dehumidification system on the main cables and are also carrying out the construction management. The RFP was advertised in the fall of 2015, for what is now only the second such project in the United States. The ongoing contract, which started in 2016, is being executed by American Bridge Company. Apart from the main cable dehumidification installation, other tasks have included, cable band bolt tightening and suspender replacements. Project completion is scheduled for June 2018.

1 INTRODUCTION

1.1 *Bridge description*

The Delaware Memorial Bridge is comprised of two separate suspension bridges each with a 2150 ft main span and 750 ft side spans (Figure 1). Structure 1 services northbound traffic into New Jersey and was opened in 1951. Structure 2 services southbound traffic into Delaware and was opened in 1968. The deck on Structure 1 was subsequently replaced with a heavier deck following the opening of the Structure 2.

The Delaware Memorial Bridge is a vital link for truck and passenger traffic in the northeastern United States. The two bridges that comprise the Delaware Memorial Bridge together carry over 30 million vehicles every year. The individual bridges carry on an average, over 45,000 vehicles a day and over 70,000 vehicles a day on some holidays when traffic flows peak. The Delaware River and Bay Authority has an aggressive and continuing program of investing in the maintenance, upkeep and long term preservation of these bridges and their structural components.

The twin main cables on both bridges are comprised of high tensile steel galvanized parallel wires, 0.196″ in diameter. There are 8,284 wires making up the cables in Structure 1 which has a compacted specified nominal diameter of 19 ½″ under the wrapping wire, and there are 9,196 wires making up the cables in Structure 2, which has a compacted specified nominal diameter of 20 ½″ under the wrapping wire.

1.2 *Project conception and development*

The Delaware River & Bay Authority (DRBA) started the process of evaluating the condition of the main cables of the two bridges in 2010. As part of this evaluation, portions of

Figure 1.   Structure 1 and 2 making up the Delaware Memorial Bridge, looking towards New Jersey. Structure 1 is on the right.

the main cables were opened at several locations along the lengths of the main cables and the extent of the corrosion in the main cables was documented. The procedure for performing these condition evaluations of the main cables followed the procedure laid out in NCHRP Report 534 – Guidelines for Inspection and Strength Evaluation of Suspension Bridge Parallel Wire Cables (TRB 2004).

While these inspections did not reveal any serious conditions such as wire breakages or extreme corrosion, there was evidence of surface rust on the parallel wires that comprise the body of the main cables. On average, for the northbound bridge, around 35% of the wires were found to be showing Stage 4 corrosion as defined in NCHRP Report 534, while for the southbound bridge, around 15% of the wires were found to be showing similar levels of corrosion.

Following technical deliberations internally within the DRBA and consultants, it was decided in 2012–2013 to move forward with the development of a main cable dehumidification project. For this purpose, the DRBA retained the services of AECOM to develop a Request for Qualifications (RFQ) followed by a Request for Proposals (RFP) – both based a two-step Design-Build project delivery methodology. The RFQ and RFP development tasks included the effort required to develop a comprehensive set of technical and performance specifications, drawings indicative of the final layout and configuration of the proposed dehumidification system, and an engineer's estimate.

While the RFP laid out the technical and performance expectations for the dehumidification system and the overall configuration of the system, there were certain elements that were clearly specified as prescriptive elements in which the contractor executing the project could not deviate. These included the location of the dehumidification plants, the dehumidification pipe blowing lengths, etc. Thus this project, while still being executed as a Design-Build project, placed certain limitations on the latitude offered to the Design-Build contractor that was to execute this contract. This methodology offered The DRBA the ability to specify the locations of the dehumidification plants, which are located inside a secure area within the bridge anchorages, and it ensured that the configuration of the system that was to be finally in place did not differ significantly from what were known limitations of dehumidification blowing lengths, etc. As such, in very broad terms, the procurement methodology adopted for this contract could be said to have followed a "Modified Design-Build" approach, rather than a "Pure Design-Build" methodology, although most seemingly pure Design-Build projects almost always have some prescribed elements, mostly based on clients' requirements and preferences.

In early 2015, the two-step procurement process was initiated with the solicitation for SOQs (Statements of Qualifications). A panel comprised of DRBA Engineers and AECOM, reviewed and ranked the submitted SOQs to build a shortlist of qualified contractors. These contractors were then invited to respond to the RFP for the project. Based on a comprehensive approach that considered both, the technical qualifications as well as the price proposal submitted by the prequalified bidders, the selection panel made the final selection of the contractor to execute this project.

The contract was governed by the DRBA's Standard Specification and the Delaware DOT Standard Specification. Proposers who had made the DRBA's Tender Short List were invited to submit competitive sealed proposals by August 28, 2015 and the contract for $33,590,000 was awarded to American Bridge (AB) on September 25, 2015.

American Bridge's main subcontractors are Safespan, Ammann and Whitney, Munters and Optimum Controls Corporation (OCC).

American Bridge received notice to proceed on February 16, 2016 and on-site construction work commenced on May 2, 2016. AECOM is also providing on-site construction management and construction phase engineering and design support services on behalf of the DRBA.

2   DESIGN OF THE DEHUMIDIFICATION SYSTEM

2.1   *System elements*

The cable dehumidification system designed for the Delaware Memorial Bridge has three main elements – plant and equipment, a control and monitoring system and an air tight cable wrapping and sealing system.

2.2   *Plant and equipment*

An individual plant room is constructed in each of the four anchorages. Each plant room contains a desiccant dehumidifier that draws air from the atmosphere through filters and fills the plant room with dried air. Two fans process the air via HEPA filters and blow dry air from the plant room through air pipes along the deck and up the suspenders to corresponding injection points on the main cables. Each fan supplies a pair of injection points at mid side span or main span quarter point locations. See Figure 2 below. The injection points are located mid-way between the exhaust points to inherently balance air flow through the cables. Plant room requirements in the contract documents were developed into fully detailed and functioning plant rooms by AB's dehumidification subcontractor Munters.

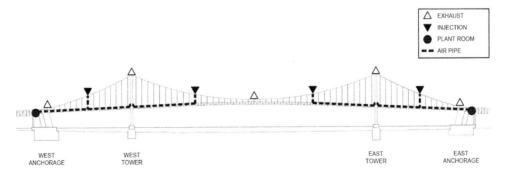

Figure 2.   Elevation showing plant room, air pipe and injection/exhaust point locations.

2.3   *Control and monitoring*

A supervisory control and data acquisition (SCADA) system is used to gather and analyze real time monitoring data. Information is gathered by the SCADA system from sensors at the injection and exhaust points on the cables, together with performance monitoring data from the dehumidifier controllers within the plant rooms and is transmitted over fiber optic links to central SCADA servers in the DRBA's offices. The run-time and historian servers give access to the recorded data and the control system over the Internet from any connected location. Control and monitoring system requirements in the contract documents were developed into a fully detailed and functioning SCADA system by AB's control system subcontractor OCC.

During the initial drying period, the system will be set to run at maximum capacity to dry the cables as quickly as possible. In the longer term the SCADA system will be used to monitor and adjust plant and equipment settings, so as to reduce energy usage and maximize long term efficiency. This will be accomplished by adjusting injection pressures, air flows and plant room humidity settings to maintain steady state non-corrosive conditions within the cables.

2.4   *Cable wrapping and sealing*

High quality rate cable wrapping and sealing is an especially important element of the cable dehumidification system as it directly affects the overall performance and efficiency of the system. The wrapping and sealing materials must be as air tight and water tight as possible, which is particularly difficult to achieve at the cable bands. DS Brown Cableguard™, a heat sealed elastomeric wrapping system, was specified for wrapping the cables; in conjunction with highly durable air tight, weather resistant end seals at cable bands. The cable bands are sealed using highly durable sealing details originally developed and used on systems in the UK (Cocksedge & Bulmer 2009; Cocksedge et al 2011). These sealing details have been shown to remain in good condition and reliably minimize water ingress and air leakage over many years. Water ingress would have a negative effect on system performance and could lead to the corrosion process within the cables being inadequately suppressed. Excessive air leakage would cause excess operation of the dehumidification plant and consequently, increased energy usage.

3   DESCRIPTION OF THE WORKS ON SITE

3.1   *Installation of the access system*

A temporary main cable access platform (MCAP) was designed and is being installed by Safespan to provide access to the main cables of Structure 1 and 2. The MCAP is being constructed beneath all four cables, full-length, in four separate field-segments per cable. The MCAP is comprised of galvanized steel deck panels that are attached and supported by wire rope support cables. The wire rope support cables run continuously from tower to anchorage and are intermediately supported by A-frame support brackets that attach to the cable band every 51 feet. The low end of each segment's wire rope support cables is tied back to the stiffening truss and the high end is connected to a custom HSS tube assemblage that is connected to the top of each tower with redundant basket cables.

To prevent uplift of each portion of the platform, wire rope vertical tie-down cables are installed from the wire rope support cables to the deck hand railing at third points. Several improvements have been made to the details of the vertical tie downs during construction to ensure connecting hardware does not loosen under wind loading.

Hand railing and kick plates are being installed along with debris mesh and anti-skid measures to create a workspace that complies with OSHA requirements, and does not require fall arrest protection for standard activities. Details are shown in Figure 3.

Figure 3.   Cable access being erected (left); completed cable access platform (right).

3.2   *Site trials of wrapping*

From the work initially carried out on the Forth Road Bridge, then adopted on Severn, Humber and Chesapeake Bay Bridges, AECOM prescribed contractual requirements for the contractor to demonstrate competency in the cable wrapping and sealing work on mock-cable test rigs before proceeding to production work on the cables (Figure 4) (Beabes et al 2016). The test rig demonstration required a pressure test to be achieved on the finished wrap and seals to demonstrate that the work had been performed to a high quality standard. The test required a target pressure be maintained for a 3-hour period with an allowable pressure loss prescribed in the specifications. The wrapping and sealing work was inspected for air leaks under pressure using soapy water (Figure 4). The purpose of the soapy water test was to inform the wrapping and sealing crews of areas which require further improvements to achieve an air-tight system. American Bridge elected to construct two mock test rigs to allow multiple crews to train on different aspects of the work simultaneously. Under close supervision and training from experienced team members, AB demonstrated proficiency in the wrapping and sealing work and passed on the air test on the first attempt.

Once the contractor successfully demonstrated the ability to wrap the cable and seal the cable bands on the test rig, contractual provisions required two experimental panels on one of the structures to be wrapped and heat sealed to the satisfaction of the Engineer prior to full wrapping production. The purpose of the experimental panels was to demonstrate that the contractor could replicate the performance witnessed on the test rig, while adapting to the actual cable conditions. This was particularly important for the heat-sealing of the wrap, as field conditions such as ambient temperature and wind affected the heating cycles required for proper heating-sealing of the cable wrap. The experimental panels also allowed the contractor to program for the production work.

3.3   *Wrapping, caulking*

Following the successful completion of the aforementioned trials, the contractor was able to commence production wrapping of the main cable and sealing of the cable band joints. To prepare the main cable for wrapping, workers cleaned and removed any sharp points in accordance with SSPC SP1 - Society for the Protective Coatings Standard SP, Solvent Cleaning (August 2106) and any loose material in accordance with SSPC SP2 - Society for the Protective Coatings Standard SP, Hand Tool Cleaning (November 2004). Existing caulking of the cable band was removed from longitudinal joints and riglets, and surfaces to receive caulking were cleaned in accordance with SSPC SP2.

Figure 4.   Trial wrapping on test rig (left); soapy water test on trial wrapping (right).

Figure 5.   Skewmaster used to wrap cableguard on main cable (left); caulking on cable bands (right).

Elastomeric cable wrapping was applied to the main cable using a purpose-made wrapper called a "Skewmaster" (Figure 5) and heated to fuse the multiple layers of overlapped seams and shrink the wrap against the underlying cable. Local heating of the wrap at edges and at roll ends was performed to ensure bonding was achieved. End seals were installed between the cable band riglets and wrapping material to inhibit air leakage. Longitudinal cable band joints and riglets were caulked using a fast cure adhesive sealant (Figure 5). A high quality of workmanship was required to ensure tight seals were achieved.

Finally the top walking surface of the wrapped cable received an anti-skid walking surface (paint with rubber granules) and the cable bands received a three coat noxyde protective system. The anti-skid surface will allow for safe access during future maintenance of the dehumidification system and periodic cable inspections.

3.4   Cable band bolt re-tensioning

Cable band bolts form a small but vital part of a suspension bridge. The suspenders support the entire deck load and are connected to the main cable via the cable bands (Alampalli & Moreau 2016). The cable bands on both Delaware Memorial Bridge structures are of typical construction being in two equal steel castings that are bolted together with cable band bolts. The suspenders loop over the cable bands and are located within grooves in the castings. A typical cable band on Structure 1 is shown in Figure 6.

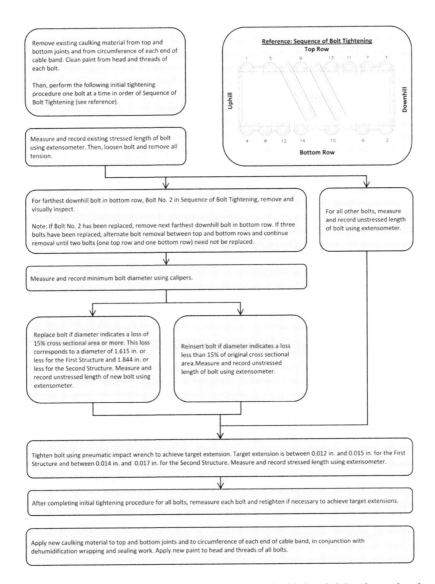

Figure 6.   Typical cable band on structure 1 and sequence of cable band tightening and replacement protocol.

The primary function of the cable band bolts is to generate sufficient friction to prevent slippage of the bands down the cable under loads applied by the suspenders. This friction is developed by pre-tensioning the cable band bolts. The slip coefficient between the wires and the inside face of the cable bands is assumed to be 0.3 on Delaware and the Factor of Safety against slippage at the critical panel point is assumed to be around 2.5.

However, over the service life of the bridge, the cable band bolts will relax and lose tension over time due to the continual slow compaction of the cables. There has been some discussion over how frequently the tension in the bolts should be checked and intervals of five, ten and twenty years have been mentioned. Whatever the interval chosen, there is little doubt that they do require to be checked a number of times over the service life of the bridge.

Given that the MCAP was going to provide access to all the cable bands, the DRBA determined that all the cable band bolts on both the Structure 1 and 2 should be checked and re-tensioned as part of the dehumidification contract. In addition, a check on the condition of the existing bolts is being carried out. The bolt in the bottom row at the low end of each cable band is being removed firstly, and examined. If the bolt had lost more than 15% of its cross sectional area, then it is replaced. The sequence of checking and replacement is shown in the flow diagram in Figure 6.

The load in the cable band bolts is being determined indirectly by measuring extension. It should be noted that the existing load in each bolt needs to be released in turn and the length of the bolt at zero load measured to establish a reference length at a known load (zero). Each bolt is then being re-tensioned, incrementally, in a pre-set sequence until the required tensile load in the bolt is achieved. The extension method assumes that all the bolts have the same value of Young's Modulus and a standard value of 29×10^6 psi was adopted. There will be a variance between bolts as they will have been manufactured from different steel batches but as the loads being applied are well within the elastic range, this is not a practical concern. It is also assumed that temperature between de-tensioning and re-tensioning a bolt does not vary.

A rigid frame extensometer with a dial gauge or electronic micrometer has been found to be best suited for measuring extension. Ultrasonic equipment can be used but the results have been found to be unsatisfactory on some bridges. The extensometers used for the construction of the original structures were refurbished and re-calibrated and are being used on this contract. Figure 7 shows a cable band bolt on Structure 1 being tightened as well as the extension of the bolt being measured via the rigid frame extensometer.

Although the Structure 1 and 2 have the same spans, at the time of design the deck on Structure 2 was heavier. Consequently, the cables on Structure 2 have a larger diameter than on Structure 1, and the suspender ropes are larger. The cable band bolts on Structure 2 required to be pre-tensioned to a higher load than Structure 1 as a result of the higher deck loading. As a consequence, the cable band bolts on Structure 2 have a larger diameter than those on Structure 1. The details of each structure, with the loads applied, are given in Table 1.

The component of the suspender loads acting parallel to the slope of the cable is a function of the slope and therefore, the cable band bolt loads could be varied at each cable band. In practice, at the cable bands in each structure, the same tensile load is applied to each bolt.

Figure 7.   Cable band bolt tightening (left); Measuring bolt extension (right).

Table 1.   Details of cable band bolts.

Description	Dia. in	Overall length ft-in	Specification	Yield strength Ksi	UTS ksi	Applied tensile load (approx.) kips	Extension in	Working stress ksi
Structure 1 orig. bolts	1¾	1-6⁵⁄₃₂	Not known	Not known	Not known	Not known	Not known	Not known
Structure 1 repl. bolts	1¾	1-5	ASTM A354 grade BC	109	125	64.6	0.012 to 0.015	26.8 (± 10%)
Structure 2 orig. bolts	2	1-10¹⁄₃₂	ASTM A354 grade BB	83	105	Not known	Not known	Not known
Structure 2 repl. bolts	2	1-9	ASTM A354 grade BC	109	125	84.2	0.014 to 0.017	26.8 (± 10%)

3.5   *Suspender rope replacement*

Some of the suspenders at socket level and the stiffeners around the suspender to deck con-nection on Structure 2 were exhibiting signs of significant corrosion. Therefore, the DRBA has included the replacement of 29 suspender sets in American Bridge's contract to dehu-midify the cables. Prior to NTP, three suspender rope sets were replaced in advance under a separate contract. The DRBA has appointed HNTB to design the new suspenders and AECOM carried out the construction support and inspection of the work as part of the contract supervision.

Replacing the suspenders is more complex when there is a main cable access system in place. However, by careful programming and collaborative working between all parties, American Bridge was able to progress the suspender replacement ahead of the cable access platform installation.

As lane closures were available for specific periods, the contractor was able to remove a set of suspenders and allow load to be transferred to the adjacent suspenders. There was a need to have some flexibility to allow a longer time period to remove and replace the existing stiff-eners at the deck socket connection. Therefore, American Bridge used a temporary jacking frame detail that transferred load from the deck, back into the ropes, above the level of the handrail. The ropes below the frame were de-tensioned, allowing the suspender sockets to be pulled clear of the deck. This arrangement meant that work could continue on the deck detail while the adjacent lane was opened. This negated the need to use temporary suspenders.

This suspender replacement work has now been completed and, in a second phase, 30 sus-penders have been added to the original contract for replacement.

3.6   *Plant room/plenum chamber testing off site (FAT)*

To facilitate the up-start of the plant rooms and verify the integration of the plant equip-ment with the control systems, the plant rooms/plenum chambers and control system were required to be fully assembled in the shop and functional trials witnessed by AECOM and the DRBA (Figure 8). Once the plant rooms were deemed operational through the shop trial process, the contractor was then allowed to install the plant rooms on the bridges. To verify the accuracy of communication between the network boxes and sensors of the SCADA sys-tem, factory acceptance testing (FAT) was required on fully assembled and wired network boxes at the vendor's site. Wiring between the control system and all network boxes mirrored the final design to be installed on the bridges. Testing was carried out to ensure signals were correctly communicated from the network box to the controls system. Additionally, alarm logic between the plant room controls software and the system software were verified. Once the SCADA and controls were deemed operational, the contractor was then allowed to install the various components on the bridges.

Figure 8.   Plant room/Plenum chamber external (left); Plant room/Plenum chamber internal (right).

4   CONCLUSIONS

After an internal inspection found corrosion within the main cables of the Delaware Memorial Bridge, the DRBA examined options for ensuring that the deterioration would not continue. The Authority concluded that dehumidification of the cables was the method that offered the promise to preserve and extend the service life of the cables. Although dehumidification systems have now been fitted to the main cables of around 20% of the global suspension bridge inventory, this is only the second application of this technology in the USA. However, over the next five years a further five major US suspension bridges are expected to have dehumidification systems retro-fitted to the main cables.

The work on the Delaware Memorial Bridges is continuing. The project is multi-disciplinary, complex and requires rigorous planning, scheduling and coordination. As on most major bridges, the environment is challenging and the safety of those working on the project and the safety of bridge users is a priority. In addition, there are operational challenges as minimizing disruption to traffic and limiting lane closures is also a priority for the client.

The DRBA as client is taking the lead role in managing the project and this pro-active approach by the client ensures that there is an over-arching view of all the multi-faceted parts of the project and other contracts being executed on the bridges.

Work is due to be completed in the fall of 2018 and there will be a two year monitoring and maintenance period following the successful commissioning of the system.

The Authors would like to thank all those who are working in an often difficult environment to bring this challenging project to completion safely and on time.

REFERENCES

Alampalli, S. & Moreau, W.J. (2016). "Inspection, Evaluation and Maintenance of Suspension Bridges." CRC Press, Boca Raton, Florida.

Beabes, S., Waldvogel, P., and Bulmer, M. (2016). "Maryland's Bay Bridge – The First Main Cable Dehumidification Project in North America." The 2016 International Bridge Conference Proceedings.

Cocksedge, C.P.E. & Bulmer, M.J. (2009). "Extending the life of the main cables of two major UK suspension bridges through dehumidification." *Bridge Structures*, 5:4, 159–172.

Cocksedge, C.P.E., Bulmer, M.J., Hill, P.G., & Cooper, J.R. (2011). "Humber Bridge Main Cable Dehumidification and Acoustic Monitoring – The World's Largest Retrofitted Systems." *Bridge Structures* 7, 103–114.

SSPC-SP1 (August 2106), Society for the Protective Coatings Standard SP, Solvent Cleaning.

SSPC-SP2 (Nov. 2004), Society for the Protective Coatings Standard SP, Hand Tool Cleaning.

TRB (Transportation Research Board). (2004). *NCHRP Report 534 Guidelines for Inspection and Strength Evaluation of Suspension Bridge Parallel Wire Cables.* Washington, D.C., USA.

Chapter 4

M48 Severn Bridge—managing the main cables since 2005

C.P.E. Cocksedge, B. Urbans & S.A. Baron
AECOM Ltd., UK

M. Maynard
Highways England, Bristol, UK

A.P. Burt
Severn River Crossing Plc, Bristol, UK

ABSTRACT: When the Severn Bridge opened to traffic 50 years ago, it was a world-lead-ing design with a very economic design for the 3240 ft. main span. In 2005 an internal cable inspection was carried out which revealed unexpectedly poor results in terms of corrosion, broken wires and strength loss. This prompted the implementation of a series of measures, starting with acoustic monitoring and followed by a cable dehumidification system. Both these systems continue to operate today, preserving the cables and providing ongoing data on their condition. Two further internal cable inspections in 2010 and 2016 have revealed that the deterioration has now slowed to a minimal rate.

This paper describes investigation and rehabilitation work carried out since 2005. It includes some special investigations into low stress/high cycle fatigue testing of cracked wires and friction tests to refine the evaluation of broken wire redevelopment length. It seeks to provide lessons learned and guidance for suspension bridge owners on how to deal with simi-larly deteriorated cables.

1   INTRODUCTION

The M48 Severn Bridge was a highly innovative design with the first use of a streamlined steel box girder deck in a suspension bridge (Figure 1). This allowed significant weight savings to be made throughout the whole structural system. With a main span of 3240 ft. (988 m), it carries a dual two lane motorway plus separate footway/cycle tracks, and each cable com-prises 8322 high tensile galvanized steel wires 0.196 in. (4.98 mm) diameter. The cables were installed using aerial spinning and after compacting they were protected in a traditional man-ner with red lead paste, wrapping wire and a conventional paint system.

This bridge opened to traffic in 1966 (Roberts et al., 1968). It soon became very heavily used, and concerns over potential extreme loading scenarios in the 1980s led to strengthening of the towers and other components (Flint, 1992). A second parallel crossing was opened in 1996, located some 3 miles to the south, and has a 456 m (1496 ft.) cable stayed bridge over the navi-gation channel with one mile long post-tensioned concrete approach viaducts on either side.

The new crossing became the main designated route across the River Severn (M4), and the original bridge remains on a lesser used, but still important, route (M48). The new bridge was procured as part of a Concession Agreement between Severn River Crossing Plc (SRC) and UK Highways England (HE), with the Government's Representative (GR) having an over-seeing role for HE. SRC employ a term maintenance contractor to carry out maintenance and inspection on both old and new crossings, however, some larger and specialist projects are let to external contractors.

Figure 1.   M48 Severn Bridge.

The bridge is located in an exposed location and receives the full force of weather systems from the Atlantic. Over the life of the bridge regular inspection and maintenance of the main cable has been undertaken, which has included periodic repainting. The regular external inspections carried out showed no significant deterioration. The cable bands are unusual in that they are split horizontally and have vertical bolts. To provide drainage of water from the cable, a small groove was provided in the lower half, however, these are inadequate and water is channeled to the cable low points, where varying quantities have been collected over the years.

Internal inspection of cables in the US prompted the owners of the Forth Road Bridge to carry out an inspection, which revealed an unexpected level of deterioration. This acted as a catalyst for SRC/HE to carry out a similar inspection of the main cables of the M48 Severn Bridge. As no UK standards were available it was decided to make use of the recommendations in the NCHRP 534 Guidelines (NCHRP, 2004) for investigating and evaluating such cables.

2   FIRST INTERNAL CABLE INSPECTION

2.1   *Planning*

AECOM was appointed to act as consultants for the inspection work in July 2005, assisted by Ron Mayrbaurl, one of the authors of NCHRP 534. Initially a scoping study was carried out including collecting information on original construction and subsequent maintenance work, followed by cable walk inspections to determine which panels should be internally inspected. The amount of water collected at mid-main span over the years suggested this should be a key area to be inspected. Balancing the budget available and the fact this was a first inspection, it was decided to inspect seven panels, four at low level and three at high level. The panels are nominally 60 ft. (18.3 m) long.

2.2   *Inspection*

The first panel inspected was at the lower end of one of the side spans and was found to be in a reasonable condition. Surprisingly the red lead paste was still pliable, as it was expected that it would have completely dried up. Wires were classified in accordance with NCHRP 534 – Stage 1 (no zinc corrosion), Stage 2 (zinc corrosion), Stage 3 (less than 25–30% steel corrosion) and Stage 4 (greater than 25–30% steel corrosion).

The second panel opened at mid main span was completely different. Here, a large number of broken wires were found, amounting to about 2% of the wires (Figure 2). Closer examination showed that many of these were Stage 3 wires, and not the normally expected Stage 4. Another mid span panel on the other cable also had a large number of broken wires, again many at Stage 3. The high level panels were found to be in a much better condition. The findings led to an additional two panels being opened on one of the cables at mid span.

2.3   *Wire testing*

The cable was originally designed with a factor of safety against tensile failure of 2.22. The wire had a specified tensile strength of 1544 to 1776 N/mm^2 (224 to 258 ksi) and a proof stress of 1158 N/mm^2 (168 ksi). The wire used on the recently completed Forth Road Bridge had been found to have an average proof stress of 1204 N/mm^2 (175 ksi) and the designers reasoned that a working stress of 1204/1.7 or about 694 N/mm^2 (101 ksi) could be justified.

Following removal of sample wires from the cable, tensile and other tests were carried out to provide statistical distributions of wire population.

Of the broken wires found during the visual inspections, it was apparent that many of these had only a relatively small amount of surface corrosion (Stage 3), and this was reinforced with the finding of cracked Stage 3 wires during the tensile testing program. Experience from the US had suggested that whilst cracked or broken Stage 3 wires are occasionally found, it is generally accepted that wire cracks were usually associated with corrosion Stage 4. Given the unusually high number of cracked and broken stage 3 wires, a further series of more detailed laboratory examinations were carried out to establish if there was a metallurgical reason for this behavior.

No specific sample wires had been removed from the cable for these tests and so broken wires removed from the cable were used. Most of these were at Stage 3, but there were some lengths of Stage 2 and 4 which were used to provide comparisons, plus a length of new wire manufactured for use as replacement for broken and sample wires.

The tests proceeded in a number of phases. The first included general items such as chemical composition, testing for contaminants, hardness measurements, and careful examination

Figure 2.   Broken wires.

of surface after cleaning for imperfections/anomalies etc. Nothing unusual was found in these tests, and in particular no high levels of contaminants were found.

The second series comprised mechanical tests in the form of tensile and tensile fatigue tests. The tensile tests included some at low temperatures (down to −50°C). All tests were carried out using specimens from the same wire sample. Interestingly, the yield stress clearly increased at lower temperatures, suggesting the steel was becoming more resistant to plastic deformation, however, failure occurred at lower strains with the fracture surfaces exhibiting a more brittle failure mechanism (Figure 3).

The intention of the fatigue testing was to determine if there was any unusual or unexpected behavior of the wire, rather than derive a stress-endurance relationship. A high constant stress range was selected to promote failure at a relatively low number of cycles. This was much higher than that experienced by the in-service cables. All fracture surfaces were examined using optical and scanning electron microscopes.

Metallographic examinations formed the third series of tests. These included checking for surface cracks in the test specimens using dye penetrant, taking sections to determine grain structure and mode of cracking and finally submerging sections in liquid nitrogen followed by breaking by hand and sectional examination.

The conclusions from these additional tests may be summarized as follows.

- The wire has a fine grained directional structure with no evidence of metallurgical defects.
- Chemical analyses showed a similar composition for all wires and no anomalies were found.
- Hardness measurements were similar for all wires with no "soft" spots.
- The zinc layer was relatively consistent and uniform and the adhesion was good.
- The crack forming mechanism follows the same process identified in other suspension bridges.
- The cracks tend to be relatively straight with little evidence of branched networks.
- The cracking will only occur within the susceptible areas, i.e. those with an adequate moisture level to cause corrosion.

Therefore it was concluded that there were no underlying metallurgical reasons for the large number of cracked and broken Stage 3 wires.

2.4   *Strength evaluation*

Strength calculations were carried out using the NCHRP Brittle Wire model and produced an appreciable strength loss. This prompted a more rigorous approach to assessing the safety

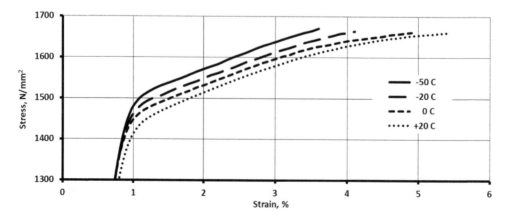

Figure 3.   Stress-strain relationship for low temperature tests.

of the cables using limit state techniques, and so bringing the assessment in line with the usual UK approach to bridge design (Young et al., 2008). The assessment showed that the bridge was safe to remain operational, although heavy trucks were restricted to the outer (slow) lanes and a traffic management strategy was developed in case the newer crossing was closed and all traffic diverted onto the suspension bridge.

3   ACOUSTIC MONITORING

As this was a first inspection, only a relatively small proportion of the cables were subject to detailed intrusive inspection. It could not of course be assumed that the worst panel had been found. Further, the only data on the long term deterioration of the cables known with certainty are the two points comprising the time of completion and the time of first inspection some 40 years later.

The rate of deterioration of a cable is essentially the rate at which wires lose their strength and finally break. The last stage can be detected using acoustic monitoring techniques, which have been used with increasing confidence on other suspension bridges.

SRC had considered installing an acoustic monitoring system even before a cable inspection was contemplated. However, the number of broken wires uncovered at mid-main span confirmed the necessity of such a system. Mindful of the time required to procure and implement a full system, it was recommended that a temporary local system be installed at mid main span, to be followed by a full system at a later date. The Temporary System was supplied by Advitam using Pure Technologies equipment and was set to monitor just six panels on the downstream cable at mid-span. The Temporary System was operational from November 2006 to November 2010. The system identified about 120 wire break events, of which over half were concentrated in one single panel.

The contract for the full system was tendered in the fall of 2006 on a quality and price basis, and was awarded to Physical Acoustics Ltd (now Mistras). Design, installation and commissioning were completed in January 2008 followed by an initial period of 5 years of monitoring and maintenance.

4   CABLE DEHUMIDIFICATION

The condition of the cable gave cause for concern and discussions were held on potential options for the future. SRC and HE were aware that AECOM had designed the cable dehumidification system being installed on the Forth Road Bridge, and they determined to appoint AECOM to carry out a study to assess the feasibility of installing a similar system at Severn. The main tower saddles had previously been strengthened against bursting and so the layout dictated that air should be exhausted here as sealing would be problematic. The main span has three injection points and four exhaust points, and the side spans each have a single centrally placed injection point. The Severn Bridge has a box girder which therefore provided a convenient location for the plant rooms. To protect the bridge in the extremely unlikely event of a fire, the plant rooms were constructed as an internal box structure using fire resistant materials.

Unfortunately the timing of the feasibility study was such that if the system were tendered as recommended, by the time a contractor had been appointed and designed and fabricated access platforms, most of the potential good weather period would have been lost delaying the project by nearly a year. An idea was developed to split the installation into two contracts. The first would tackle the area of the cable believed to be in the worst condition at mid main span. To enable an early start to be made the contract was offered to the Term Maintenance Contractor for the bridge, Laing O'Rourke. A cost reimbursable type of contract was adopted, which was measured against a pre-determined target cost. This was known as the Advance Works (AW) Contract.

Whilst the arrangement of the injection and exhaust points suited the complete system, it did not suit the proposed partial system. However, it was realized that as the design of the injection and exhaust sleeves was similar and could readily be adapted from one to the other. In the AW system an exhaust is placed at mid-span, with injection points about 150 m away and a further set of exhaust points 150 m beyond. An advantage of this arrangement was that it would assist any water in the cable run down to the low point, where a drain point had been provided in the sleeve for this purpose. The layout of the partial and full systems is shown in Figure 4.

The main effort required in the installation of a cable dehumidification system is making the cable as airtight as possible. This requires over-wrapping the cable with a proprietary material, with re-caulking of the cable bands with airtight sealants. Access is therefore required to the full length of the cables, and the contractor's method of achieving this is critical to the success of the project. For the AW contract, access to the cable was limited to the lower levels above the deck and the cable wrapping was installed from proprietary self-powered access cradles (Figure 5). This method suited UK working practice where the labor is non-unionized and operatives can carry out work of several trades.

Like most suspension bridge cables, it was not perfectly circular and hence the stiff cylindrical heating blankets did not evenly bond the two layers of the wrap together, and often required reworking with hand held heat guns and rollers. A sub-contractor working on the wrapping came up with an innovative solution to this problem using an inflatable blanket that naturally assumed the shape of the cable. This proved to form a much more consistent and uniform bond than the rigid blanket, and the concept is now universally used.

In spite of best intentions, poor summer weather followed by a poor fall led to delays in installing the wrapping and by December 2007 only half of the wrapping had been completed, which was the central section between injection points. However, the plant room had been commissioned and following a brief test run just prior to Christmas 2007, it started injecting

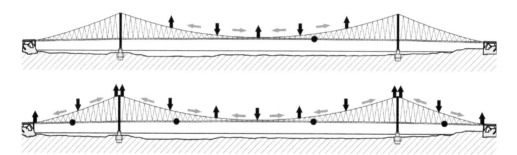

Figure 4.   Layout of partial (top) and full (bottom) dehumidification systems.

Figure 5.   Access systems for advance works contract (left) and full system contract (right).

dry air continuously from February 2008. The contractor remobilized his wrapping team in April 2008 and the remainder was completed by July 2008.

Having made a start using the AW Contract, attention turned to procuring the second (main) contract. As will be appreciated, the work to be done under the Full System (FS) Contract included wrapping the side span and upper sections of the main span cables, installing further plant rooms and converting the AW system to blow in the opposite direction. A target cost contract was devised which included a financial pain/gain element. There was an aim of completing the work by the end of 2008. C Spencer was the successful contractor.

Part of their offer included the use of three cable-walking gantries (Figure 5) which were considered to offer somewhat better protection against the weather than the cradles used on the AW Contract. This proved to be a wise decision as the weather during summer 2008 turned out to be the worst for many years. By the end of July 2008 there was an apparent risk of not completing all of the cable wrapping before winter, and at the suggestion of the AW contractor, who were just finishing their work, the wrapping of 10 panels was transferred between contracts. Again this proved to be a good choice and completion of wrapping was achieved in October 2008.

Commissioning of the three additional plant rooms was achieved in November 2008 and the changeover of the Advance Works system was executed to a carefully drawn up plan so as to minimize the period of system down time (Cocksedge et al, 2010).

As an additional measure to protect the cable wires, HE decided to use the cable dehumidification airflow to introduce a vapor phase corrosion inhibitor (VpCI). These work by forming a thin molecular protective layer on the metal surface and are used in many fields including for electronics and military equipment protection. It is not believed to have been used to protect a suspension bridge cable, and so a series of tests were carried out. These included examining if it would affect the inter-wire friction (important for redeveloping the strength of a broken wire), and employed a jig with one fixed wire and one moving wire at right angles. Friction was measured with the wires dry, wet with water, coated with VpCI and with water plus VpCI, and it was found to consistently increase in the presence of VpCI, a satisfactory result. Tests were also carried out on the cable wrapping material and sealants. It was found that there was a marginal change in tensile strength and failure strain, but these were not considered significant. Finally a relative humidity/temperature sensor was exposed to VpCI, but there was no difference in output to a reference probe. VpCI was introduced into the dehumidified air supply from late 2009, almost a year after the dehumidification system had been fully commissioned. SRC continue to replace the VpCI sachets three times a year.

In addition to the normal relative humidity/temperature, flow and pressure sensors required for system operation and monitoring, corrosion sensors were fitted in the exhaust sleeves to provide numerical data on the likely internal corrosion rate as the humidity reduced during drying,

5   SECOND INTERNAL INSPECTION

The poor results of the first inspection, coupled with the data emerging from the Temporary Acoustic monitoring system gave rise to an impetus for a second internal inspection just 4 years after the first in 2010. The consultant role was tendered by SRC and AECOM was engaged to carry out this work, again working with Ron Mayrbaurl. There were three main aims for the inspection; firstly to provide more information on the overall condition of the cable by opening up more panels, secondly to provide information on the deterioration rate by re-opening previously inspected panels, and finally to confirm the performance of the acoustic monitoring systems.

A total of nine panels were selected, including two that had been previously inspected. These included three contiguous panels at mid-span on the downstream cable, and the contractor, C Spencer, erected a long scaffolding platform. Six panels were at low level and three at high level.

The inspection was carried out in a similar manner to the first inspection with broadly similar findings of many broken wires at mid-span and lesser deterioration at higher levels.

One unexpected discovery was the presence of mold on the underside of the dehumidification wrapping in a few of the panels. This was tested and found to be caused by common harmless mold spores probably deposited by rain just before wrapping. Further wires were removed for testing to increase the sample database and refine the proportion of cracked wires.

The calculated strengths were similar to those computed for the first inspection. Both of the panels re-inspected showed a very small reduction in theoretical strength.

This inspection generated two pieces of research, with the University of Manchester trialing a magnetoresistive sensor to detect the extent of corrosion inside the fully wrapped cable, and a series of low stress/high cycle fatigue tests.

5.1   *Low stress/high cycle fatigue tests*

The investigations to date had shown that there were a large number of cracked wires existing in the cable. The number of cracks is not known, but an estimate of the potential number can be assessed from the tensile tests of the intact sample wires. Of the 100 sixteen foot long Stage 3 and 4 wires tested in the first and second inspections, about 20% were found to contain at least one crack. The average crack depth was just less than 1 mm. Of the approximately 1100 broken wires ends examined after removal from the cable the average crack depth to cause fracture was just less than 2 mm.

The results from the acoustic monitoring system suggest that following dehumidification the rate of wire fracture has been dramatically reduced. However, the question remains as to what might happen to the cracked wires in the longer term, and an obvious subject of concern would be whether the cracks could grow as a result of fatigue caused by traffic crossing the bridge.

To address this concern, two investigations were carried out: the actual stress ranges in the cable wires were measured, and fatigue tests were undertaken on cracked wires.

For convenience one of the cables at mid-main span was strain gauged. Data was collected for a period of nearly two weeks with continuous sampling at 20 Hz. Strain gauges were attached to individual wires at the top, bottom and two sides of the cable. Adjacent to each strain gauge a thermocouple was mounted to permit temperature adjustments to be made, and another thermocouple was provided to record the ambient temperature. The data collected was sorted into "bins" of 0.5 N/mm^2 stress ranges, and grossed up to the equivalent of annual cycles. The data is shown in Figure 6. It is clear that the data splits into two distinct groups, towards the left the count is primarily due to vehicles and to the right it is due to diurnal temperature changes. It is interesting to note that the data from 1 to 6 N/mm^2 fits a straight line (on semi-log graph). A check was carried out using weigh-in-motion data collected over the same period, which showed generally good correlation.

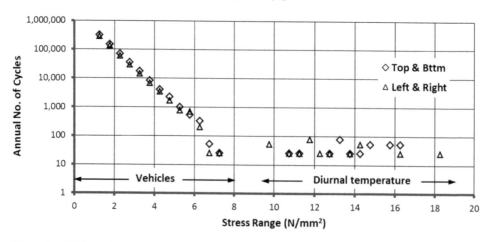

Figure 6.   Cable strain measurements.

In the laboratory it proved difficult to confirm the presence of any pre-existing cracks in the wires. The wires selected were broken wires, so it was known that they were susceptible to cracking. Initially a stereo microscope was used, paying particular attention to the inside of the cast curve, where it was more likely that cracks would be found. Dye penetrant was then tried, but none of the specimens showed evidence of transverse cracks. Finally, a micro focus radiography technique was tried which has a focal spot of about 1 mm, but it was thought that the cracks were too narrow to be resolved. There remained a concern that the wires did not contain cracks, so some short lengths were immersed in liquid nitrogen followed by shock loading to reveal any pre-existing cracks. Of the seven wires immersed three were found to contain cracks, thus confirming the presence of cracks.

The specimen wires were subject to 10 million cycles, commencing with a stress range of between E6 and E30 N/mm^2, around a static stress of 580 N/mm^2. These were deliberately chosen to be at the high end of the stress ranges normally produced by traffic to act as a worst case.

A total of 18 specimens were fatigue tested, but only two failed below 10 million cycles; these had pre-existing cracks about 0.9 mm and 1.45 mm deep. The surviving specimens were then tensile tested to failure and another two were found to have pre-existing cracks, although these were smaller at 0.2 mm and 0.75 mm. This meant that the majority of the test specimens unfortunately were not cracked. Figure 7 shows SEM images of the fracture surfaces of two wires with pre-existing cracks, one that propagated before failure and one that did not propagate (crack found after subsequent tensile testing).

However, it was possible to perform some analysis on the test results and draw some tentative conclusions. The wires failed in order of pre-existing crack size and it was possible to infer a likely crack depth/endurance curve (Figure 8). It is tentatively suggested that wires with cracks of less than 0.75 mm will not fail in fatigue at the stress ranges present in the cable. An analysis

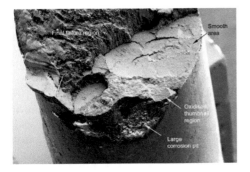 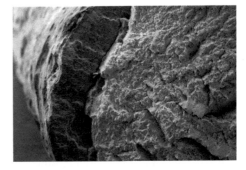

Figure 7. SEM images of fracture surfaces—left 0.9 mm pre-existing crack + fatigue growth; right 0.75 mm pre-existing crack—no fatigue growth.

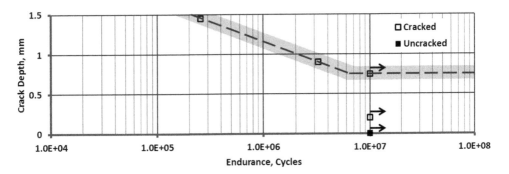

Figure 8. Possible crack depth—endurance relationship.

of the size of cracks found in broken wires and in tensile tested specimens suggests that there are very few wires present in the cable with crack depths greater than 0.75 to 1.0 mm, with reasons for this postulated in Mayrbaurl (2010). Given the very low level of stress fluctuation measured it is further tentatively concluded that even the few wires with crack depths greater than 0.75 to 1.0 mm will not fail through fatigue caused by normal traffic flow.

5.2   *Use of acoustic monitoring system*

When panels were unwrapped during the first inspection it was possible to hear some wires breaking. Of course it was very difficult to positively locate the breaks once the cable had been fully unwrapped and cleaned, so it was suggested that the acoustic monitoring system could be harnessed to provide an accurate break location.

In the event, it proved difficult for the full system to provide the location data as it struggled to cope with noise associated with cutting the wrapping wire. It should be noted that the system had been designed for continuous background monitoring rather than dealing with persistent extreme noise. Nevertheless, it was possible to find audible events through back-analysis of the data permanently recorded.

An attempt was made to carry out a blind test on one of the high level panels, but as one sensor had to be removed to accommodate the platform rigging, the wire cut did not register as the noise had attenuated such it was below the trigger level at the first sensor encountered.

6   THIRD INTERNAL INSPECTION

6.1   *Inspection*

SRC and HE had determined that a third cable inspection should be undertaken in 2016. This was in part to maintain the regular inspection cycle that was considered warranted from the poor findings of the earlier inspections, and to form part of the handover of the crossings back to the Government at the end of the concession in 2017/18. AECOM were again appointed to carry out the inspection.

As previously a Scoping Study was carried out to determine which panels should be inspected. The overall aims of the inspection were the same as the second inspection, to increase the knowledge of the condition of the cables by opening up more panels, to provide information on the deterioration rate of the cables by re-opening previously inspected panels and to confirm the performance of the acoustic monitoring and dehumidification systems. All relevant information was assembled together and it was apparent that mid-main span formed the area of greatest interest and panels were selected so that a continuous line of six panels would have been inspected on each cable, with the re-inspection of one panel previously inspected once and one previously inspected twice. Just two high level locations were selected to cover two previously uninspected quadrants, a decision strongly influenced by the increased risks of working at height above the carriageway.

Following a competitive tender, American Bridge International was appointed as contractor to facilitate the inspection. They programmed their work to start on one cable and subsequently move to the other cable. Access to the low level platforms at mid span was provided by scaffolding and one new high level access platform was designed and fabricated for use in the two high level locations. Somewhat fewer broken wires were encountered than in previous inspections, plus the contractor was asked to accelerate the work due to a planned closure of the Severn rail tunnel leading to increased traffic crossing the bridge and likely disruption. This meant the inspection was completed early.

The inspection procedure followed that adopted previously, wedging in the usual pattern and extracting sample wires for testing. The contractor had suggested using a hydraulically driven wedge to open the cable, which had been used on the Bosporus Bridge in Turkey, but the idea was eventually rejected. There were two additional facets to the investigation, namely to carry out tensile tests on long wire specimens, and an attempt to measure the friction generated by the wrapping wire.

6.2   Long wire tests

In addition to the normal tensile tests on 12 inch long specimens it was decided to carry out some tensile tests on long specimens. The purpose of carrying out these tests was two-fold, firstly to confirm if a longer wire behaved in a similar way to a standard short specimen and to assist in the detection of cracked wires. To facilitate this slightly longer wire samples were removed from the cable (7 m rather than 6 m), with the long specimen being taken from the same wire subject to the normal NCHRP tests. The test laboratory assembled a special rig to undertake the tests, with a 1.5 m long specimen held horizontally. A 1 m long extensometer was fitted to the wire which enabled strains to be measured up to failure. The arrangement of the test rig is shown in Figure 9.

The stress-strain curves produced were very similar to those generated from the standard specimens, as were the failure loads. Several of the wires were found to have pre-existing cracks, these corresponded to standard wire test findings except for one for which no cracks had been found in the standard specimen.

6.3   Use of acoustic monitoring system

As with the second inspection, early discussions were held with the acoustic monitoring supplier with the aim of getting the system to detect wire breaks during the course of the inspection. It was hoped that the upgrade would enable it to continue working even through the noisiest activities.

The system had been set up such that when a potential wire break was detected (as automatically classified) an email was distributed. Over the course of the inspection some 50,000 such emails were sent, the majority of these associated with wrapping wire removal and driving wedges into the cable. This presented a huge data analysis challenge. To assist with this analysis, a detailed record was taken of all inspection activities, including when wires were heard to break during removal of wire wrapping, so that the manual analysis tasks would focus on specific time periods and locations. The majority of audible wire breaks were located in the data.

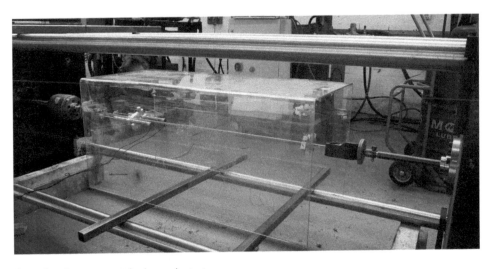

Figure 9.   Arrangement for long wire tests.

6.4   *Review of NCHRP as applied in the UK*

In view of the growing experience of internal cable inspections to NCHRP 534 Guidelines on the three major UK bridges, HE considered that it was an appropriate time to review the strict applicably of the guidelines to the UK bridges, particularly given that they all had acoustic monitoring and cable dehumidification systems fitted. Accordingly the HE convened a group of UK experts to have open discussions with the intention of gaining a consensus of opinion that might ultimately lead to the publication of a standard outlining how NCHRP 534 would be applied in the UK. This could include modifications, as necessary, to the Guidelines.

Three areas were initially considered for discussion, these were (1) extending the interval between inspections, (2) reducing the amount of each inspection, and (3) improving the calculation of remaining strength. It was considered that the interval between inspections could be increased up to 10 years provided the humidity inside the cable was keep low for the majority of time, few wire breaks detected, and there were no significant load changes. Regarding the extent of each inspection, this could be reduced to as little as one panel per cable at low level, unless the findings were worse. For improving the strength calculation there were different views on where the work of the group should concentrate its efforts, but initially it was agreed to examine the effective redevelopment length. The led to a series of tests being carried out, as described in the next section.

6.5   *Wire friction tests*

An important input into the NCHRP strength calculations is the effective development length of broken wires. Practical observations have suggested that this might be over-conservative given the finding of wires broken in several places, sometimes less than 500 mm apart, suggesting that having broken once there was tension rapidly regained to enable the wire to break again nearby. A series of tests was devised which were carried out during removal of the wire wrapping. In all cases a 1 m length of wrapping was removed in the middle of the panel so that two wires could be teased out and gently cut with a micro-disc cutter. Various 50 mm strips of wrapping were removed elsewhere to permit any wire slippage to be seen. Spray paint and scribe marks were made to help identify movement, see Figure 10.

In the first test, following cutting the wire, the wrapping wire was progressively removed from the center outwards, but no slippage was found. In the second test, following the wire cut, the wrapping wire was removed from the cable bands working inwards in steps of 1 m, and there was no slippage until the last 750 mm remaining was removed.

The third test was essentially a repeat of the first, but was on a panel previously inspected and was coated with zinc paste which had not dried. As soon as the wire was cut, slippage

Figure 10.   Wire cutting and wire slip measurement.

was immediately apparent along the whole length, suggesting that the zinc paste was acting as a lubricant. The fourth test was a repeat of the second, but with the last section of wrap removed in small increments. The four half-length wires behaved differently, with the major slip occurring when there was between 300 and 800 mm of wrapping remaining, which tied in well with the broken wire observations.

These tests provided a useful insight into the effect of wrapping wire, but of course it was only possible to use outer wires, and the presence of the dried red lead paste clearly acting as "glue". To provide some information on the behavior of internal wires, HE commissioned some tests on a small mock-up cable.

The mock-up cable comprised about 400 wires, which were almost all new apart from a number of corroded Stage 3 and 4 wires extracted from the Severn Bridge cable, which were arranged towards the center of the cable. The Stage 3 and 4 wires protruded from the end so they could be gripped individually and pulled to establish the frictional restraint. The bundle of wires was clamped a close intervals and the clamping force could be varied (Figure 11). Tests were carried out with the clamping bolts slack and with a torque of 10 and 25 Nm, with the highest value approximating to the equivalent force applied by the Severn Bridge wrapping wire. There was an inevitable spread of the individual results, but the average values followed a clear relationship, as shown in Figure 12. The Stage 4 wires had consistently greater friction values.

If these results are scaled up to the full cable it would appear that the actual effective redevelopment length is much less than calculated using NCHRP 534. For the Severn Bridge, the NCHRP effective development length is between 5 and 7 panels, but it could be as low as

Figure 11.   Mock-up cable friction test.

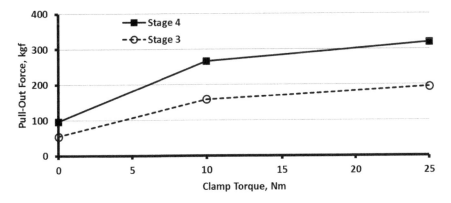

Figure 12.   Mock-up cable friction test results.

2 to 3 panels. For a deteriorated cable this would considerably improve the estimated cable strength. However, more work needs to be undertaken before it can be developed into clauses for incorporation into guidance for use in the UK.

7 LONG TERM PERFORMANCE OF ACOUSTIC MONITORING & DEHUMIDIFICATION SYSTEMS

7.1 *Acoustic monitoring system*

At the end of the initial 5 year period Mistras was approached with a view to extending the monitoring period for a further 5 years. To extend the life of the system a number of improvements were made including replacing the computer hardware and upgrading the monitoring software; removing and re-fixing each sensor with new adhesive with selective replacement as required; and replacement of some of the cables running down the suspender ropes.

The upgrade has improved its reliability and it now averages about 97% uptime, with the shortfall comprising planned and occasional reactive maintenance.

7.2 *Dehumidification system*

The dehumidification system has continued to perform well since 2008. Following the initial drying of the cables which achieved a reduction of relative humidity to 40% within 18 months, the system has maintained an average exhaust air humidity of about 30%. Regular maintenance of the plant is carried out which helps keep the amount of downtime to a minimum. Reliability of the sensors was improved following the fitting of protective cover boxes to reduce the impact of wind, rain and UV sunlight.

The system is controlled via SCADA and it can be viewed (and adjusted) on a web interface. Data is regularly downloaded and summary monthly reports are produced.

The effectiveness of the system is best illustrated in Figure 13, which plots the relative humidity at a typical exhaust with the cumulative number of wire breaks recorded by the acoustic monitoring system. As the amount of water has reduced inside the cable, the rate of wire breaks has also slowed. In the last 6 years there have been just 6 wire breaks identified.

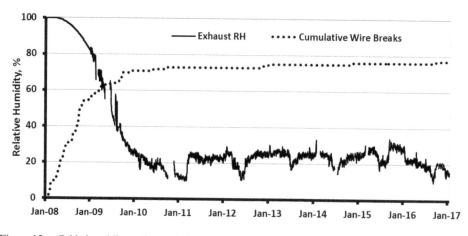

Figure 13.   Cable humidity and cumulative acoustic monitoring results.

8   ANCHORAGE DEHUMIDIFICATION IMPROVEMENTS

Dehumidification had been installed in the anchorages in the 1980s. Initially it was planned to dehumidify the entire void within the gravity chambers, but it was soon realized that moisture seeping in through the concrete walls was preventing a satisfactorily low humidity being achieved. SRC then installed shrouds over the splaying strands and small dehumidifier units were able to keep the relative humidity below 40%.

The control system was very simple, with the dehumidifiers being switched on and off by a humidistat, but the only way of checking the system was correctly functioning was to go into each of the anchorage chambers and carry out a physical check.

In 2015 SRC wanted to eliminate the need for the weekly check and AECOM were asked to review the existing system and propose an upgrade, which would include integrating it into the existing dehumidification SCADA system to allow control and monitoring from the same website as the cable dehumidification system.

The upgrade was procured as an extension to the third cable inspection contract. Each shroud dehumidification arrangement now includes a combined dehumidifier and condenser unit and circulation fan to improve system efficiency and air flow through the shroud. Jet diffusers distribute the dry air into the splayed strands at the bottom of the shroud. The new equipment is located in a more convenient location for maintenance. RHT sensors generate the RH set point, and this allows remote control of the RH setting and air flow via the website to maintain 40% RH within the shroud.

9   CONCLUSIONS

It is now twelve years since the first internal investigation commenced and much has been learned about the cable condition and the 8322 wires that make up each cable. Extensive laboratory testing concluded that the wire had been manufactured to a good standard. The reasons for the large number of cracked and broken wires are believed to be associated with a combination of a large amount of water retained inside the cable plus a somewhat higher stress from a low design safety factor.

The poor results from the first inspection dictated that urgent action was required. Fortunately the bridge owners appreciated the seriousness of the issues, realized that dehumidification offered the best solution, and committed to installing a system as soon as practicable. Analysis of data generated by the acoustic monitoring system, coupled with two further intrusive inspections have confirmed that the decision to install cable dehumidification had been very prudent.

ACKNOWLEDGEMENT

The views expressed in this paper are those of the authors alone. The authors would like to thank Severn River Crossing plc and the Highways England for agreeing to publication.

REFERENCES

Cocksedge, C., Hudson, T., Urbans, B., Baron, S., 2010, M48 Severn Bridge—Cable Inspection and Rehabilitation, ICE Proceedings, Bridge Engineering 163, December 2010.

Flint, A.R., 1992, Strengthening and Refurbishment of Severn Crossing, Part 2: Design. ICE Proceedings, Structures and Buildings, February 1992.

Mayrbaurl, R.M. 2010, The Strange Behavior of Cracked Wires in a Cable, *AISC SEI Congress*, Orlando, May 2010.

National Cooperative Highway Research Program (NCHRP), Report 534 – Guidelines for Inspection and Strength Evaluation of Suspension Bridge Parallel-Wire Cables, Transportation Research Board, 2004.

Roberts, Sir G, Gowring, G., Hardie, A., Hyatt, K., 1968. Severn Bridge, 1, Design and contract arrangements, 2, Foundations and substructure, 3, Fabrication and erection, Proceedings of Institution of Civil Engineers, Volume 41, September 1968.

Young, J., Lynch, M., Lambert, P., Fisher, J., 2008, Assessment of the Suspension Cables of the Severn Bridge, UK, 17th Congress of IABSE, Chicago, 17–19 September 2008.

Chapter 5

Superstructure replacement works for the Macdonald Suspension Bridge, Canada

D. Radojevic & K.F. Kirkwood
COWI Bridge North America, North Vancouver, BC, Canada

ABSTRACT: The Angus L. Macdonald Bridge, completed in 1955, connects Dartmouth and Halifax, NS, Canada. The suspension bridge is 762 m long, with a 441 m long main span. The bridge deck reached the end of its functional life and was replaced segment-by-segment during bridge weekend closures, with traffic using the bridge during the weekdays. New deck segments were fully prefabricated, including an initial layer of wearing surface, and erected in a way that allowed traffic to use the bridge immediately following replacement. Fabrication began in early 2015 and erection in late 2015. The last segment was replaced in 2017 February. Concurrently with the segment replacement, the entire deck was raised to its final profile (3 m at midspan) to allow for the increased shipping clearance. The hangers were then replaced. This paper describes some aspects of the superstructure fabrication, segment erection, and challenges encountered during erection.

1 INTRODUCTION

The Angus L. Macdonald (ALM) Bridge, shown in Figure 1 and Figure 2, opened in 1955 to connect Dartmouth and downtown Halifax, Nova Scotia, Canada. The suspension bridge is 762 m long and the overall length, including approaches, is 1347 m. The main span is 441 m long. The deck of the suspended spans had reached the end of its functional life and the Contractor (American Bridge Canada Company, ABCC) replaced it segment-by-segment. Cross-sections of the old (pre-replacement) and new superstructures are shown in Figure 3.

Buckland & Taylor (B&T), now COWI Bridge North America (COWI), assisted by local subconsultant Harbourside Engineering Consultants (HEC), completed the design of the bridge and its erection sequences in 2014. Subsequently, the Owner, Halifax Harbour Bridges (HHB), awarded the contract for construction to ABCC.

Figure 1.   Angus L. Macdonald Bridge—pre-superstructure replacement photo.

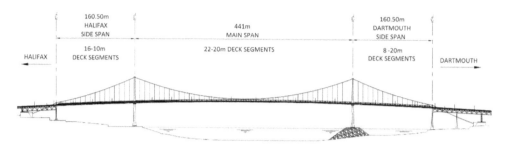

Figure 2.   Bridge general arrangement.

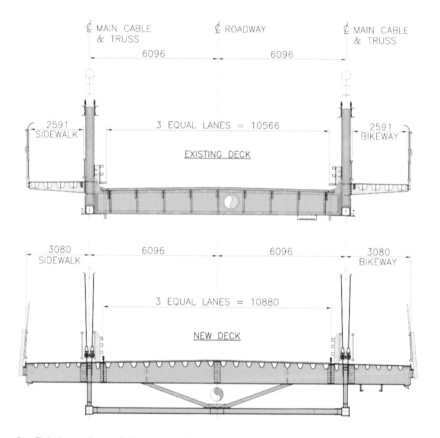

Figure 3.   Existing and new deck cross-sections.

Fabrication of deck segments began in 2015 April, and was completed in 2016 August. Deck replacement work started in 2015 October and the contractor performed it during night closures (with traffic running during the day) and full weekend closures. Because the public was using the bridge during the daytime, the contractor was allowed to close the bridge at 7:00 pm, but was required to open it no later than 5:30 am following the night of segment replacement.

This paper describes main fabrication and erection aspects of the project. Background information regarding the general scope of the project is in Radojevic & Kirkwood (2016) and Radojevic et al. (2014).

2   FABRICATION

2.1   *General*

Fabrication was completed by Cherubini Metal Works (CMW) of Dartmouth, Nova Scotia, in two separate fabrication shops. One shop fabricated the trusses. The other shop fabricated the deck plates and assembled the trusses and the transverse bracing system between the trusses to the deck plates/transverse floorbeam system. This assembly was carried out with the deck segments inverted. As each fabrication operation described below was completed, the segments were moved along an assembly line series of work stations, as shown in Figure 4.

2.2   *Orthotropic deck system*

The orthotropic deck system consists of plates of three thicknesses (14 mm in the roadway, 12mm over the trusses and 10 mm in the sidewalk and bikeway, SW/BW) and U-shaped longitudinal stiffening troughs. Deck plates forming half of each segment were cut to size and longitudinally welded. Handling these large, relatively flexible built-up systems was challenging because the plates required flipping to provide access to the back-side of complete joint penetration longitudinal welds.

Next, the fabricator developed a special press and used it to form 20 m long longitudinal troughs. The deck plate sub-assembly was laid on a specially shaped (hyperbolic paraboloid) pre-cambering welding table and the 300 mm wide troughs were positioned at 600 mm centres. Welding gantries were used to place four concurrent partial penetration longitudinal welds to join two troughs at a time to the deck plate. This process was repeated on all 30 deck troughs. The special shape of the welding table accounted for distortions due to welding shrinkage and the resulting deck plate assembly was flat and ready for mating with floorbeams and trusses.

2.3   *Trusses and floorbeams*

Trusses and floorbeams were fabricated in the same shop. The trusses comprise a combination of built-up top and bottom chords, I-shaped verticals and Hollow Structural Section (HSS) diagonals. The outer web of the bottom chords extended above the top flange in order to provide a convenient running surface for under-deck inspection travellers. A great deal

Figure 4.   Deck segment assembly line.

of fabrication effort went into ensuring that the trusses were fabricated within contractually required dimensional tolerances. The trusses were placed in position and tack-welded to the deck plates. Floorbeams (scalloped to mate with the orthotropic deck plate troughs) were inserted transversely through the trusses and welded to the truss diagonals, and the undersides of the deck troughs and deck plates. The lateral bracing system was then installed between the trusses.

2.4   *Adjoining deck segment match drilling*

The deck segments comprised two halves temporarily joined at the deck plate mid-line. Adjoining deck segments were "match-drilled" to each other with the proper fabrication profile to ensure that the deck was geometrically correct in the final condition.

After pre-assembly, the deck segments were split and flipped to the "right side up" position, as shown in Figure 5. They were then re-joined with a permanent longitudinal deck plate weld and bolts at the joints of the floor beams and transverse bracing members. Deck segments then exited the fabrication shop and were stored in CMW's yard until they were painted and paved.

2.5   *Painting*

A specially fabricated, temporary shelter was used to protect the deck segments during sand blasting of all steel prior to application of a zinc primer. Mid, strip and top coats were then applied and the coating allowed to cure.

2.6   *Pre-paving*

A thin layer of pre-paving was applied in another specially fabricated temporary shelter, as shown in Figure 6. This pre-paving served two purposes: First, it provided traction for vehicles (driving on bare steel deck plate would have been dangerously slippery). Second,

Figure 5.   Deck segment flipping in the shop.

Figure 6. Application of pre-paving.

Figure 7. Loading of a new deck segment with appurtenances onto the barge.

it protected the steel deck plate from damage. However, it had to be as thin as possible to minimize the weight of deck segments during installation. The pre-paving is described in Section 6.

2.7 *Installation of deck appurtenances*

After the application and curing of pre-paving, deck segments were stored on CMW's ocean-side wharf where the inner and outer barrier sections, longitudinal catwalks, and water pipe were added, as shown in Figure 7. Also, some temporary installation equipment (Temporary Hanger Extensions, THEs) were installed to speed up deck segment replacement.

3   REPLACEMENT OF TYPICAL DECK SEGMENTS

The general principles of the replacement of the suspended superstructure and the sequences of replacement of bridge deck segments are described in Radojevic & Kirkwood (2016) and Radojevic et al. (2014). The typical deck segment replacement works described in this chapter apply to typical, 20 m long, segments in the Dartmouth side span and main span.

The replacement of the suspended superstructure began in 2015 October, with the first deck segment in the Dartmouth side span.

The process of the segment replacements started with closing the bridge to traffic. Once the existing and new decks were disconnected and spans restrained longitudinally as needed, the existing roadway deck and stringers were cut at the segment cut line. Then, the existing hanger pins were removed, and the existing trusses relieved of global loading to facilitate the existing segment replacement. The existing trusses were then cut using thermic lances as shown in Figure 8.

Upon cutting of the existing trusses, the existing segment was "free" from the bridge and was lowered onto a barge using strand jacks located on the Erection Gantry (previously positioned above the segment being replaced, as shown in Figure 9 and Figure 12). The Erection Gantry was also used to raise new deck segments into position.

Figure 9 shows lowering of the first existing deck segment in the Dartmouth side span. A barge onto which the existing deck segment was lowered, and a barge which delivered the new deck segment, are shown at the bottom of the photo in Figure 9.

Figure 10 shows the gap in the superstructure during lowering of an existing segment (view from deck level). Also shown (in the background) is the existing bridge cross section in a view that is typically not seen during the life of a bridge.

During lowering of an existing segment, the geometry of the new installed deck segments on the bridge was adjusted for proper alignment with the new deck segment that would replace the existing one that was being lowered. These adjustments were made using THEs on the new deck, see Figure 11. The THEs supported the new deck from the existing hangers until the hangers were replaced with new ones.

Figure 8.   Existing truss bottom chord cutting.

Figure 9.   Lowering of the first existing deck segment in the Dartmouth side span.

Figure 10.   Gap in the superstructure during segment replacement.

Once an existing segment was lowered onto a barge, a new deck segment was brought on a barge under the gap in the deck, and then raised into position using strand jacks supported on the Erection Gantry, see Figure 12. Because of the need to open the bridge for traffic immediately following segment replacement, the new deck segments were delivered and installed fully prefabricated, including barriers and a thin initial layer of wearing surface to accommodate the traffic, as described in Section 2. Each lowering and lifting operation of a deck segment typically took about 60 minutes.

After lifting of a new deck segment into position, it was connected to the existing deck with a Temporary Deck Connection (TDC), shown in Figure 13.

Once the existing and new decks were connected with the TDC, the new deck splice was aligned and bolted. New THEs were then installed on the newly installed segment, and strand

Figure 11.   Temporary hanger extensions (installed between the existing hangers and new deck).

Figure 12.   Raising of a new deck segment.

jacks were disconnected from the segment. The newly installed segment was at this time fully integrated into the bridge superstructure and free from the Erection Gantry.

The hangers near the erection front were then adjusted using THEs to the required forces for running traffic on the bridge. After the hangers had been adjusted to run traffic,

Figure 13. TDC in place between the new and existing segments (brown colored steel under the deck).

Figure 14. Bridge during construction—elevation following a typical segment replacement.

temporary transition barriers and traffic plates were installed, and the bridge was opened for traffic.

Figure 14 shows the bridge elevation following replacement of a typical (14th) segment in the main span. A characteristic dip in the bridge deck profile is visible due to the presence of heavy construction weights (Erection Gantry and the TDC) at the erection front.

4   REPLACEMENT OF DECK SEGMENTS IN THE HALIFAX SIDE SPAN

There is no water access under the Halifax side span and access for erection below the bridge is restricted by the Department of National Defence (DND) buildings. Therefore, the segments had to be transported on and off the bridge and installed from above the deck. To facilitate this process, the segments in this side span were typically fabricated in 10 m lengths. This enabled just enough clearance for the contractor to transport the deck segments across the bridge when they were rotated 90° from their final orientation. Figure 15 shows the transportation of a new deck segment in its rotated orientation (with the lifting frame on top) using Self Propelled Modular Transporters (SPMTs). As shown in the photo, the clearances during segment transport were very tight, and very careful segment manipulation was required.

As shown in Figure 16, the contractor used an Erection Gantry equipped with a swivel to perform the rotation of the segment and lowering/lifting operations. Powered chainfalls were used for segment lowering and raising operations (in contrast with strand jacks used in the other two spans).

Figure 15.   Transport of a new deck segment in the Halifax side span.

Figure 16.   Halifax side span erection gantry equipped with a swivel and lifting frame.

The removal of an existing segment generally followed a sequence similar to the typical segment removal sequence, with the exception that it had to be rotated 90° before it was placed on the SPMTs to transport it to the Halifax side for dismantling and disposal (see Figure 17).

Figure 17.   Existing Halifax side span segment rotation for segment removal.

A new segment was transported from the Halifax side sideways to clear the hangers (Figure 15). Once the new segment was positioned above the gap between the decks, it was rotated 90° before it was lowered into place. After these steps, the erection sequence was similar to the typical segment erection sequence.

Figure 18 and Figure 19 show a new deck segment during installation. The clearances between the segment and the bridge elements were tight, and this required careful segment manipulation during installation. Aside from segment rotation, longitudinal and transverse segment movements were required for the segments to avoid conflicts with the bridge. These movements were achieved by movable "cathead" support beams which were located on the Erection Gantry.

The last bridge deck segment (the last segment in the Halifax side span) was replaced in 2017 February.

5   RAISING OF THE SUSPENDED SPANS

As described in Radojevic & Kirkwood (2016), it was originally planned to increase the shipping clearance in the main span by raising the suspended spans 3.0 m at midspan after replacement of the entire superstructure.

However, due to schedule constraints, the contractor's preference was to raise the suspended superstructure simultaneously with replacement of deck segments. This change was a significant deviation from the original plan, so COWI re-engineered the span raising sequences to allow for raising of the spans to be concurrent with segment replacement. An additional constraint was that this new method was not to interfere with segment replacement works. However, it allowed the contractor a great degree of freedom to perform span raising essentially independently of segment replacement works. The re-engineering accomplished

Figure 18.   New Halifax side span segment rotation during installation (roadway view).

Figure 19.   New Halifax side span segment rotation during installation (view from the erection gantry).

Figure 20.   Bridge superstructure elevation after raising of the main span.

this by requiring the span raising to lag behind the segment replacement erection front a prescribed amount, and by limiting values of raising so that would the erection front would not be affected. The raising of the spans was accomplished using THEs, shown in Figure 11.

Figure 20 shows the bridge superstructure profile following replacement of all segments in the main span and full main span raising at midspan. At this stage, only a minimum amount of main span raising near the Halifax tower was not completed. The contractor completed the remainder of span raising after having replaced all deck segments (including Halifax side span) in 2017 April.

After completion of the raising of the superstructure, the existing hangers and the THEs were replaced with new permanent hangers. Galfan was chosen as the corrosion protection on the new hangers. To the best of the authors' knowledge, this is the first application of Galfan coating on long span bridge hangers in Canada.

6   PAVING

6.1   *General*

The need to have a safe driving surface as soon as each deck segment was installed into the bridge necessitated use of a unique two-layer paving system (Epoxy Asphalt, EA, provided by Chemco Systems). EA comprises a two part epoxy (resin and binder) combined with carefully graded aggregate. The first shop applied layer, known as pre-paving, was briefly discussed in Section 2.

6.2   *Pre-paving*

The pre-paving comprised a thin (nominal 10 mm) shop applied layer to minimize segment weight for lifting. The mix design needed to be robust to resist wear of winter tires in the Halifax climate. The deck was sandblasted and then primed with an inorganic zinc-rich primer. A layer of hot EA was spray-applied and then heated aggregate chips were embedded into liquid epoxy with rollers. Excess aggregate was swept off and the resulting pre-paving layer was cured approximately one week to ensure adequate adhesion.

6.3   *Final paving*

The pre-paving described above was integrated into the final paving layer, which was applied after completion of deck segment installation and field welding of the deck plates at transverse joints between segments.

The final layer of paving is nominally 35 mm thick (on top of the 10 mm thick pre-paving) and was field-applied on the Dartmouth side span and main span in 2016 October (prior to the inclement weather associated with the winter of 2016/17). The final paving was applied to the Halifax side span in 2017 May.

7 DESIGN FOR WIND EFFECTS

Halifax is situated in an area where many weather systems collide, resulting in high winds, including hurricane winds which have caused major damage in Halifax in the past. A site-specific wind climate study was conducted by wind specialists RWDI. Based on this study, the mean hourly design wind speed at deck level is 36.1 m/s, while the deck must remain aerodynamically stable for 10-minute wind speeds of 49.2 m/s. During construction, the bridge had to withstand reduced mean hourly wind speeds of 30.5 m/s and remain aerodynamically stable for 10-minute wind speeds of 44 m/s.

Dynamic wind-structure interaction is of high importance on long, flexible structures such as suspension bridges, and was a concern for the Macdonald Bridge. Therefore, three circumstances of interest have been investigated:

1. Wind on the existing bridge;
2. Construction wind on the partially-completed bridge, for checking of all erection stages and for the design of the temporary deck connection; and
3. Design wind on the completed bridge.

Extensive physical modeling in the wind tunnel was conducted to adequately cover concerns related to the bridge design and during erection. Several models of the existing and new bridges were built and tested in the wind tunnel (by RWDI) in order to confirm the aerodynamic stability of the completed bridge and during erection, and to measure aerodynamic coefficients for input to the dynamic buffeting analysis. The following models were built and tested:

1. Sectional model of the existing deck;
2. Sectional model of the new deck;
3. Aeroelastic model in the final condition; and
4. Three aeroelastic models during erection. An aeroelastic model with the temporary deck connection between the existing and new decks at midspan is shown in Figure 21.

An interesting case with a gap in the deck (at main span quarter point) during erection was investigated. This aeroelastic model is shown in Figure 22 and was used to investigate the wind effects on the deck segment during lifting, effects on the bridge with a gap between the existing and new decks, and effects on the erection gantry for operational gust wind speeds of 15 m/s.

The wind tunnel testing indicated that the bridge deck is aerodynamically stable during erection and in its final condition.

Figure 21. Aeroelastic model of the bridge during erection—erection front at midspan.

Figure 22. Aeroelastic model of the bridge during erection—gap in the deck at main span quarter point.

In addition to the physical testing in the wind tunnel, several dynamic buffeting analyses of the bridge in its various configurations were performed to calculate the force effects in the bridge structure. The force effects were then used for final design of the new bridge elements, checking of the elements to remain and design of some erection equipment. Buffeting analysis of the bridge in its existing and final condition, along with the six buffeting analyses runs for the erection stages, were performed to adequately cover various wind loads on the bridge throughout its transformation from existing to new.

8 CONCLUSIONS

The following conclusions are drawn from this paper:

1. A major bridge can be renovated or even replaced while the public is using it. However, careful planning, engineering, fabrication, erection and management are required to execute a complex project like the Macdonald superstructure replacement. Superstructure replacement works must be executed in a timely manner to open the bridge to public traffic on time.
2. Complex fabrication of Macdonald deck segments required a skilled fabricator to carry out the fabrication. Rigorous QC/QA is a requirement for achieving a quality, long lasting product.
3. Temporary works and major equipment, that were specifically designed by the consultant for this project (Erection Gantry, Temporary Deck Connection and Temporary Hanger Extensions), played a key role and contributed to the success of the project.
4. Final paving had to be applied during a good weather window necessitating that the Contractor be on schedule.
5. Because the probability of failure of a bridge is greatest during erection, utmost attention of the owner, consultant and the contractor is required when the public is using the bridge during construction, as in the case of Macdonald Bridge.
6. Wind effects are critical during construction of major suspension bridges and for the replacement works as described in this paper. Careful assessments of wind loads and aerodynamic stability, along with performing dynamic buffeting analysis, are required to maintain safety of the bridge during construction.

ACKNOWLEDGMENTS

The authors acknowledge HHB for providing photos for Figure 8, Figure 9, Figure 10 and Figure 12.

REFERENCES

Radojevic, D., Eppell, J., Kirkwood, K.F., and Buckland, P.G. 2014, Deck Replacement of the Angus L. Macdonald Suspension Bridge, *ASCE Structural Engineers International Conference 2014*, Boston, USA: ASCE.

Radojevic, D., Kirkwood, K.F. 2016, Macdonald Bridge Suspended Spans Deck Replacement: Construction Engineering Challenges and Solutions, *IABSE Symposium Proceedings 2016*, Stockholm, Sweden: IABSE.

Chapter 6

Fabricating orthotropic deck panels for the Macdonald Bridge, Halifax, Canada

S. Ross
Cherubini Metal Works Limited, Dartmouth, NS, Canada

ABSTRACT: Over the course of approximately two years, the Macdonald Bridge had the suspended span deck and stiffening trusses replaced with new sections while remaining open to traffic. Cherubini fabricated new deck sections for the project. Deck segments were either 10 m or 20 m long and approximately 18 m wide. The fabrication process is an important phase in any project, but the nature of this work was particularly demanding. The new segments had to be matched to adjacent replacement segments as well as matched to temporary deck connectors to provide continuity with the existing, soon to be replaced, sections. A high level of dimensional control was required. The fatigue sensitive nature of orthotropic decks raised the quality control requirements of the project beyond what is typically expected of bridge construction. The amount of welding required to fabricate the orthotropic panels made the dimensional and quality control an exceptional challenge.

1   INTRODUCTION

1.1   *Orthotropic steel decks*

The details of orthotropic decks vary from project to project; however, the basic components of the system remain. Orthotropic decks consist of a steel deck plate located just below the wearing surface, the deck plate is stiffened by longitudinal ribs placed at a regular spacing, and the longitudinal ribs span between transverse floor beams, which in turn span between the main support system of the bridge (girder or truss). The support conditions of the orthotropic deck can vary; typically, a longitudinal stringer will be located in-line with the main support system of the bridge. The stringer would directly support the deck plate as well as the transverse floor beams (FHWA 2012). Refer to Figure 1 for a typical cross section of an orthotropic steel deck. In the case of the Macdonald bridge, the top chord of the supporting truss was welded directly to the deck plate.

1.2   *The Macdonald Bridge*

The Macdonald Bridge was originally opened on April 2, 1955. The bridge has a suspended span of approximately 762 m, approach spans of approximately 148 m and 437 m, for a total

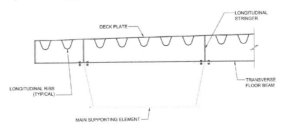

Figure 1.   Generic orthotropic steel deck cross-section.

span of approximately 1.3 km (Radojevic et al. 2014). The original roadway was two lanes. Refer to Figure 2 for an aerial view of the Macdonald Bridge. The view is from the Dartmouth side of the bridge looking towards Halifax; the deck replacement had almost reached mid-span of the bridge in this photograph.

The main cables of the bridge are 12.2 m center to center. The original roadway, sidewalk, and ductway were all contained within the space between the two main cables. The work completed by Cherubini in 1999 moved the walkways and a new bike lane to the outside of the cables, and a third lane added to the roadway. The additional width was provided by adding orthotropic deck sections which cantilevered from the existing bridge deck. The 585 m of approach spans were replaced with orthotropic deck sections. The bridge deck had been increased to a width of about 18.3 m. The current redecking project has maintained the width of about 18.3 m while replacing the 762 m main span deck with orthotropic deck sections. Refer to Figure 3 for a typical cross section of the orthotropic deck located on the Macdonald Bridge.

Figure 2.   Aerial view of the Macdonald Bridge; part way through redecking.

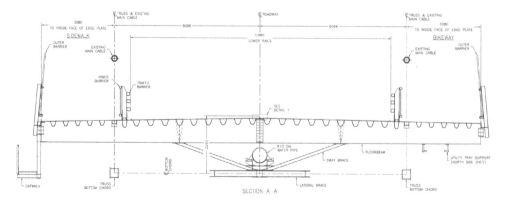

Figure 3.   Typical cross-section of the new Macdonald Bridge orthotropic bridge deck (HHB 2014a).

1.3 *Consultants, contractors, and governing codes and standards*

Halifax Harbour Bridges retained Buckland & Taylor (now COWI North America Ltd.), along with local consultant Harbourside Engineering Consultants, for the design of the retrofit. American Bridge Canada Company was selected as the general contractor for the project, and Cherubini Metal Works were the steel fabricators. The design was carried out per the Canadian Highway Bridge Design Code, CAN/CSA-S6-06. All components of the bridge were fabricated and inspected in accordance with the Canadian Highway Bridge Design Code, CAN/CSA-S6-06, and CAN/CSA-W59-13.

2 CONTROLLING WELD SHRINKAGE AND DISTORTION

2.1 *Fabrication and erection tolerances*

In order to understand the challenges of fabrication and the level of precision required, some basic project specifications are needed for context. It should be noted that the specified tolerances were more stringent than typical tolerances in CAN/CSA-S6-06. Relevant specified tolerances are as follows (HHB 2014b):

1. Side span overall running dimension cumulative length; plus/minus 10 mm
2. Center span overall running dimension cumulative length; plus/minus 20 mm
3. The work point of the vertical hanger pin centerline, deck segment width, deck segment length, and deck segment diagonal; plus/minus 5 mm
4. Deck flatness; plus/minus 6 mm over 3 meters for roadway
5. Deck plate welded splice maximum vertical offset; 10% of the thinner plate thickness
6. Trusses
   a. Depth of truss at bolted splice locations; deviation shall not be greater than 10 mm through its length, with 3 mm at ends
   b. Truss panel points; plus/minus 5 mm of theoretical position
   c. Alignment of field splices; Vertical alignment not to exceed 10% of thinner part or 3 mm
   d. Traveller rail outer chord plates centerline to centerline; plus/minus 3 mm and shall not deviate more than 5 mm from the design profile

2.2 *Weld distortion and shrinkage*

The welding process occurs at very high temperatures. Because of this, the deposited weld metal has a much higher temperature than the base metal. As the weld metal cools to ambient temperature, it will shrink. The cooler base metal restrains and limits the shrinkage, which in turn causes distortion in the assembled component.

Shrinkage occurs longitudinally and transversely. Distortion will occur locally in the form of angular distortion; distortion will also occur globally in the form of camber or sweep in the assembled component. Refer to Figure 4a and 4b for an illustration of angular distortion and Figure 4c for an illustration of distortion creating camber.

2.3 *Controlling weld distortion and shrinkage*

The most effective method of controlling the distortion and shrinkage of a welded component is to make allowances for the anticipated change in geometry when cutting and fitting the individual components. Pieces may be cut longer than their finished dimension so that they will shrink to the required length; components may also be assembled in a cambered position to counter the expected camber created during the welding process.

In the case of the orthotropic deck sections for the Macdonald Bridge, a full-size prototype section was fabricated while closely monitoring the geometry of the assembly before and

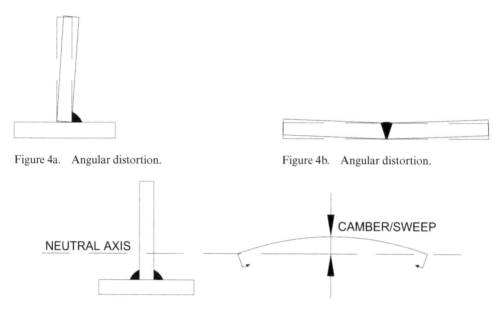

Figure 4a.   Angular distortion.   Figure 4b.   Angular distortion.

NEUTRAL AXIS   CAMBER/SWEEP

Figure 4c.   Camber/sweep caused by welding.

Figure 5.   Prototype section of Macdonald Bridge deck section.

after the welding process. The section measured approximately 20 m long by 9 m wide; the 9 m direction being one half of the bridge deck width. The prototype section consisted of a deck plate, longitudinal ribs, transverse floor beams, and a section of stiffening truss. Refer to Figure 5 for a picture of the prototype section.

Based on the prototype assembly, it was determined that the deck section had to be fabricated approximately 10 mm longer than the finished dimension. In order to account for the distortion of the finished assembly, the deck plate was cambered both longitudinally and transversely before welding the ribs, floor beams, and truss sections. Shown in Figure 6 is a partially fabricated deck section; it can be seen that the deck plate has been cambered in two directions, the ribs have been tack welded in place, and the welding gantry shown in the foreground of the picture is ready to complete the rib to deck weld.

Figure 6.   Partially fabricated deck section.

3   FABRICATION PROCEDURE

The new deck system consists of 46 segments measuring approximately 18 m by 20 m, except on the Halifax side of the bridge where the segments were 10 m in length. The 18 m wide deck segments were fabricated in 9 m half width sections which were joined near the end of the fabrication process via full penetration deck welds and bolted floor beam splices. The fabrication procedure can be broken down into the following basic steps:

1. Preparation and fabrication of sub-assemblies.
2. Stage 1 – Fitting and welding the ribs to the deck. This entire process happens while the deck assembly is upside down.
3. Stage 2 – Weld inspection and testing. The deck assembly remains upside down.
4. Stage 3 – While still in the upside-down position the half width deck sections were trial fitted to form the full width bridge deck and at the same time, the full width segment was trial fitted with adjacent segments. Also at this stage is the welding of the transverse floor beams to the deck and ribs, welding the truss to the deck plate, and welding miscellaneous components such as edge plates.
5. Stage 4 – The segments are reduced to half width deck assemblies again and then flipped upright. The half width deck sections were again trial fitted and then permanently joined. Final dimensional verification takes place at this stage as well as a profile check.
6. Stage 5 – This stage consists of painting the assembled segments and applying the temporary wearing surface on the deck plate. Two semi-permanent tent structures were erected outside of the main production facility in order to accommodate the painting and surfacing processes.

3.1   *Preparation and fabrication of sub-assemblies*

The preparation and fabrication of the sub-assemblies included bending the ribs, preparing the deck plates, fabricating the truss, truss bracing, and preparing the miscellaneous plates for fixture attachments.

3.1.1 *Rib preparations*

The project specifications allowed for a maximum of one splice per panel for each rib which allowed the fabricator to bend a more manageable length rib. However, splicing the ribs via full penetration welds created additional expense, added time to the fabrication schedule, increased the required quality control, and most importantly, created another potential location for a weld defect. The potential weld defect is not only in the rib splice location, but also where the rib to deck weld pass by the rib splice. As will be discussed later, the rib to deck weld is difficult and critical to future performance of the bridge deck. Taking all of this into consideration, it was determined that the most effective means of forming the ribs was to purchase full length raw plates and bend full 20 m sections of rib; a purpose-built press was fabricated. The press is capable of bending plate just over 20 m in length. The large press used to form the ribs is shown in Figure 7.

In addition to bending the ribs, the edges were also prepared for welding to the deck. Between deck segments, the ribs have a bolted splice. Bolt holes for the rib splices were pre-drilled on every second deck segment, while the adjacent deck segments had the rib splice bolt holes match drilled during Stage 3 of construction.

3.1.2 *Deck plate preparation*

A typical half segment of bridge deck consisted of 14 mm, 12 mm, and 10 mm thick deck plate. The raw plate was purchased in full 20 m sections, eliminating the need for transverse butt splice. With the deck plate in the upside-down position, the longitudinal butt weld was made between adjacent plates. The butt weld was made from one side only, with the remaining portion of the weld being completed on the top side of the deck plate once the segment is flipped into its upright position at Stage 4.

3.1.3 *Truss fabrication*

The truss top and bottom chords are built up box sections fabricated from plate; vertical and diagonal members consisted of HSS and built up sections. Tolerances related to the truss fabrication were critical particularly related to straightness and alignment on the bottom chord outer plate that forms the traveler rail. Extra care was taken to monitor and control fabrication methods to minimize weld distortion. Subcomponents of the truss top chords were monitored for longitudinal shrinkage during the welding process and adjustments made to maintain the design lengths and gusset plate location.

Survey and dimensional checks of the truss construction were carried out during fit up and after welding to monitor and control dimensional tolerance. Techniques such as pre-cambering of chord members to account for distortion, additional internal stiffeners, and welding sequencing were used to control critical dimensions.

Figure 7.   Purpose-built press for bending 20 m long plate.

The truss ends were match drilled to adjacent segments. The lead end of the trusses, the direction of site installation, had to be attached to the temporary deck connection. The temporary deck connection provided a connection between the newly installed panel and the adjacent, soon to be replaced, existing panel. The lead edge of each deck segment had to be matched to the adjacent panels as well as the temporary deck connection. A temporary deck connection jig was fabricated in order to match drill the lead edge of the deck segments. Match drilling took place at Stage 3 of fabrication.

3.1.4   *Floor beam fabrication*
The floor beams were fabricated in approximately 9 m lengths and spanned the entire width of the half-deck segment. The web of the floor beam was cut to fit around the ribs and the truss, but at this stage were only rough cut. The cutouts were trimmed when the floor beams were fitted onto the deck assembly. The project specifications allowed a maximum gap of 5 mm between the floor beam web and the ribs and trusses. The floor beam to rib weld detail changed depending on the gap achieved.

3.2   *Stage 1—assembly*

The deck plate was first placed on a floor jig that had been built to camber the deck plate in order to counter the weld distortion; the amount of camber was determined based on past experience and confirmed during the fabrication of the prototype.

Once the deck plate was firmly clamped to the floor jig, the ribs were placed on the deck plate. A secondary jig, similar in shape and profile to the floor beams, was used to press the ribs firmly down to the deck plate. Once the ribs were in firm contact with the deck plate, the ribs were tack welded to the deck plate. The tack welds were minimum 50 mm long and spaced at a maximum of 600 mm. Tack welds were made carefully, inspected, and marked for future reference. An improper tack weld had the potential to create a discontinuity in the final rib to deck weld; it was important for the tack weld to be completely consumed by the final rib to deck weld. The welding process was closely monitored from the inside of the rib via a track mounted camera. The rib to deck weld was required to be an 80% partial penetration weld, with a tolerance of plus or minus 5%. Making a partial penetration weld from one side only with that level of accuracy is a difficult task. If there is not enough penetration, the weld does not reach the required strength. If there is too much penetration, the weld will blow through the backside of the rib plate. Weld blow through is considered a serious defect and had to be avoided. The penetration of the weld was controlled by the included angle of the joint, the edge preparation of the rib, the position angle of the weld consumable, and by varying the voltage, amperage, weld speed, and wire feed rate.

Refer to Figure 8; in this figure, the ribs are being welded to the deck plate. The welding gantry is in the process of welding both sides of the rib simultaneously. In the foreground of the picture, the track for the internal camera can be seen. Also in the foreground, on each side of each rib is a run-off tab; these tabs were used to move the inherent discontinuity created by starting and stopping a weld, off of the finished product. Once the weld is complete, the run-off tabs were removed; the cut-off tabs were periodically used to check the cross-section of the weld. To the right of the picture, a weld inspector is performing an ultrasonic weld inspection of the finished weld. Also, visible in the picture is the longitudinal and transverse camber provided by the floor jig.

3.3   *Stage 2—inspection and testing*

Some weld inspection had started during Stage 1 and the remainder of the required weld inspection was completed during Stage 2.

To reduce the potential for corrosion, end sealer plates were required at the ends of the ribs. The end sealer plates were added at this stage and periodically pressure tested to ensure

Figure 8.   Rib to deck weld process.

a tight seal of the ribs. At this stage, longitudinal and transverse shrinkage was measured. Longitudinal shrinkage was typically in the 10 mm range as expected. The transverse shrinkage was not significant, being in the range of 1–2 mm. The transverse shrinkage, however, continued to be monitored.

3.4   *Stage 3—initial trial fit*

At this stage, the half-width deck segments are trial fitted in their upside-down positions. Adjacent half segments were aligned and longitudinal and transverse pop marks are established on the underside of the deck plate. These pop marks were the control lines for geometric control and setting all attachments to the deck. Adjacent segments were aligned and the fabrication profile was checked. The was verified by independent surveyors using total station survey to verify control lines.

The control lines were critical to the overall running dimension control and longitudinal alignment. A control line, with a corresponding control line on the existing bridge structure, was established on the first panel fabricated. The first panel remained at Stage 3 until the second panel was moved to Stage 3 and fitted to the first panel. At this point the control line from the first panel was transferred to the second panel; this process was repeated for the second and third panel, the third and fourth panel, and so on. The accuracy of the survey was in the range of 1–2 mm, the accuracy of placing the pop marks is comparable. Therefore, the accuracy in transferring the control lines from one segment to the next could only be completed with plus or minus a few millimeters. Referring back to the allowable tolerances for the total running dimensions, the tolerance for the side spans was 10 mm and for the main span was 20 mm. The Dartmouth side span consisted of eight segments, the main span consisted of twenty-two segments, and the Halifax side span consisted of 16 segments. The number of segments, and the overall allowable tolerance illustrate the required level of precision in the fabrication of each segment as well as the assembly of adjacent segments. Since the control lines can only be transferred with an accuracy of a few millimeters, the cumulative error could easily exceed the total allowable tolerance; it was expected however, that statistically the plus and minus tolerance for each control line transfer would correct itself and based on the feedback from the on-site survey, this was in fact the case.

Trusses were positioned on the segments, fixed to a temporary deck connection jig as well as adjacent segments, and tack welded to the deck. Dimensional checks were completed and the truss was welded to the deck. Floor beams were fitted to the deck with cut outs for ribs and truss top chords. The process of fitting the floor beams was a trial fit with scribing and trimming to ensure a satisfactory fit up for the subsequent welding process of floor beams to ribs, deck and truss top chord. A pressing jig was used to press the floor beams to the mating surface. Sway bracing and lateral bracing were installed as well, although at this point they are only attached as a trial fit-up and would be later removed. The ends of the trusses, bracing, floor beams, and ribs were then match drilled to adjacent sections using the splice plates as templates.

Dimensional checks were repeated after final welding of all components at Stage 3. The lead segment was then flipped upright and moved to Stage 4. The trailing segment in Stage 3 became the new lead segment and the next segment moved from Stage 2 to Stage 3 to be fitted and matched to the new lead segment. Refer to Figure 9; this picture shows the transition from Stage 3 to Stage 4. In the background of the picture, a full width deck section is assembled and ready to be trial fitted to the panel being flipped. In the foreground, there is a half width segment which had just been matched to the other half of the full width segment. The main focus of the picture is the panel which is partially flipped.

Figure 9.   Partially flipped panel.

3.5   *Stage 4—final deck assembly*

At this stage, trailing and lead panels are set on benches to the fabrication profile, then matched and bolted together with sufficient bolts and full-size pins to maintain the profile. Pinned locations were identified for correlation to site fit up.

Cross bracing and sway frames were re-assembled. A detailed survey was carried out to re-establish the control survey on the top of the segment and re-confirm fabrication tolerances.

The longitudinal deck plate seam weld for the half segment was then completed. Top surface details are added; lift hitches, traction rod brackets, deck drains, paving stops, barrier post connections etc.

After completion of all welding and attachments of components to the segment, a final survey and inspection of the segment was carried out. After the final survey, the lead segment is removed from the shop and the trailing segment moves forward to become the new lead segment; a new trailing segment was then introduced and matched to the new lead segment and the temporary deck connection jig.

3.6   *Stage 5—painting and surfacing*

At this stage, the completed deck segments are moved to semi-permanent tent structures for the application of the specified paint system and the deck wearing surface. The application of the wearing surface and the paint both have sensitive environmental restrictions. Figure 10 shows the process of applying the wearing surface. The painted deck plate has been partially covered with an epoxy resin, and the aggregate will be distributed over the surface shortly after. Once the aggregate is distributed over the bridge deck, a rubber tire roller makes multiple passes over the bridge deck. The deck plate preparation and the application of the wearing surface was done in accordance with the manufacturer's specifications. Surface preparation and painting was done in accordance with the Society of Protective Coatings (SSPC) standards. Also of interest in Figure 10 are the miscellaneous items attached to the topside of the bridge deck during Stage 4.

3.7   *Storage and shipping*

All of the major assembly work, the painting, and surfacing were all completed at the same location. Figure 11 shows a schematic of assembly line inside the shop, and Figure 12 shows the location of the painting and surfacing tents, as well as the storage location for the panels.

Figure 10.   Segment in the painting tent.

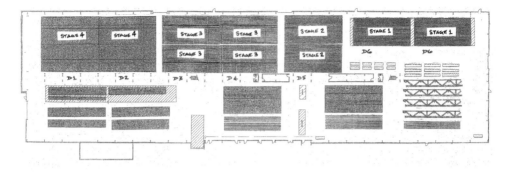

Figure 11. Shop floor layout showing location of each stage of fabrication.

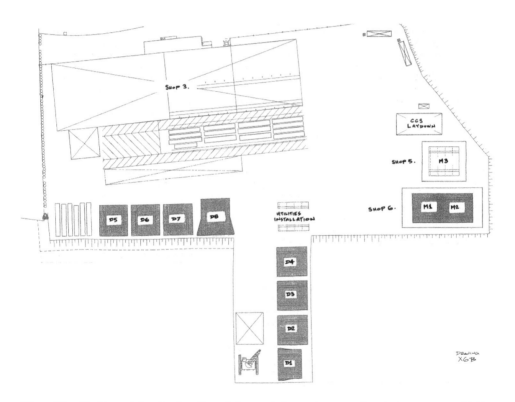

Figure 12. Yard layout showing location of shop, painting and wear surface tents, storage, and jetty.

At the bottom of Figure 12, there is a jetty and a crane shown; at this location, trial fit items that had been removed are reattached; the completed deck segments were loaded to a barge and shipped to the bridge. The fabrication shop is located approximately 6 km from the Macdonald Bridge by way of water. Figure 13 shows a panel being lifted from the barge into its final position on the bridge.

Figure 13.   New deck segment being lifted in place.

4   CONCLUSIONS

Fabricating orthotropic steel decks is a difficult process. It is especially challenging in a retrofit project with additional geometric constraints. The process itself is inherently difficult, but is exacerbated by the fact that there are so few orthotropic deck projects available to gain experience. General fabrication experience is critical in the success of projects such as this, however, there is still a steep learning curve associated with fabricating orthotropic decks and the specialized experience in fabricating this system is invaluable.

Cherubini successfully supplied the orthotropic deck segments for the Macdonald Bridge. The segments were provided in a timely manner such that the fabrication process never delayed the project schedule. The segments were fabricated with the accuracy required for the erection contractor to proceed efficiently.

Although the fabrication process was complex, the keys to success were simple; ensure accuracy and precision at each stage, be consistent with the process, and provide continuous quality control and dimensional checks.

REFERENCES

CAN/CSA-S6-06 Canadian Highway Bridge Design Code, *CSA Group,* Toronto, Ontario
CAN/CSA-W59-13 Welded Steel Construction, *CSA Group,* Toronto, Ontario
Halifax Harbour Bridges 2014a, Suspended Structure Typical Cross Sections, Drawing No. 1960-107, *Construction Drawings.* Halifax, Nova Scotia
Halifax Harbour Bridges 2014b, General Project Specifications. Halifax, Nova Scotia
Radojevic, D., Buckland, P., Kirkwood, K. & Eppell, J. 2014, Extending the Life of the Angus L. Macdonald Suspension Bridge, *2014 Conference of the Transportation Association of Canada.* Montreal, Quebec.
US Department of Transportation FHWA 2012, *Manual for Design, Construction, and Maintenance of Orthotropic Steel Deck Bridges,* Report No. FHWA-IF-12-027, Washington, D.C.

Chapter 7

Design and construction of the New Champlain Bridge, Montreal, Canada

M. Nader
T.Y. Lin International, San Francisco, California, USA

Z. McGain
International Bridge Technologies, Laval, Quebec, Canada

S. Demirdjian
SNC Lavalin, Inc., Montreal, Quebec, Canada

J. Rogerson
Flatiron Construction Corp, Richmond, British Columbia, Canada

G. Mailhot
Infrastructure Canada, Montreal, Quebec, Canada

ABSTRACT: The deteriorating condition and associated high maintenance costs of the existing Champlain Bridge (opened to traffic in 1962) prompted the accelerated need to design, build, operate, maintain, and finance the New Champlain Bridge Corridor Project.

Part of the largest transportation infrastructure currently underway in North America, the 3.4 km New Champlain Bridge is comprised of a signature cable-stayed bridge, an east approach, and a west approach. Performance and design criteria of this life-line structure must meet the design-life requirement of 125 years. The severe winters in Montreal, along with the requirement to maintain navigation channel traffic during construction present unique challenges to design and erection. The design-build team provided innovative use of pre-casting, modular segments, and non-tradition erection techniques and sequencing to meet the fast-track project schedule.

1 INTRODUCTION

Spanning the St. Lawrence River between Île des Sœurs and the Brossard shore in Montreal, Quebec, the New Champlain Bridge replacement is a part of a larger New Champlain Bridge Corridor Project. With the rapidly increasing cost of maintaining the existing bridge, building a replacement was considered to be more economically and socially beneficial for the region.

While the focus of this paper is on the New Champlain Bridge, the overall work scope includes a new Île des Soeurs Bridge, highway reconstruction, and widening of the federal portion of Autoroute 15, making this project one of the largest in North America.

The new 3.4 km bridge consists of three independent superstructures supported by common piers along the 2044 m West Approach with a typical span of 80.4 m, a 529 m asymmetric Cable-Stayed Bridge with a main span of 240 m; and a 762 m East Approach with a maximum span of 109 m (Figure 1). The design includes up to four highway lanes in each direction, a central transit corridor for mass public transportation with provision for future light rail transit and a multi-use path for cyclists and pedestrians.

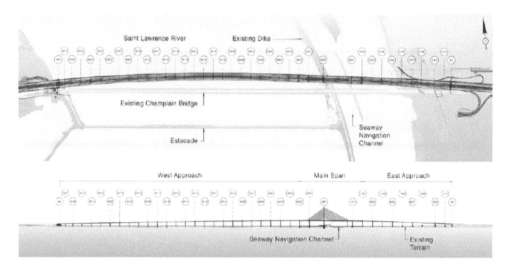

Figure 1.   Plan and elevation views of the New Champlain Bridge.

Figure 2.   Rendering of the New Champlain Bridge.

The CSB creates a signature element. The 160 m high single-pylon is comprised of a tuning fork configuration of twin masts. Inclined lower tower legs echo the inclined approach pier legs. The pier caps throughout the approaches form "W" shapes, defining the unique aesthetics of the bridge (Figure 2).

The project has been advanced through a Design Build Finance Operate and Maintain Public–Private Partnership model. A competitive selection process was undertaken with three teams being shortlisted from six registered parties. The winning bid presented the lowest net present value after meeting a set of mandatory technical principles which addressed architectural, durability and schedule criteria among others.

2   BACKGROUND

2.1   *Need to intervene*

The existing Champlain Bridge is one of the busiest bridges in Canada and plays a vital role in the Canadian economy with some 40 to 60 million vehicles and 11 million transit users travelling on the structure each year. Moreover, it is estimated that $20 billion in Canada-US trade crosses the bridge every year.

Figure 3.   Support trusses installed below several of the 100 exterior girders of the existing Bridge.

At the time the structure was designed and first opened to traffic in 1962, it was not intended for the bridge to be subjected to de-icing salts and was therefore not designed to accommodate these aggressive agents. A few years after the bridge was opened, the use of de-icing salts for winter operation became widespread in the province of Quebec and greatly contributed to the structure's accelerated aging. Prestressed concrete, an innovative technology at the time, had recently been introduced in North America. Whereas structural aspects supporting this new technology were sufficiently mastered at the time, it is apparent today that knowledge about the techniques required to ensure the durability of precast, post-tensioned concrete structures subjected to a salt environment were lacking. This is also evident from the lack of details developed with respect to durability, particularly as it relates to deck drainage. Deck drainage was replaced in the early 1990's, unfortunately only after the passing of some 30-years of aggression by de-icing salts which by then had allowed chloride ions to penetrate the concrete cover and reach the steel post-tensioning ducts.

Because of its condition and importance, the existing bridge has undergone extensive major structural repairs over the years by The Jacques Cartier and Champlain Bridges Inc. (JCCBI), the owner and operator of the existing crossing. Pending the bridge's replacement, monitoring, inspection and major structural interventions over the past few years have increased substantially in order to maintain the bridge in a safe operating condition. In fact, costs to maintain the existing bridge have totaled more than $300 million CDN over the past six years (2011 to 2016 period).

These expenditures have been used to undertake repairs and strengthening of pier caps including structural reinforcement by the addition of external longitudinal post-tensioning, the implementation of a multitude of girder reinforcement strategies applied to the 100 edge girders including extensive concrete repairs to delaminated concrete, the addition of exterior longitudinal post-tensioning, deviated post-tensioning strands anchored to the girder web faces, the installation of a below deck queen-post system incorporating deviated post-tensioning cable and twin steel posts, reinforcement of webs in shear by the addition of external stirrups, reinforcement of webs by the use of GFRP (glass fiber reinforced polymers) materials, installation of steel trusses to fully support the edge girders (Figure 3) and steel bents to support all seven girders at four spans crossing above land near the east and west abutments.

In light of its condition, in December 2013 the Government of Canada announced that it would strive to replace the existing bridge under an accelerated timeframe by 2018.

2.2   *Architectural considerations*

An important facet of the project is architectural quality. This led the Government of Canada to adopt a directives approach resulting in a precise definition of the most prominent and visually significant features of the bridge crossing.

Architectural guidelines were developed regarding structural form and architectural lighting. These guidelines were framed by a "definition design" such that the government could guarantee to the community that what it displayed during its public announcements would in fact be delivered.

Key architectural features of the new bridge which stemmed from this process included the requirement for a curved bridge for its alignment in-plan (hence offering a changing view of Montreal's skyline for its eventual users), an asymmetrical cable stayed bridge equipped with a tall slender center tower, characteristic piers and pier caps ("W" shaped as shown in Figure 4) and the desire to limit the number of columns in the river so as to avoid creating a forest of columns.

Measures were incorporated in the Request for Proposal as well as the Project Agreement's technical requirements to ensure that the architectural vision set out in the development phase would be preserved in the delivered bridge.

2.3   *Durability objectives*

In addition to schedule and meeting architectural expectations, one of the Government of Canada's other principal objectives for the project was to ensure the delivery of a new bridge of a very high quality and endowed with an extended design life. To this end, the project specifications and performance objectives imposed by the Authority included among others the following design requirements:

- Design life of 125-years for all non-replaceable elements.
- Mandatory use of stainless steel reinforcement in 100% of all deck slabs and at other strategic locations subjected to salt spray or salt leakage.
- Good deck drainage system including longitudinal carrier pipes and vertical drain pipes extended so as to discharge close to water level.
- Requirement to develop a Durability Plan that demonstrates that the durability objectives set out in the Project Agreement can be met for all components.
- Requirement to undertake time-to-corrosion modelling for concrete components using state-of-the art modelling techniques.
- Fatigue resistance of components to be considered over the extended design life.
- Reserve capacity for structure design which allows for the replacement of a cable stay with traffic and which also accounts for the potential loss of multiple stays in an extreme event.
- Limitation on the number of expansion joints; a maximum of only 8 expansion joints is permitted, including the expansion joints at the abutments. This is in strong contrast to the existing bridge which incorporates 57 expansion joints.

Figure 4.   Rendering of distinctive pier shape.

- Incorporation of an efficient system for maintenance access and inspection, including shuttles within box girders, elevators within the main span tower shafts, supply of under-bridge-inspection-vehicle and access devices within the interior of all hollow pier columns as well as maintenance travellers for the main span and back span of the cable stayed bridge.
- Incorporation of a comprehensive Structural Health Monitoring System.
- Requirements to mitigate stray currents and induced currents, particularly in light of the eventual implementation of an electrified mass transit system.

3 DESIGN CRITERIA

The bridge design was performed in accordance with the principal standards of CAN/CSA-S6-06 (R2013) Canadian Highway Bridge Design Code (CSA, 2012), Ministère des Transports du Québec (MTQ) Manuel de Conception des Structures, Volumes 1 and 2 (MTQ, 2009), MTQ Collection Normes—Ouvrages Routiers, Volumes I to VII (MTQ, 2010), and Eurocode (EN, 2008), 2 (EN, 2005), & 3 (EN, 2008) with UK National Annexes. Additional design references for specific applications were stipulated by the Project Agreement.

The local geological conditions, harsh climate and seismic hazards of this location present unique challenges to the design and construction of the bridge. Specialized studies on wind, seismicity, scour potential, vessel collision and ice loading were performed as technical inputs to the design criteria. Addressing durability, the design adopts a comprehensive approach to corrosion protection of the bridge components, taking into account the environment, design detailing, materials selection, construction quality and accessibility for maintenance, inspection, repair and replacement. The bridge is designed to ensure 125 years of service life under the site conditions addressed below.

3.1 *Wind*

The design wind speed criteria are based on one-hour (3600-second) average wind speeds resulting from wind climate analysis. At the bridge deck elevation, design wind speeds were derived for return periods representing construction design and final design, based on site data corrected for approach terrain and surface roughness. 10-minutes (600-second) averaged wind speed apply to aeroelastic stability during construction and final Bridge.

Wind tunnel testing analysis on sectional and full bridge models identified any potential for vortex shedding induced vibration and onset of flutter instability associated with the completed bridge and critical intermediate construction stages.

3.2 *Seismic*

The site is within the stable, yet seismically active continental interior of the North American Plate. The sub-surface conditions of the bridge consist of artificial fill or native clay, overlaying glacial tills and overlaying shale rock with various degree of weathering.

Seismic analysis used the essentially elastic design approach assuming minor inelastic behavior, and considered non-synchronous ground motions in three dimensions due to wave passage effects, site effects and incoherence. The analysis included time-history and geometrically non-linear response (P-delta effects).

The artificial fill along much of the bridge alignment is susceptible to liquefaction when submerged. Accordingly, the west abutment, main span tower (MST), Pier E01 and the east approach foundations have been designed assuming liquefaction for the largest 2475 year event and the intermediate 975 year event.

The liquefaction potentials in specific areas were captured by modelling soil springs with reduced to zero stiffness in the non-linear time-history analysis for the entire NBSL. Ground motions input included imposed lateral spreading due to liquefaction.

3.3   *Scour*

Scour depths and final scour elevations were computed for each pier. As the design calls for all foundations to be founded in the layer of competent rock, the bridge is not vulnerable to scour for 100 year floods stipulated by the design criteria.

3.4   *Vessel collision*

The New Champlain Bridge is designated as a Class I critical bridge. The vessel collision study considers the vessel types and sizes that navigate through and adjacent to the St. Lawrence River. The vessel collision analysis determined the design demands for a maximum annual frequency of collapse of 1/10 000.

3.5   *Ice loading*

The bridge piers are designed to account for dynamic ice forces due to collisions of moving ice floes, static ice forces caused by thermal movement of continuously stationary ice sheets and ice adhesion forces. In addition, the pier concrete mix design accounts for ice abrasion effects due to ice thicknesses of 0.9 m during the 125 year design life.

3.6   *Transit corridor loading*

In addition to the highway corridors, the design takes into account two non-concurrent phases on the transit corridor: (a) the RBL phase (reserved bus lane) and (b) the SLR phase (light rail transit). Structures were designed in accordance with the Canadian Highway Bridge Design Code (CSA, 2012) load requirements and load combinations for the RBL phase and the load requirements of Eurocode4 for the SLR phase.

4   SUBSTRUCTURE

4.1   *Foundations*

The West Approach extends ~2044 m and spans across 25 spread footings bearing on rock. The East Approach extends ~762 m and includes two types of foundations: spread footings bearing on rock and foundations founded on drilled shafts socketed in rock. The foundations of the CSB consist of cast-in-place (CIP) footings built on drilled shafts, with the exception of Pier E02 which is CIP spread footing on rock.

Examples of drilled shafts and spread footings for piers not constructed in water are shown in Figure 5. The foundations' structural designs are based on strength and stability (resistance to sliding, overturning, and uplift) under the governing load combinations.

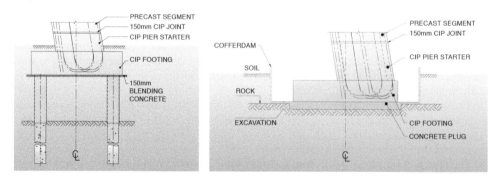

Figure 5.   Drilled shafts (left) and spread footing (right).

4.2 *Abutments*

The east abutment geometry consists of three independent structures supporting the north and south-bound highway superstructures as well as the central transit corridor. Each pile cap rests on two sets of cast-in-drilled-hole (CIDH) piles anchored to rock. A series of mechanically stabilized earth walls are placed below the pile cap.

The west abutment geometry consists of two multiple-column bents linked by a concrete superstructure cast monolithically with the abutment backwall. The multiple-column bents rests on a pile supporting footing that is in combination with both the New Champlain Bridge abutment and the neighboring highway bridge abutment supporting overpass P11. The abutment behaves like a rigid frame in both principal directions (Figure 6).

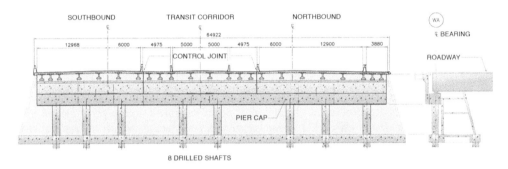

Figure 6.   West abutment.

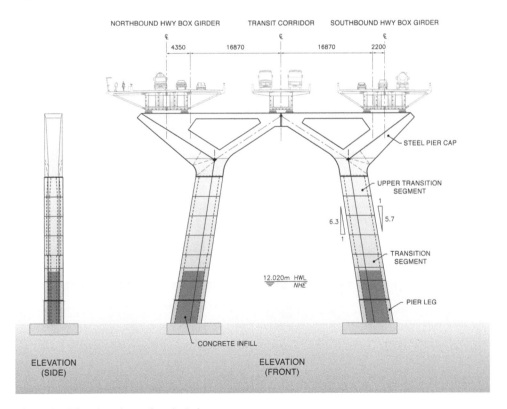

Figure 7.   Elevation views of typical pier.

4.3   *Piers*

The segmental pier leg substructures consist of hollow concrete box-sections stacked one on top of another and joined together using post-tensioning. Pier legs are variable in height to obtain the required superstructure profile.

The upper 11.4 m of the piers have pier caps consisting of steel box-sections with an interior matrix of diaphragms and stiffeners. The pier caps form two triangles that rest on the more prismatic pier legs and join in the center. The top members of the triangles are slender tension members (Figure 7).

5   APPROACH SUPERSTRUCTURE

The approach superstructure makes up almost 3 km of the 3.4 km crossing. The geometry of the superstructure was largely dictated by fabrication and transportation requirements. Segments were limited in weight to facilitate shop handling and transportation. To facilitate transportation to the site without detours, the twin tubs of the highway girders and the transit corridor boxes are limited to a depth of 3250 mm and a width of 4400 mm (Figure 8).

The deck consists of precast panels with in situ stitch pours to create composite action with the steel box girders. Reinforcement of the deck is entirely stainless steel as per Project Agreement to ensure the design life of 125 years. The precast deck panels include integral highway barriers, including relief joints to minimize cracking and extend their service life. These measures reduce the amount of finishing works required, which will aid in delivering the bridge on time.

To offset the cost of stainless steel reinforcement in the deck slab, the girder design includes larger top flanges that reduce deck reinforcement.

The design of the longitudinally stiffened box-girders was required to comply with provisions of Eurocode 3: EN1993-1-5 (EN, 2008) while the remainder of the steel and concrete elements were designed according to the Canadian Highway Bridge Design Code (CSA, 2012). This mixing of codes was challenging at times and required careful review to ensure consistent designs, and to avoid over- or under-design. The use of the EN1993-1-5 (EN, 2008) for the longitudinal design, combined with preferred fabricator processes, resulted in box girders without longitudinal stiffeners (albeit with marginally thicker webs).

Double composite action was used to maximize section efficiency and limit the weight of the steel in the generally heavier segments over the piers. Strain compatibility and local buckling between lines of shear studs were checked using Eurocode provisions and verified against the recommendations of Florida DOT reports (Potter & Ansley, 2010 and Sen et al., 2010) on recent developments. The addition of bottom flange concrete at the piers lowered the

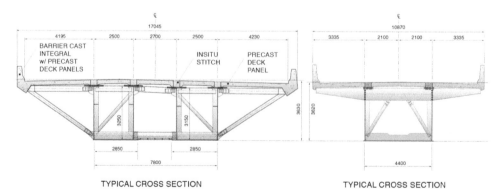

Figure 8.   Cross sections of southbound highway corridor (left) and central transit corridor (right)/ northbound corridor supporting multi-use path (not shown).

elastic neutral axis of the section, raising service stresses in the deck reinforcement. A careful balance was struck to optimize the girder section design, minimizing stainless steel reinforcement for crack width control and the total weight of structural steel.

The typical erection sequence consists of assembling a full span of girder segments and lifting it into place adjacent to the previously erected deck. Once in place, the girder lines are spliced and prepared to receive the precast deck panels. Deck panels are placed and stitches poured in a sequence that minimizes reinforcement stresses at the piers.

6 CABLE-STAYED BRIDGE

The CSB consists of a four span superstructure, a MST, supporting piers and a cable-stay system.

6.1 *Superstructure and crossbeams*

The superstructure consists of three longitudinal girders, supporting north-bound and south-bound roadways, as well as a center transit corridor. The girders are lightweight composite members with precast deck panels, designed to support all the design demands, including earthquake, wind and extreme events. Unbalanced spans and stays, important to the overall architecture of the bridge, require use of concrete counterweights in the shorter backspan to achieve overall balance at the MST.

A crossbeam at each pair of stay cables supports the three girders into a two-dimensional grid of steel box girders. The crossbeams transfer the weight of the girders to the stay cables, distributing the stay forces to mitigate twisting of the upper tower shaft. The crossbeams and the three girder segments form the basic assembly unit for the erection of the mainspan (Figure 9).

6.2 *Tower*

The tower consists of two shafts built of precast and cast-in-place concrete segments on a CIP footing with piles. The tower shafts are hollow to provide passageways for elevators, ladders and utility and are connected by a lower crossbeam and an upper crossbeam resembling a bowtie. The lower crossbeam is framed into the superstructure and the "bowtie" is above the clearance envelope of the transit corridor (Figure 10).

The lower portions of the shafts up to the bowtie are sloped at 1:7 from the vertical, while the upper portions are vertical and free-standing. This upper vertical portion, standing on the rigid A-frame of the lower shafts and cross—beams, supports the stay-cable anchorages. The architectural requirements of the shape of the shaft and the location of the stays in the shaft section result in an eccentricity of the downward component of the stay force onto the shaft. This produces a permanent moment in the shaft about the bridge longitudinal axis, requiring

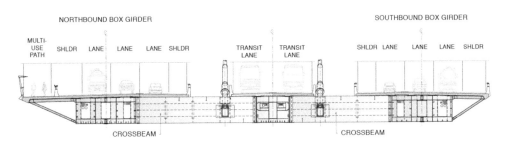

Figure 9.   Superstructure at crossbeam.

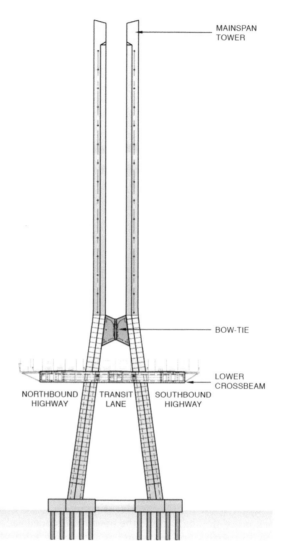

Figure 10.   Elevation of tower.

an initial transverse camber of the shafts, bowing inward to offset the permanent dead load deflections outward.

In addition to forming the backbone of the lower A-frame of the MST, the lower cross-beam supports about 60 m of the back span and main span superstructures, and resists any twisting due to differential loads in the back and main spans. Structurally, it is one of the most rigid components of the entire CSB.

Functionally, the lower crossbeam serves as a major cross-passage between the three longitudinal girders; as a center for the coordination and distribution of utility lines in the superstructure and MST; as the chief elevator service landing; and as a base station for the under-bridge maintenance gantry.

6.3   *Erection techniques*

The major challenge in the CSB erection is crossing the St. Lawrence Seaway, the major water-way of eastern Canada and the Great Lakes region. No temporary structures are permitted

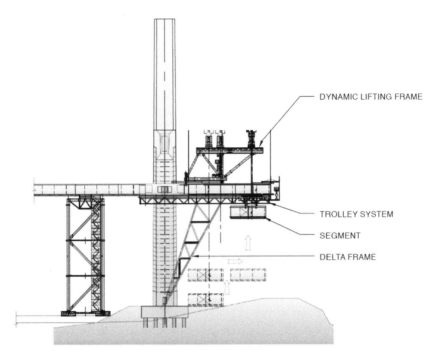

DYNAMIC LIFTING FRAME

TROLLEY SYSTEM

SEGMENT

DELTA FRAME

Figure 11.   Erection of first segments of main span.

in the channel and over-channel clearances must be maintained with limited impact to shipping. The seaway potentially freezes from December to March with few restrictions to working over the channel; however, the severity of winter makes placing concrete difficult and overall productivity is reduced. Therefore, the main span erection will proceed through the spring and summer of 2017, after obtaining from the Seaway management corporation permission for passage of bridge segments through the upper portion of the shipping clearance. Each segment will be lifted to a gantry, which will transport it under the main span soffit to the erection front over the Seaway (Figure 11). There, another gantry will lift the segment into position for connection to the previously erected girders. The transit over the Seaway will occur over a period of several hours. Once lifted into place, the segment will no longer obstruct clearance. During the segment erection cycle, restrictions to shipping will be limited to several meters of vertical clearance over a few hours per month.

One traditional method for constructing CSBs is cantilever construction: first erecting the steel box girders, then stay cables and finally placing the concrete deck slabs. To accelerate construction, this procedure will be modified, and the steel segment will be erected with most of the concrete deck panels already in place. By increasing the construction cantilever moments at the tip of the girders, the time spent per segment will be reduced.

7   KEY DESIGN FEATURES TO ACCELERATE CONSTRUCTION

Other key time-saving features implemented in the design are highlighted as follows:

7.1   *Precast footings*

Concrete spread footings are pre-cast on site in a certified temporary pre-casting facility, which allowed work to continue through cold winter periods—when pouring concrete would otherwise be extremely difficult. Precasting allows foundation preparation and footing casting to be carried out in parallel, rather than in series—as in conventional construction.

7.2   *Precast segmental pier legs*

The precast segmental construction of each pier leg is cast in parallel. This facilitated the incorporation of the required architectural features and provides a high quality product.

7.3   *Steel pier caps*

Although constructing the pier caps in concrete is possible, there are several issues that increase the risk of schedule delays. The complex shape requires difficult formwork and long curing times in large staging areas. The top tension members require a high concentration of post-tensioning (PT); placing such PT includes several additional construction steps for the placing, tensioning, grouting and final concrete casting to restore the architectural shape. Finally, the lifting of such a heavy concrete element is challenging.

   The steel pier caps reduce construction time and effort by being fabricated offsite, requiring fewer assembly stages during erection and smaller cranes.

7.4   *CSB substructure construction*

The pier and tower shafts of the sub-structures of the CSB are precast concrete segments, except for transitions at the superstructure and the change of slope in the tower shaft.

   The foundation designs were started at-risk during the early works phase (prior to financial closure), and issued for construction early in the design phase to allow construction to proceed during fall 2015. Footings and starter segments of the foundation shafts are CIP.

8   CONCLUSIONS

The New Champlain Bridge is one of the high-profile infrastructure projects in North America. With unique challenges of environmental constraints, durability requirements and architectural considerations, special solutions needed to be developed with respect to the design and construction of key elements of the project. This design-build project is subject to a fast-track schedule of 42 months from design to construction. Through the innovative use of precasting, prefabrication, and erection techniques, the design is tailored to give the most accelerated construction in order to achieve the targeted opening date of December 2018.

REFERENCES

CSA. 2012. Canadian Highway Bridge Design Code, *CAN/CSA-S6.*
EN. 2008. Eurocode 1, Actions on Structures, Part 2 Traffic Loads on Bridges, *BS EN 1991-2 UK National Annex (NA).*
EN. 2005. Eurocode 2, Design of Concrete Structures: Part 1-1 General Rules and Rules for Buildings, *BS EN 1992-1-1 UK National Annex (NA).*
EN. 2008. Eurocode 3, Design of Steel Structures: Part 1-1 General Rules and Rules for Buildings (with document S.), *BS EN 1993-1-1.*
MTQ. 2009. Manuel de Conception des Structures. Vol. 1 & 2, *Ministère des Transports du Québec.*
MTQ. 2010. Normes – Ouvrages d'Art Tome III, *Ministère des Transports du Québec.*
Potter W, Ansley M. 2010. Investigation of the Double Composite Box Girder Failure Criteria, *FDOT Structures Research Centre.*
Sen, R., Stroh, S., Pai, N., Patel, P., and Golabek, D. 2010. Design and Evaluation of Steel Bridges with Double Composite Action, *University of Southern Florida.*

Bridge construction

Chapter 8

Design-Build replacement of the I-278 Kosciuszko Bridge Phase 1—approaches and connectors

P. D'Ambrosio, G. Decorges, C. Lauzon & B. Sivakumar
HNTB Corporation, New York, USA

ABSTRACT: The Kosciuszko Bridge is a vital transportation link carrying the Brooklyn-Queens Expressway (I-278) over Newtown Creek. With a construction cost of $554,700,000, the Kosciuszko Bridge Phase 1 bridge replacement project is NYSDOT's largest Design-Build Project to date. Choosing the Design-Build delivery method allowed NYSDOT to reduce the number of bid packages and the duration of construction by several years. With Notice to Proceed on May 23, 2014 for design and construction, the new bridge is scheduled to open in the spring of 2017.

The existing bridge is being replaced with two new parallel bridges, and the work will be performed in two phases. Phase 1 represents the design and construction of the eastbound structure and westbound Brooklyn connector, with the remainder of the westbound structure in Phase 2.

The Brooklyn and Queens Connectors and Approaches carry traffic to the new Cable-Stayed Main Span, spanning over Newtown Creek. This paper highlights the design challenges and solutions for the Approaches and Connector structures built in Phase 1 of the project.

1 INTRODUCTION

1.1 *Existing bridge*

The Kosciuszko Bridge, named after the Polish-born Revolutionary War hero Thaddeus Kosciuszko, opened to traffic on August 24, 1939. The bridge is part of Phase I, which carries a 1.1-mile segment of the Brooklyn-Queens Expressway (BQE, Interstate 278 over Newtown Creek) between Morgan Avenue in Brooklyn and the Long Island Expressway (LIE, Interstate 495) interchange in Queens. The bridge is a vital link to the region's transportation network, serving commuter, local traffic, and a significant amount of commercial traffic. It has three lanes of traffic in each direction (eastbound and westbound) and is heavily traveled with an estimated 160,000 vehicles per day.

The existing approach span structure between Varick Avenue in Brooklyn and 54th Road in Queens is made of steel trusses with spans varying between 120 foot and 230 foot in length. The main span is a steel through truss and spans 300 feet over Newtown Creek. The existing Brooklyn Connector runs adjacent to Meeker Avenue and is comprised of steel girder spans flanked with brick facade walls along Meeker, with openings at the three underpassing streets: Morgan Avenue, Vandervoort Avenue, and Varick Avenue. The existing Queens Connector consists of two steel girder bridge spans, approximately 100 feet each, between 54th Road and 54th Avenue in Queens. The existing bridge with its main truss over Newtown Creek is shown in Figure 1 below looking east.

1.2 *Project background*

The bridge replacement project began in November 2001 with a public scoping process. After this initial phase, 26 alternatives were developed, taking into consideration maintenance of access, highway connections, number of lanes, property impact, local businesses, residential neighborhoods, etc. An Environmental Impact Statement (EIS) was prepared to review the

Figure 1.   Existing Kosciuszko Bridge looking east into Queens with new Queens Approach substructure.

potential impacts of the proposed project. The record of decision to replace the bridge was made in March 2009. In 2011, legislation was modified allowing the New York State Department of Transportation (NYSDOT) to make the project Design-Build.

There were several key factors which led to contracting the project as Design-Build as opposed to traditional Design-Bid-Build. First, having the Engineer and Contractor teamed up from the start would lead to an innovative design with improved constructability. Next, the project risks could be shared among the Owner and Contractor. Finally, going to Design-Build would reduce the construction duration from over five years to three and a half years.

1.3   *Project details and goals*

The existing bridge is being replaced with two new parallel bridges, with the project divided into two phases. Phase 1 consists of the design and construction of the new eastbound (EB) Main Span, EB Brooklyn and Queens Approaches, EB and westbound Brooklyn Connector, EB Queens Connector, and the Laurel Hill Boulevard Pedestrian Bridge over the LIE interchange ramps. The new approaches, main span, and Queens Connector are built south of the existing bridge, while the Brooklyn Connectors are replaced in place of the footprint of the existing structure.

The sequencing of the traffic is as follows. After Phase 1 work is completed, the eastbound and westbound traffic will shift from the existing bridge to the new EB bridge, which can accommodate six lanes of traffic. The existing bridge will then be demolished and Phase 2 of the project, the new westbound bridge, will be constructed in its place. Phase 2 will consist of the design and construction of the westbound (WB) Main Span, WB Brooklyn and Queens Approaches, and WB Queens Connectors. When both bridges are complete, five lanes on the eastbound bridge and four lanes on the westbound bridge with standard shoulders will be available. The westbound bridge will also accommodate a bikeway/walkway on the north side.

The existing structure shows severe deterioration and is vulnerable to fatigue. The new structures will provide superior safety, operational and structural improvements. There will also be significantly improved grades across the entire facility, since the demand for 125 feet of vertical clearance over Newtown Creek has been lessened to a 90-foot minimum vertical clearance. The proposed grades will be reduced from more than 4% to less than 3%.

1.4   *Design-Builder*

The Design-Build Project for Phase 1, which this paper focuses on, was awarded to the Joint Venture SKE (Skanska, Kiewit, Ecco III) with HNTB, the designer of record, in April 2014.

The team's proposal was selected for best value, having received the highest technical score. The design and construction approach balanced cost and schedule, minimized traffic and environmental impacts, and achieve NYSDOT's transportation goals.

Some of the challenges of the project site are that it is in an urban area, has a main span over a waterway, and has contaminated ground conditions within its footprint. After assessing these conditions, the Design-Builder developed a design that accounted for environmental constraints, hazardous and contaminated materials, right-of-way constraints and requirements, construction staging and work zone traffic control, avoidance and minimization of utility relocation, constructability, and minimization of impacts to businesses and the community.

There were several key features, innovations, and benefits of the design and construction approach. For example, the use of retained earth walls eliminates approximately 30% of the existing bridge structure within the site, reducing long-term maintenance costs. Alignment shifts reduce long-term maintenance and provide a more durable infrastructure. Profile corrections improve driver safety and reduce bridge height, enhancing visual quality. Also, by optimizing the limits of construction, lane shifts and construction phases were reduced. This culminated in the elimination of temporary structures throughout the entire project limits. Figure 2 shows an overall plan view of the project limits highlighting the different areas of the project where new construction is taking place. A rendering of the completed project from the Design-Builder's winning technical proposal is found in Figure 3 below.

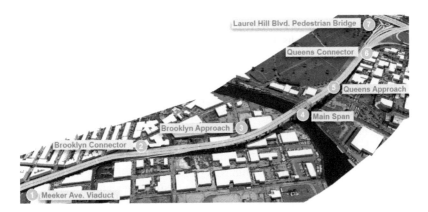

Figure 2.   Overall site plan view highlighting different areas of the project.

Figure 3.   Rendering of new Kosciuszko Bridge after Phase 2 work is completed (only the EB bridge is included in Phase 1).

1.5   *Task force meetings*

As the design progressed, the Owner, Contractor, and Engineer attended weekly Task Force Meetings. There were separate task force meetings for different disciplines, including structural, civil, and electrical. At these meetings, the Design-Builder presented the concepts that he was working through and received direct feedback from NYSDOT and their Owner's Engineer. Through this process, issues were worked out before the designer started to develop final calculations and plans, and this helped to streamline Owner approval on many design items such as joint locations, bearing types, continuity diaphragms, pier, and abutment details.

2   OVERALL PROJECT REQUIREMENTS

2.1   *Design requirements*

The structures were designed in accordance with AASHTO LRFD Bridge Design Specifications, Sixth Edition, referred herein as AASHTO, and NYSDOT LRFD Blue Pages. The live loads consist of HL-93 (AASHTO 2013) and NYSDOT Permit Vehicle (NYSDOT LRFD 2013) shown in Figure 4 below.

All the non-replaceable major structural components needed to be designed to meet a 100-year service life. The required design service life could be provided either by selecting materials with reduced corrosion potential, by selecting materials and details which resist degradation, or by other means acceptable to NYSDOT. The bridge is in a severe corrosive environment due to the extensive use of de-icing salts. It was required that stainless steel reinforcement be used for all bridge decks. For reinforced concrete elements, the service life was to be determined using the STADIUM® (Software for Transport and Degradation in Unsaturated Materials) model (SIMCO 2014–2015). Early in the design phase, the contractor developed a detailed Corrosion Protection Plan showing how to satisfy the design requirements.

The main span was required to be a single tower, cable stayed bridge. The solution developed by the Design-Builder consists of two vertical legs with no horizontal struts, avoiding the risk of falling snow and ice on the deck below. The proposed cable stayed spans consist of a 624-foot-long main span and a 377-foot-long back span.

As with any new bridge structure within New York City, NYSDOT's Region 11, the bridge had to be designed to meet the seismic performance requirements for the region (NYSDOT LRFD 2013). Due to its importance, the Kosciusko Bridge main span and approaches are classified as critical structures, and the seismic criteria was to meet the requirements of minimal damage under the functional evaluation event (FEE) corresponding to a 1000-year return period and repairable damage for the safety evaluation event (SEE) corresponding to a 2500-year return period. A site-specific response spectra was developed for the project site as mandated by the project requirements. For the single span connector bridges, no formal seismic analysis took place. The minimum support widths of the bridge seat required based on AASHTO were provided.

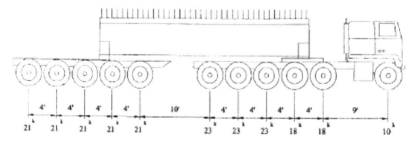

Figure 4.   NYSDOT permit vehicle axle loads and spacing.

3 CONNECTORS

The existing low level concrete viaduct with closure walls between Morgan and Varick Avenue is called the Brooklyn connector. Both Eastbound and Westbound are included in Phase 1. The structures located between 55th Avenue and the LIE Interchange are called the Queens connector. During phase 1, only the eastbound Queens Connector structures will be built.

A 100-year service life requirement was stipulated for structures and their abutments (including the walls in the abutment regions) and a 75-year service life requirement for the retaining walls and moment slabs. Various combinations of concrete mixes, concrete cover, and reinforcement (uncoated, epoxy, and stainless) were analyzed with STADIUM® software, and the results were discussed at the various task force meetings (SIMCO 2014–2015). Integral and semi-integral bridges were the optimal solutions for these single span connector bridges. Stainless steel reinforcement was used for the bridge decks. Galvanization of the steel superstructure was found to be the most cost effective solution for construction and durability.

Compared to the existing viaducts, the amount of elevated bridge structures was greatly reduced not only to optimize construction cost but also to facilitate stage construction and minimize maintenance. Between local streets, the new roadway is built on embankments retained by T-WALL® units (Neel 2014–2015).

To make the spans less vulnerable to vehicular impact, vertical clearance over the under-passing streets was also improved in several areas. The vertical clearance underneath the Queens connectors was increased from 13¢-4² to 14¢-6² at 54th Road and from 14¢-0² to 14¢-6² at 54th Avenue. Underneath the Brooklyn connector, where even longer spans were installed, standard vertical clearances of 14¢-6² were provided.

3.1 *Bridges*

In its original configuration, the Meeker Ave Viaduct, the elevated structure above Meeker Avenue, carried four lanes of traffic. During the 1960s, the structure was modified to accommodate six lanes of traffic. The width of the existing lanes and shoulders were non-standard. The Preliminary Design and indicative plans provided with the RFP showed a temporary structure to be constructed along the Meeker Avenue viaduct at the same location where another temporary structure was installed in the 1960s. This concept is shown in Figure 5 below. The construction of a temporary structure was not only costly but also presented high construction risk, since the foundation of the 1960s structure was left in place. The Design-Builder intensively researched another alternative and found that the proposed roadway

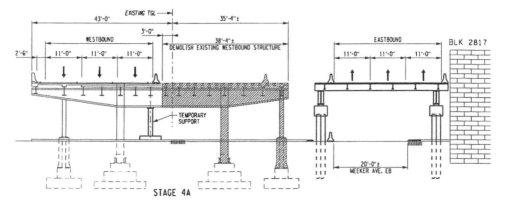

Figure 5.   Proposed cross section by owner during preliminary design with bridge demolished in two stages and use of a temporary bridge.

alignment could be shifted to the west creating sufficient space to keep six lanes opened to traffic. An additional benefit of shifting the roadway alignment was the portion of the roadway overhanging the new embankment along eastbound was reduced and became equal to the overhang along westbound.

A total of five connector bridges were built in Phase 1. The structures are designated below by the underpassing street with span length in parentheses (from West to East):

- Morgan Ave (96¢-8²),
- Vandervoort Ave (86¢-9²),
- Varick Ave (57¢-0²),
- 54th Road (62¢-0²), and
- 54th Ave (62¢-3²).

The connector bridges are all integral bridges except for Morgan Ave., which is a semi-integral structure. The vertical piles of the integral abutments resist the contraction and expansion of the superstructure under temperature changes. Therefore, they are subjected to bending stresses, and NYSDOT only accepts HP piles in this application. With the abutments restrained by embankment and the deck connected to both abutments, the abutments need not be designed for longitudinal loads due to temperature and live load. Since the length of these bridges is relatively short, the bending stresses did not exceed the yield limit of the steel used for the pile.

The foundation of the existing concrete frame bridge at Morgan Ave. had to be cored to allow for construction of the abutment under the existing viaduct. The large thickness of the existing foundations and limited overhead clearance would not have allowed the construction of a typical stub abutment founded on H piles. Instead, the abutment at Morgan is supported on micropiles, and the bridge is semi-integral. NYSDOT standardized details were used for the piles, abutments, and superstructure abutment connection.

Due to the proximity of the tie-in with the remaining Meeker Avenue viaduct, the structure at Morgan had to be built in five stages. To minimize construction time during the weekend outages and weather dependency, full depth precast deck panels with ultrahigh performance concrete (UHPC) closure joints were used. UHPC closure joints require minimal development length of the overlapping reinforcing bars in the closure pour allowing the width of the pour to be only 8 inches. The composite connection between steel girders and the precast panels was realized with short shear studs fitting within hidden pockets formed in the precast panels. The hidden pockets provide a uniform and continuous connection between the steel girders and the concrete deck. However, the numbers of rows of stud is limited due to the narrowness of the hidden pockets, as illustrated in Figure 6 below.

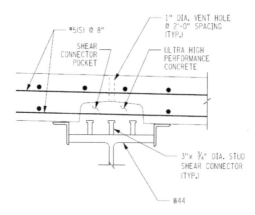

Figure 6.   Precast panel with hidden pockets for shear connectors.

The large overhang required at Morgan increased the loading on the exterior girders. To obtain a more uniform transverse load distribution, the first three girders were braced with top and bottom lateral members and diaphragms, increasing the torsional rigidity similar to a box girder. A 3D model was created using MIDAS Civil software to obtain an accurate distribution of loads among the outside girders (MIDAS 2014). When the HL93 tandem (AASHTO 2013) loading is placed along the barrier, the first exterior girder receives 65 percent of the load, the second exterior 30 percent of the load, the third interior girder 5 percent of the load.

In contrast to Morgan Avenue, the decks of all the other connector bridges were cast in-place, and the number of construction stages was much reduced. The bridges over Vandervoort (shown in Figure 7 below) and Varick Avenue, were constructed in two stages and three stages respectively.

The structure at Varick Avenue is over 200 feet wide. To accommodate thermal movement, a longitudinal joint was introduced between the eastbound and westbound roadway. During staging, traffic traversed the joint.

3.2   *Embankments and walls*

A large portion of the Meeker Avenue viaduct was replaced with fill contained by retaining walls. The T-WALL® retaining walls solution offered a durable solution able to resist the vertical pressure generated by both the typical and cantilever moment slabs. In contrast to the typical moment slabs which terminate at the fascia barrier, the moment slabs between Morgan Avenue and Vandervoort Avenue feature an overhang, cantilevering beyond the barrier up to $4¢-8^2$. This special moment slab not only has to resist TL-5 impact loading (NYSDOT BM 2014) on the barrier, but also the overturning due to the overhang. The stability against overturning is achieved by the weight of the thickened concrete slab poured within the fill behind the walls. The design provides a safety factor of 2.0 against overturning under ultimate loading. The vertical stress behind the walls was calculated assuming a uniformly distributed pressure over an effective base area (B-2e) in accordance with AASHTO Article 11.6.3. The bearing stresses at the strength limit state are typically between 2 and 2.5 ksf, and the T-WALL® units were designed for the additional loading. A 2^2 thick pre-molded resilient joint filler is installed on top of the T-WALL® preventing it from receiving direct loading from the moment slab. See Figure 8 below for a section of the T-WALL® with the special moment slab (Neel 2014–2015).

A plan was developed based on staged construction to abandon the substructure of the existing Meeker viaduct in place. The layout of the T-WALL® units had to be carefully planned to avoid conflict with the columns supporting the existing viaduct (Neel 2014–2015). Fill was placed underneath the existing concrete deck while the bridge was still in service. The amount

Figure 7.   Vandervoort first stage completed and pile installation for the second stage in the back.

Figure 8.   Section across embankment capped with moment slab overhanging T-WALL®.

Figure 9.   Fill placement under Meeker Avenue Viaduct with isolation sleeves around columns.

of fill placed while the bridge was in service was only limited by the vertical clearance required by the equipment used for compaction. The remaining fill was installed in stages within each weekend lane closure corresponding to the removal of a portion of the existing viaduct.

The effects of filling the spans on the substructure elements of the existing viaduct were carefully evaluated by the Design-Builder to capture the effects of construction. Reinforced concrete columns sitting on large spread footings support the existing viaduct. A detailed differential settlement analysis was performed to evaluate the influence of the surcharge due to the fill placement under the existing viaduct. The results of this analysis indicated that the fill could only be placed in 1 foot lift increments and with maximum differentials of 2¢-6² across the width of the viaduct and 10¢ over 120¢ along the length of the viaduct. The cap beams of the existing bents that could not withstand limited differential settlement due to their poor condition, were reinforced with fiber polymer wrapping. The existing concrete columns were wrapped with corrugated pipes to isolate them from down drag effects. Figure 9 shows existing columns wrapped with corrugated pipes as the T-WALL® units were being constructed (Neel 2014–2015).

A comprehensive monitoring system was installed to measure settlement continuously during placement of the fill. Thresholds were established and the recordings of the monitoring system were reviewed daily.

As an aesthetic treatment, the retaining wall panel faces were finished with a vertical steeped wood plank finish. Vertical orientation was provided to prevent potential climbers from escalating the walls. At the East end of the Brooklyn connectors, the walls supporting the embankment reach a height of 50 feet.

4 APPROACHES

4.1 *Superstructure*

The new 995-foot-long continuous Brooklyn approach spans extend from the anchor pier of the Main Span to the abutment west of Stewart Avenue. The new 970-foot-long continuous Queens approach spans extend from the anchor pier of the main span to the abutment east of 55th Avenue. In the indicative plans provided with the RFP, preliminary design showed an 11¢-6² deep concrete box girder superstructure with approximately 180 foot spans. Based on the subsurface conditions, the type of foundations and access to the site, the Design-Builder found that 125 foot spans resulted in an optimal span arrangement. For this span length, composite prestressed girders were a viable option and were selected for the superstructure type. These shorter spans need two more foundations per approach as compared to the design provided with the RFP plans but allowed for smaller foundation elements. The key would be the maximum girder lengths, approximately 135 feet, that the shorter spans allowed. This enabled the transportation of all girders to the site by truck. Also, by using prestressed gird-ers, prestressing would be done at the plant during the fabrication of the girders as opposed to a field splice with post-tensioning on site in a less controlled environment. See Figure 10 below for the Approach cross section developed by the Design-Builder.

The new span layout called for eight spans in Brooklyn and another eight in Queens, accounting for approximately 2000 feet of bridge structure. A multi-girder composite super-structure type was used to support the proposed roadway, which has a complex geometry following a curved path with large superelevation (up to 5.8 percent). The overall deck width varies from approximately 94 feet to 108 feet and 9 to 10 girders were used per span. The girders are 71² deep precast PCEF (Prestressed Concrete Committee for Economical Fab-rication) bulb-tee girders (NYSDOT BM 2014). Although prestressed concrete girders are seldom used in the northeast, for this project they offered the advantage of material cost and speed of fabrication over steel superstructure alternates. The minimum concrete compressive strengths of the precast girders were 10 ksi at final. The forces in the strands were released to the girders once the concrete strength had reached 7 ksi.

The deck followed a curved path throughout much of the Approaches. The piers were set radial to the curve in most spans, but there were two skewed spans, approximately 20 degree skews in Queens Approaches, after the roadway transition off the main span. To cut down on fabrication cost and construction complexity, the girders are chorded not curved. With straight girders, intermediate diaphragms could be removed and the girders were designed for load sharing under live load due to the deck alone. The girder ends in each span are embedded into a constant width diaphragm which provides continuity between girders in adjacent spans and transfers the loads into the substructure. Using chorded girders to follow a curved path and constant width diaphragms resulted in varying overhangs and varying girder lengths over the entire approaches.

The girders were made composite with an 8-1/2 concrete deck with stay-in-place forms reinforced with stainless steel reinforcement offering superior corrosion resistance. The number of girders and their spacing, 11¢-0² maximum, was set to approach the limits of isotropic deck design as outlined in NYSDOT's Bridge Manual. When the parameters of isotropic deck design are met, it has been proven by testing that the failure mode for the deck will be punching shear as opposed to flexure and much less reinforcement is required

Figure 10.   Proposed Approach cross section from the Design-Builder.

(NYSDOT BM 2014). Therefore, setting the girder spacing based on the limits of the iso-tropic deck design resulted in significant cost savings by reducing the amount of stainless steel reinforcement. See Figure 11 below for an aerial view of girders on the piers along the Brooklyn Approach during construction.

The prestressed girders meet the 100-year service life criteria via the type of concrete used and protective measures against corrosion. Type HP concrete was used for the prestressed girders and a calcium nitrate corrosion inhibitor and penetrating protective silane sealers were applied to all exposed faces. Also, due to the absence of joints in the deck, a main inno-vation of the superstructure discussed below, the ends of the girders are not exposed to leaks and chlorides. The prestressing strands and all mild reinforcement in the girders are uncoated, except for the composite bars which extended into the deck and are epoxy coated.

With the girder spacing set at $11¢\text{-}0^2$ maximum, the varying deck overhangs approach up to a $6¢\text{-}0^2$ maximum and the exterior girders govern over the interior girders in each span. The NYSDOT Bridge Manual requires the girders to be designed to meet Service III Tension limits for the heavy NYSDOT Permit vehicles in addition to the HL93 truck and tandem loadings required by AASHTO. For the exterior girders, the Permit Vehicle always governed. The girder strand layouts consisted of either all straight or draped and straight patterns. The use of draped strands was limited mainly to exterior girders and girders in the end spans. Prestressed girders in their casting beds are shown in Figure 12 below.

Figure 11.   Aerial view of new Brooklyn Approaches with stay-in-place formwork being laid out over the prestressed girder superstructure.

Figure 12.   Prestressed girder with draped and straight strand pattern in casting bed.

The anchorage zone at the ends of the prestressed girders are subject to large localized forces when the prestressing forces are released into the concrete. To combat the introduction of this force, AASHTO requires a check of the resistance of the anchorage zone due to bursting forces to ensure that the concrete does not split when the load is transferred. The webs of the girders were typically reinforced with #4 stirrups at 12 inches. At the ends of the girders they were increased to #6 bars and spaced at 3 inches for the first two feet of the beam due to the increased shear demand, but mainly to resist against the bursting forces. The bottom flange reinforcement which confines the steel was kept at size #4 but followed the stirrups with 3 inch spacing over the last 2 feet as well. During the casting of the girders, no major issues were reported in terms of cracking or spalling and the additional closely spaced reinforcement paid dividends during fabrication.

The main innovation of the superstructure is that it is 8 span continuous, jointless, and bearing-less. Continuous behavior, as opposed to simply supported behavior, was achieved through the continuity diaphragms at the piers, 3¢-0² wide at typical piers and 4¢-0² wide at skewed piers, and resulted in optimized girder designs. The continuous superstructure is restrained against longitudinal movement at all intermediate piers, and movement is accommodated via modular deck joints at the ends: the abutment and the anchor pier. The total movement range of each modular joint is 9 inches at the abutments and 21 and 24 inches at the Brooklyn and Queens anchor piers respectively, which included the movements from the cable-stayed main span. By reducing the number of deck joints, long-term maintenance of the structure was minimized with fewer joints to maintain and a watertight seal (continuity) between the superstructure and substructure. Permanent bearings are provided only at the abutments and the main span anchor piers. Due to the large longitudinal movements, multi-rotational uni-directional disc bearings are used. Bearings are replaceable components and have been designed to last 40 years per the service life requirements. Fewer bearings on the bridge provide better long-term maintenance with less items to inspect and maintain.

How the Design-Builder was able to remove permanent bearings at the intermediate piers was tied directly into how the diaphragm was constructed. The girders were designed for simply supported behavior under their own self-weight and during the pouring of the deck. Once the deck hardened, the girders were designed to be continuous and composite with the concrete deck. The construction sequence for the cast-in-place decks and continuity diaphragms is summarized below.

The girders were placed on temporary supports during construction, and in the final configuration those supports were removed. The girders were lifted on top of the piers and placed on temporary sand jacks. To reduce the risk of cracks forming above the pier area under negative moment induced by the deck dead load, the diaphragms were poured in two stages. A half depth diaphragm was used for pouring of the deck. The girders remained supported on sand jacks until the portion of the deck and full depth diaphragm were poured and hardened. The Contractor first poured the partial depth diaphragm and then began pouring the deck with a pour over the positive moment section of a particular span. The subsequent pours captured the remaining depth of the diaphragm and the area over the pier. No notable cracks were observed in the deck or diaphragms after placement of the deck. This pattern was repeated until all the spans were poured. Once the full-depth diaphragms had hardened, the sand jacks were removed.

By design, the load is transferred from the superstructure through the continuity diaphragm and into the substructure. The girders are embedded into the diaphragm approximately 2 to 4 inches. The shear force at the end of the girder is transferred to the diaphragm via shear friction (AASHTO 2013), and the strength of the connection is provided through a combination of cohesion between the two concrete surfaces and the hairpin reinforcement. The ends of the girders received an exposed aggregate finish to create a strong bond between the precast and fresh concrete. The vertical reaction is transferred from the continuity diaphragm to the pier cap through a 13-inch-wide shear key. Horizontal reactions are transferred from the diaphragm to the pier via dowels between the diaphragm and pier cap beam. The rounded shear key between the two concrete interfaces allows for rotation of the superstructure releasing moments from being transferred from the superstructure into the substructure. See Figure 13 for section of continuity diaphragm at girders.

A traditional "multiplier" method based on the PCI Bridge Design Manual is often used in practice to predict prestressed girder camber at erection. First, the girder self-weight and camber due to prestress force at release are computed using structural analysis. Next, the camber growth due to creep and shrinkage is predicted by multiplying the camber due to prestress force by 1.85 and the camber and the deflection due to the self-weight of the girder by 1.80 and taking the difference between these two products (PCI 2011). The result is used as an estimate of the camber at the time of erection of the deck. Based on previous project experience with similar girder size and span length, a straight multiplier of 2.0 was placed on computed girder prestress at release less self-weight to predict the final camber growth. Whether the strand pattern consisted of all straight strands or draped and straight strands did not seem to have too much of an impact.

The prestressed girder camber was directly related to the setting of the haunches for the deck. By design, the haunches were set with a 2-inch minimum haunch at centerline of girder and 1-1/2-inch minimum haunch at the girder edge to account for the cross slope of the deck which ranged from 2 percent up to 5.8 percent. To achieve these minimum haunches and account for the predicted camber growth, the haunches at the piers were typically set at 5 inches at the centerline of girder. However, they varied up to 9 inches at the end of the Approaches in Queens where the roadway follows a convex vertical profile.

4.2 *Piers*

For wide superstructures, with tall piers, multi-column piers with cap beams provide the optimal structural solution in terms of material efficiency. For the approach spans, two column piers were used (see Figure 14). The same column spacing (66¢-0²) and column diameter (6¢-0²) were used throughout the approach piers to facilitate and accelerate construction. The uniform pier geometry allowed the contractor to use the same formwork and column layout for the entire Approaches. Column heights ranged from 30¢-0² to 70¢-0². This resulted in the short pier columns with heavy demands under seismic loading. The constant spacing of the columns resulted in varying overhangs at each cap beam. The positive bending moment

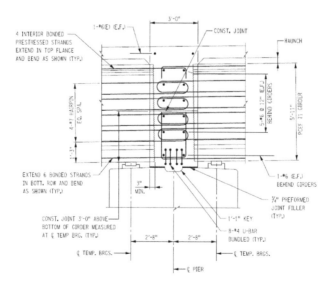

Figure 13. Continuity diaphragm section at girders.

Figure 14.   As-constructed approach pier elevation.

section dominated the design over the negative bending moment region. An $8\cent \times 8\cent$ pier cap beam section was used at each approach pier.

The amount of reinforcement in the substructure was optimized using Grade 75 reinforcement in all cap beams, columns and footings. The Designer took advantage of this higher strength to reduce the amount of reinforcement.

The service life criteria were met by selecting materials with reduced corrosion potential, the use of corrosion resistant reinforcement, and increased concrete cover. As the design and detailing of the pier elements was performed, these three important parameters were put into the STADIUM® model to be sure that 100-year service life requirement was satisfied. All concrete for the piers was high-performance concrete. By eliminating all intermediate deck joints throughout the approaches, pier cap reinforcement could be uncoated but an increased cover, 2.5 inches, was provided to meet the 100-year service life (SIMCO 2014–2015). The columns also were detailed with 2.5-inch cover and uncoated reinforcement except for splash zones, where the piers could be sprayed by deicing salts from adjacent roadways. For columns, in the splash zone, the reinforcement was epoxy coated to satisfy service life criteria. The footings were buried $1\cent$-0^2 minimum below the final grade and a 3-inch cover with uncoated reinforcement was provided.

For seismic design of the piers, a multi-modal response spectrum analysis and inelastic push-over analysis were performed, and the piers were designed based on the traditional force-based "R" factor method (AASHTO 2013). For the FEE, the structure was designed to remain elastic and R, the response modification factor, was taken as 1.0. Per AASHTO, when members in structures undergo significant lateral deflection under applied loads, force effects should be determined using a second-order analysis. An inelastic pushover analysis was performed to compute displacement ductility ratios under the SEE in the transverse and longitudinal directions from which the R factors for the SEE were established. The analysis was performed using MIDAS Civil Software with strain limits set per *AASHTO Guide Specifications for LRFD Seismic Bridge Design* performed for individual piers (AASHTO Seismic 2012). Moment curvature diagrams (shown in Figure 15 below) were developed under the SEE seismic loads, and the R factor was determined to be 3.5 in the transverse direction and 2.0 in the longitudinal direction (MIDAS 2014).

The R factor was applied to the seismic demands resulting from the elastic response spectrum analysis, for the design of each of the substructure elements. The R factor for the SEE was applied by dividing the column demands under the SEE by R and taking those results as the demand on the column. For the pier cap design, no response modification factors were applied (R = 1.0) for the SEE, and the negative moment section was designed for the full demands of the elastic analysis limited by the plastic moment developed by the columns.

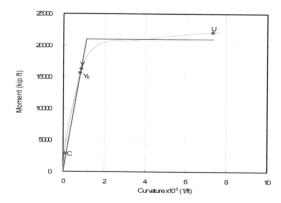

Figure 15.   Moment curvature diagram for individual pier under SEE demands generated with MIDAS.

When designing the pile cap and foundations, similarly, a response modification factor of 1.0 was used, and these elements were designed considering the full effects of the elastic analysis for the SEE, again limited by the plastic moment capacity of the columns. This design methodology ensures ductile behavior of the piers under the SEE.

4.3   *Foundations*

The foundation design for the approach piers was that of a footing supported on piles. The pier columns each have their own reinforced concrete footing. The footings are supported on deep pile foundations consisting of 18-inch outside diameter tapered steel piles between 35 and 45 feet long. Each footing is comprised of 16 tapered steel piles. The required maximum nominal compressive resistance of the piles are 590 and 545 kips for Brooklyn and Queens Approach Spans respectively. The piles also can resist uplift demands of 155 kips and 200 kips at Brooklyn and Queens Approach spans respectively.

Tapered steel piles were used for their advantages as a structural solution as well as their ability to minimize environmental impacts. The advantage of this pile type is the ability to achieve excellent end-bearing resistances at shallow depths. A project requirement was that all foundations must terminate 10 feet above the raritan clay layer of soil. With the tapered steel piles, this was easily attained. Since the soils beneath the Kosciuszko Bridge were highly contaminated, the steel casing around each of the piles was designed to be sacrificial and the reinforced concrete section alone provides the required capacity. Figure 16 below is a rendering of the tapered steel piles taken from the Design-Builder's winning proposal.

4.4   *Abutments*

The abutments at the ends of the approaches are conventionally reinforced concrete backwalls and bridge stems supported on tapered steel piles. Prefabricated T-WALL® units (shown in Figure 17 below) were used at the front and sides of the abutment and were designed to resist lateral earth pressures only (Neel 2014–2015). The bridge vertical reactions and horizontal forces such as wind and seismic loads are taken by the piles. When compared to the existing bridge, the Brooklyn abutment was moved to the east reducing the length of the Brooklyn Approach by approximately 200 ft.

The abutment backwalls and bridge seats are located beneath modular expansion joints and could not meet the 100-year service life requirements without special corrosion protection. The front face of the backwall and the top layers and front face of the bridge seat were reinforced with stainless steel bars. The STADIUM® analysis indicated that epoxy coated

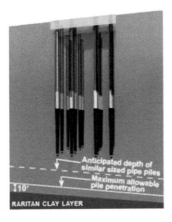

Figure 16.   Tapered steel piles with minimum tip above other similarly sized piles.

Figure 17.   Front face of abutment with T-WALL® units.

reinforcement in the other (rear) faces and bottom layer of the bridge seat would meet the 100-year service life requirements (SIMCO 2014–2015).

5   CONCLUSIONS

The Kosciuszko Bridge Design-Build Replacement Project is NYSDOT's largest Design-Build project to date and will set precedents for future Design-Build work in New York. The new bridge will provide improvements from the standpoints of safety, operations, and maintenance. The improved vertical profile and wider lanes and shoulders should reduce the number of accidents. Fewer accidents and a safer bridge will improve operations, reducing congestions and delays. The existing deteriorating structure will be replaced with a durable structure made from high quality materials designed to provide 100 years of service. This project highlights the benefits of early teaming between the Contractor and Engineer. The Contractor and Engineer worked hand-in-hand throughout the entire process: Pre-Award, Design, and Construction. Issues were resolved in an expedited fashion, as the Engineer and Contractor were in constant communication. The weekly Task Force Meetings with the Owner and Owner's Engineer also proved invaluable. Staged construction is particularly challenging and requires the collaboration of all parties (Owner, Contractor and Engineer). To be most effective, new or rehabilitated structures must be designed to facilitate construction especially in New York City metropolitan area where the infrastructure is so congested. In the Northeast, where there are severe corrosive conditions due to extensive use of deicing salts, the elimination of deck joints and permanent bearings will create structures that are more durable and offer better long-term maintenance. Prestressed concrete superstructures, although seldom used in New York State, were advantageous both in terms of cost and constructability. As the new bridge opens in the spring of 2017, the lessons learned will surely be applied to future Design-Build and bridge work throughout the region.

REFERENCES

American Association of State Highway and Transportation Officials (AASHTO) (2nd Edition). 2011 with 2012 Interim Revisions. *AASHTO Guide Specifications for LRFD Seismic Bridge Design*. Washington, D.C.: American Association of State Highway and Transportation Officials.

American Association of State Highway and Transportation Officials (AASHTO) (6th Edition). 2012 with 2013 Interim Revisions. *AASHTO LRFD Bridge Design Specifications*. Washington, D.C.: American Association of State Highway and Transportation Officials.

MIDAS Information Technology Co., Ltd. *MIDAS Civil 2014 v2.3*. Computer Software. 2014.

New York State Department of Transportation (NYSDOT). July 8, 2013. *NYSDOT LRFD Bridge Design Specifications*.

New York State Department of Transportation (NYSDOT) Office of Structures (4th Edition). 2006 with Addendum #3, 2014. *Bridge Manual New York State Department of Transportation (BM)*.

Precast/Prestressed Concrete Institute (PCI) (3rd Edition). 2011. *Precast Prestressed Concrete Bridge Design Manual*. Chicago, IL.

SIMCO Technologies Inc. *STADIUM®: A SIMCO Solution*. Computer Software. 2014–2015. Quebec, Canada.

The Neel Company, *T-WALL® Retaining Wall System*. 2014–2015. Springfield, VA.

Chapter 9

"A new Belt for Brooklyn"—the five mile Belt Parkway reconstruction project

D. Hom
New York City Department of Transportation, USA

W. Ferdinandsen
Greenman Pedersen, Inc., USA

P. Dombrowski
AECOM, USA

ABSTRACT: In 2009, the New York City Department of Transportation (NYCDOT) embarked on an ambitious, 12 year, $750 million reconstruction program of the historic Belt Parkway between Exit 9 (Knapp Street) and Exit 14 (Pennsylvania Avenue) in Brooklyn, New York. Included within the five (5) mile construction limits are six (6) bridges of varying span lengths, configurations and design features. All but one (1) of the structures span navigable waterways that flow into Jamaica Bay, which is a sensitive marine environment and a protected habitat for a multitude of fish, wildlife and plants. Each bridge was independently designed, and the program was advanced and packaged in three (3) phases to reconstruct the parkway in the most efficient manner, while minimizing impacts to the travelling public and neighboring communities. The program is highly complex, with the combined constraints of a coastal national park/recreation area, a dense urban location, varying site conditions, and a myriad of stakeholder permits and approvals required.

1 INTRODUCTION

In early 1930, the Belt Parkway was first proposed by master builder Robert Moses to provide modern highway access to Manhattan and to connect to similar parkways constructed on Long Island and Westchester County, New York. First described as a "Marginal Boulevard", the construction was advanced by the New York City Parks Department as a "Circumferential Parkway" and ultimately as the Belt Parkway (New York City Department of Parks and Recreation, 1937). Similar to other parkways, Moses and the Parks Department planned a series of "ribbon" parks along the route to control access and to encourage residential development.

The 36 mile long Belt Parkway, which includes the 11 ½ mile long Cross Island Parkway, began construction in 1934 Originally, each section of the Belt Parkway was designated independently, as the Shore Parkway, Southern Parkway, Laurelton Parkway and Cross Island Parkway. Today, only the Cross Island Parkway section remains separately distinguished from the Belt Parkway (www.nycroad.com/roads/belt).

The entire parkway construction included 47 road bridges, six (6) pedestrian overpasses, five (5) railroad bridges, and six (6) crossings over water bodies. There were 26 individual park areas that are adjacent to the parkway, which totaled more than 3,500 acres. Included among the remaining park areas are beaches, a golf course, fishing piers, and a horse riding academy (New York City Department of Parks and Recreation, 1940).

The parkway was dedicated on June 29, 1940 (New York Times, 6/29/1940), and construction was completed in May 1941. Originally opened with two (2) travel lanes in each direction, the parkway was widened to three (3) travel lanes (Figure 1) by the end of the decade, due to the rapid increase in traffic volumes. Since opening 75 years ago, the demands on the Belt Parkway have increased exponentially. In 1941, the average daily traffic was approximately 20,000 vehicles per day in total for both directions. Today, approximately 150,000 vehicles use the parkway each day (New York City Department of Transportation, 2016). Several events led to the tremendous growth, most notably the opening of New York International Airport (now JFK Airport) in 1948, and the construction and opening of the Verrazano Narrows Bridge in 1964. Additionally, the steady increase in vehicular ownership and the population shifts to the suburban communities on Long Island, after World War II, have greatly contributed to the rise in traffic volume on the parkway.

In 1971, New York Governor Nelson Rockefeller announced plans to widen and rebuild the parkway, between the Verrazano Narrows Bridge and JFK Airport, to the standards set for an Interstate highway. The concept was advanced as a means to reduce truck congestion on local Brooklyn streets and to improve overall mobility for goods. The plan met immediate opposition from local elected officials and civic leaders, with the many concerns leveled at anticipated increases in air pollution and widespread damages to the Jamaica Bay ecosystem (New York Times, 4/25/1971). The Governor quickly withdrew support of the plan, within months of the announcement (New York Times, 5/21/1971).

Over the decades, the parkway has benefited from numerous safety and maintenance projects aimed at keeping the bridges and roadways in a state of good repair. Resurfacing, of the roadways and overlaying/replacing bridge decks were commonplace, as were the installation of enhanced median barrier/guide railing systems throughout the length of the corridor. However, in the early 1990's, NYCDOT programmed several of the bridges for rehabilitation/replacement.

Figure 1.   Original Belt Parkway (Late 1940s).

2   REHABILITATION OR REPLACEMENT

The NYCDOT Division of Bridges is responsible for the design, construction, rehabilitation and reconstruction, maintenance, operation and administration of the 789 bridges (including 5 tunnels) and 53 culverts, Citywide, that are under the jurisdiction of the agency. Included in the inventory are the majority of the original Belt Parkway bridges that were built and opened in the early 1940s.

Each bridge is inspected on a biennial basis and given a condition rating based on a series of evaluated parameters. Typically, bridges are programmed for repair based on the condition ratings received during the inspections, and initially, the five (5) bridges that carry the Belt Parkway between Exit 9 (Knapp Street) and Exit 14 (Pennsylvania Avenue), had been scheduled for rehabilitation over the course of time.

However, as these bridges were in close proximity, in similar condition and in need of upgrade to bring them into compliance with current state and federal standards, the decision was made to package them together in a five (5) mile, corridor-wide replacement program (Figure 2). Also, included in the program were two (2) bridges outside of the corridor that are not the subject of this paper.

Within the five (5) mile construction corridor are four (4) waterways and one (1) local street that are crossed by the Belt Parkway as shown in Table 1.

Included in the construction contracts is the complete replacement of the connecting roadways and ramps with Portland cement concrete.

Reconstruction of these bridges and their approach roadways will bring the five (5) mile corridor into full compliance with current state and federal standards, including:

- Standard width travel lanes and safety shoulders;
- Super-elevated roadway/bridge segments around curves;
- Realignment of approaches to improve sight distances;
- Concrete median barriers;

However, the program not only brought the bridges up to current standards, it harkened to the history of the parkway by incorporating several, traditional elements to enhance the character of the parkway, including:

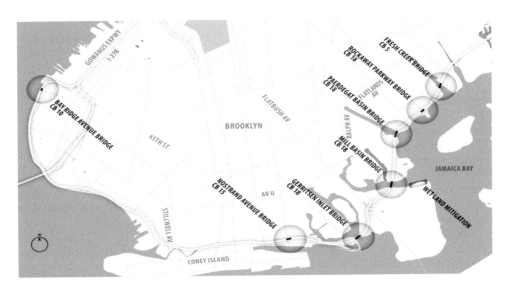

Figure 2.   Belt Parkway Bridge replacement program.

Table 1. Belt Parkway Bridges.

Bridge identification number	Feature carried	Feature crossed	Year built	Year replaced
2-23145-0	Belt Parkway	Gerritsen Inlet	1941	2017
2-23147-9	Belt Parkway	Mill Basin	1940	2019
2-23148-9	Belt Parkway	Paerdegat Basin	1940	2013
2-23149-9	Belt Parkway	Rockaway Parkway	1940	2013
2-23150-9	Belt Parkway	Fresh Creek Basin	1940	2013

- Art Deco bridge railings that mirror the original design;
- Stonework detailing on the bridge abutments;
- Relief detailing on the bridge parapet walls;
- Timber rail fences adjacent to the bicycle / pedestrian paths;

In July 2006, the Art Commission (now the Public Design Commission) awarded the program an Excellence in Design award.

3 INDIVIDUAL BRIDGE DESCRIPTIONS

3.1 *Belt 1*

The first construction project in the Belt Parkway program was the "Belt 1" package of the bridges over Paerdegat Basin, Rockaway Parkway and Fresh Creek Basin. This 2.12 mile section of the parkway between the Pennsylvania Avenue interchange (Exit 14) and the approximate midpoint between Paerdegat Basin and Mill Basin, included the following elements common to all the bridges in the program:

- Complete removal of the existing bridge superstructures and substructures, including the abutments and wingwall, and construction of new bridge superstructures and substructures;
- Construction of new approach slabs, and new approach and connecting roadways between the bridges;
- Installation of concrete median barrier along the entire project limits and new timber guide railing on the approach roadways;
- Installation of temporary and permanent bicycle/pedestrian paths, including timber rail fence;
- Installation of new street lighting, drainage facilities and overhead sign structures;

The construction project was awarded to the joint venture of Tully Construction and Posillico Civil in October 2009.

3.1.1 *Paerdegat Basin*

The former Belt Parkway Bridge over Paerdegat Basin (Figure 3) was a 692 foot long, 13 span, multi-girder bridge that was rated Fair prior to construction. It had numerous structural deficiencies that were only exacerbated by a significant barge accident in 2005, which severely impacted one of the piers adjacent to the navigational channel. Although emergency repairs were performed to keep the bridge operational, the damage accelerated the need for full replacement of the structure.

The design of the replacement structure was performed by NYCDOT—In House Design, which opted for two (2) independent, replacement structures (Figure 4). The Eastbound bridge is 1,227 feet long with five (5) spans, and the Westbound bridge is 825 feet long with three (3) spans. Both bridges carry three (3), 12 foot lanes of traffic with a 12 foot wide safety shoulder on the right side. The Eastbound bridge also carries a dedicated, 12 foot wide, barrier protected, bicycle/pedestrian path.

Figure 3.   Original Belt Parkway Bridge over Paerdegat Basin (2005).

Figure 4.   Replacement Belt Parkway Bridges over Paerdegat Basin (2013).

Included in the design were the following, unique elements, some of which presented challenges to the project (New York City Department of Transportation, Contract Number HBK 1024, 2009):

- Construction of new reinforced concrete abutments, wingwalls and piers founded on 24 inch diameter concrete filled steel piles;
- Installation of isolation bearings conforming to seismic requirements;
- Installation of High Performance Steel (HPS) Grade 485 W box girders and diaphragms;
- Installation of modular type bridge expansion joints at the abutments;
- Installation of swale BMP grading, drainage and plantings in accordance with NYCDEP standards and design;
- Relocation of a NYCDEP sludge force main;

The new Eastbound bridge was opened to traffic on December 19, 2011, and the new Westbound bridge was opened on December 19, 2012. The former bridge was permanently closed to traffic on December 20, 2012, and the demolition was completed in May 2013. The joint venture of GPI/CTE was selected to provide resident engineering and construction inspection services, and URS was selected as the project construction support services (CSS) consultant.

3.1.2 *Rockaway Parkway*

The previous Belt Parkway Bridge over Rockaway Parkway (Figure 5) measured 150 feet in length and had 4 spans. The structure had a multi-stringer, simple supported framing configuration on steel cap beams. The Rockaway Parkway bridge was also rated Fair prior to the start of construction, with numerous structural flags that required repair.

The replacement structure (Figure 6) was designed by Hardesty and Hanover, and consists of a single span that greatly improves the visibility along Rockaway Parkway. The bridge deck was widened to 109 feet, from 84 feet, and accommodates three (3), 12 foot lanes of traffic and

Figure 5.   Original Belt Parkway Bridge over Rockaway Parkway (2005).

Figure 6.   Replacement Belt Parkway Bridge over Rockaway Parkway (2013).

a 12 foot wide safety shoulder, in both the Eastbound and Westbound directions. The Rockaway Parkway Eastbound bridge is the only bridge in the corridor without a bicycle/pedestrian path.

The Rockaway Parkway design specifically included the following features (New York City Department of Transportation, Contract Number HBK 1091, 2009):

- Replacement of four (4) access ramps within the project limits;
- Construction of new reinforced concrete abutments, wingwalls and piers founded on 16 inch diameter concrete filled steel piles;
- Installation of elastomeric bearings, and ASTM A709 Grade 345 multi-girder steel superstructure;
- Installation of swale BMP grading, drainage and plantings in accordance with NYCDEP standards and design;

The bridge was completed in four (4) main stages. The southern stage was completed on December 5, 2011, followed by center stage completion on October 18, 2012. The northern stage was completed in July 2, 2013, and the center median was completed on August 22, 2013. The joint venture of GPI/CTE provided resident engineering and construction inspection services, and Hardesty and Hanover was the project construction support services (CSS) consultant.

3.1.3 *Fresh Creek Basin*

The old Belt Parkway Bridge over Fresh Creek Basin (Figure 7) was a five (5) span, 264 foot long multi-girder bridge that had been previously rated Fair in biennial inspections. Many years prior to construction, steel shorings had been placed at each of the abutments to alleviate the concerns of the bearings and pedestals. The underdeck had a significant amount of timber shielding to protect the waterway from spalling and loose concrete.

The new bridge (Figure 8) was designed by Earth Tech, and measures 316 feet in length with a reduction to three (3) spans to allow for a wider navigational channel.

The new bridge is 126 feet wide, and carries three (3), 12 foot lanes of traffic with a 12 foot wide safety shoulder on the right side, in each direction, along with a dedicated, protected, bicycle/pedestrian path. The profiles of both approach roadways and the bridge structure were raised to remedy the insufficient stopping sight distance conditions and to accommodate a 60 mile per hour design speed. The design also included several components that were unique to Fresh Creek Basin (New York City Department of Transportation, Contract Number HBK 1072, 2009):

- Construction of new reinforced concrete abutments, wingwalls and piers founded on 18 inch diameter concrete filled, tapertube, steel piles;
- Installation of elastomeric bearings, and ASTM A709 Grade 345 steel stringers and diaphragms;
- Installation of armorless bridge expansion joints;

Figure 7. Original Belt Parkway Bridge over Fresh Creek Basin (2005).

Figure 8.   Replacement Belt Parkway Bridge over Fresh Creek Basin (2013).

- Installation of inspection platforms on each of the piers;
- Installation of dolphins and pier protection at both piers;

As a cost and time savings proposal, the contractor initiated a Value Engineering proposal to construct a temporary bridge on the north side of the existing bridge (Figures 9 and 10).

Upon approval of the temporary bridge proposal, the project was able to be completed in three (3) main stages. The temporary bridge was completed and opened to traffic on March 4, 2011. The northern stage was completed on March 24, 2012, and the southern stage was completed on February 13, 2013. Finally, the center median was completed on August 22, 2013. Weidlinger Associates provided resident engineering and construction inspection services, and Earth Tech was the construction support services (CSS) consultant.

The Belt 1 project contained a daily incentive of \$35,000 per day for early completion of the project, including demolition of all bridge. By attaining substantial completion on August 22, 2013, the contractor completed the project 428 days early and earned the full incentive of \$14.98M.

3.2   *Gerritsen Inlet*

The Belt Parkway Bridge over Gerritsen Inlet was placed into service in 1941, thereby completing the original parkway. The original structure (Figure 11) consisted of 11 spans with a pair of pin and hanger connections on each of the three (3) original through girders. In the early 1980's, the pin and hanger connections were rehabilitated and fixed, and the number of spans was reduced to nine (9).

The original bridge was 520 feet long and was comprised of a steel girder superstructure, supported on reinforced concrete piers and abutments, supported by timber piles. The bridge was rated Fair in previous biennial inspections. Originally intended for construction with Mill Basin, the design of the new Gerritsen Inlet bridge (Figure 12) was performed by URS.

The new bridge will have a total of three (3) spans and a length of 496 feet. The width of the new bridge will be 121 feet, and carry three (3), 12 foot lanes of traffic with a 12 foot wide safety shoulder, in each direction. The dedicated, protected, bicycle/pedestrian path will be constructed adjacent to the Eastbound roadway. Similar to the Fresh Creek Basin bridge, the profiles of both approach roadways and the bridge structure were raised to improve mobility and safety of the parkway.

At Gerritsen Inlet, the site conditions and design parameters of the project led to the inclusion of the following elements (New York City Department of Transportation, Contract Number HBK 643, 2012):

Figure 9. Belt Parkway Bridge over Fresh Creek Basin temporary bridge (2012).

Figure 10. Belt Parkway Bridge over Fresh Creek Basin temporary bridge to the left of the existing bridge (2012).

Figure 11. Original Belt Parkway Bridge over Gerritsen Inlet (2010).

- Construction of new reinforced concrete abutments, wingwalls and piers founded on 24 inch diameter concrete filled, steel piles;
- Installation of multi rotational bearings, and ASTM A709 Grade 345 steel girders and diaphragms;
- Installation of armorless bridge expansion joints;
- Installation of inspection platforms on each of the piers;
- Installation of dolphins and pier protection at both piers;
- Installation of swale BMP grading, drainage and plantings in accordance with NYCDEP standards and design;

The Gerritsen Inlet project was awarded to China Construction of America—Civil, Inc. (CCA) in December 2012, and constructing started in February 2013. The project will be performed in four (4) main stages. The northern side of the new bridge was completed on

Figure 12.   Replacement of Belt Parkway Bridge over Gerritsen Inlet—Stage 3 (January 2017).

August 15, 2015, and the center portion was completed on November 18, 2016. The last two (2) major stages are scheduled to be completed during 2017, and the project is scheduled to be completed in Spring 2018. GPI/CTE provides resident engineering and construction inspection services, and URS is the construction support services (CSS) consultant.

3.3   *Mill Basin*

The Belt Parkway Bridge over Mill Basin (Figure 13), also known as the Mill Basin draw-bridge, was opened on June 29, 1940, as the only movable bridge on the Belt Parkway. When it first opened, the New York City waterways were among the most heavily navigated thoroughfares in the country. In 1941, the bridge was opened 3,100 times. By 1953, the number of openings had decreased to 2,173, and, in 2015, the bridge opened a total of only 210 times. The existing bridge is 864 feet long and consists of 14 spans. The bridge was rated Fair in previous biennial inspections.

The replacement Mill Basin bridge (Figure 14) was designed by HNTB. The new bridge will be the longest of the replacement Belt Parkway bridges with a total of 17 spans and an overall length of over ½ mile. The new bridge will be 121 feet wide, and at a height of 60 feet above mean high water, the new bridge will not need to be movable.

Like the other Belt Parkway bridges over waterways, the new Mill Basin bridge will have three (3), 12 foot lanes of traffic with a 12 foot wide safety shoulder, in each direction, and a dedicated, protected, bicycle/pedestrian path adjacent to the Eastbound roadway.

The new Mill Basin bridge design includes the following elements (New York City Department of Transportation, Contract Number HBK 1023, 2015):

- Construction of new reinforced concrete abutments, wingwalls and piers founded on 24 inch diameter concrete filled, tapertube, steel piles;
- Installation of multi rotational bearings, and ASTM A709 Grade 345 steel girders and diaphragms;
- Installation of modular type bridge expansion joints at the abutments, Pier 5 and Pier 12;
- Installation of inspection platforms on each of the waterway piers;
- Installation of dolphins and pier protection at the main channel piers;

Figure 13.   Original/existing Belt Parkway Bridge over Mill Basin (2012).

Figure 14.   Replacement of Belt Parkway Bridge over Mill Basin—Stage 2 (December 2016).

- Relocation of a NYCDEP sludge force main;
- Installation of swale BMP grading, drainage and plantings;
- Relocation of existing watermain, electrical and telephone services to the Jamaica Bay Riding Academy;

The contractor for the Mill Basin project is Mill Basin Bridge Constructors, which is a joint venture of Halmar International and Michels. Construction began in June 2015, and Stage 2 is currently in progress. The project has a contractual duration of 5 ½ years, but, similar to the Belt 1 project, an incentive program is included that will reward the contractor for finishing the project early. The project is currently on schedule to be completed in early 2019. GPI/CTE provides resident engineering and construction inspection services, and HNTB is the construction support services (CSS) consultant.

4   OTHER CONSIDERATIONS

4.1   *Environmental protection and tidal wetland mitigation*

The five (5) bridges in the Belt Parkway corridor between Knapp Street and Pennsylvania Avenue are all located adjacent to the Gateway National Recreation Area (GNRA), which is a division of the National Park Service, and all the new drainage systems outfall to Jamaica Bay. During construction, a full-time Environmental Engineer is on staff to inspect and monitor all Storm Water Pollution Protection Plans (SWPPP) in accordance with standards of the New York State Department of Environmental Conservation (NYSDEC). All of the project designs included the installation of erosion and sediment controls, such as silt fencing and turbidity curtain.

As all of the new bridges are wider than the bridges being replaced, and some of the new bridges are constructed off line, the construction unavoidably removed a significant number of trees and impacted wetland areas beneath the bridges. NYCDOT, in partnership with the New York City Department of Parks and Recreation (NYCDPR), has established a schedule and budget for independent upland mitigation/tree planting contracts to begin construction immediately following the bridge replacement projects. NYCDPR let the first upland mitigation, within the Belt 1 project limits, in 2016 and is schedule to complete all replacement tree plantings by the spring of 2018. Similar upland mitigation projects are planned for Gerritsen Inlet and Mill Basin.

The wetland impacts were mitigated through a separate Wetlands Mitigation contract that was administered by NYCDOT at Floyd Bennett Field (Figure 15). The NYSDEC mandated that the wetland losses were mitigated at a single location, in close proximity to the areas of impact, which is why nearby Floyd Bennett Field was selected (New York City Department of Transportation, Contract Number HBK 1072TW, 2011).

The project was awarded to the joint venture of Tully Construction/Posillico Civil, and began construction in March of 2011. In 2011, approximately 2.3 acres of land was cleaned of rubbish and debris (Figure 16).

In 2012, the area was converted to tidal wetlands (Figure 17).

Following Hurricane Sandy in 2012, the area was devastated and in need of replanting. The replanting operations were performed in 2015, and monitoring of the area was completed in January 2017. GPI/CTE provided resident engineering and construction inspection services, and HNTB was the design and construction support services (CSS) consultant.

Figure 15.   Tidal wetland mitigation at Floyd Bennett Field.

4.2   *Multi-modal coordination*

Through extensive coordination with all project stakeholders before and during construction, the impacts to the local communities, including all modes of transportation, were minimized:

- Three (3) lanes of traffic are maintained in each direction during all peak hours. Lane closures are permitted during non-peak hours only. One (1) lane, in each direction, can be closed during weekdays from 10 AM to 2 PM. Two (2) lanes, in each directions, can be closed during weeknights from 1 AM to 5 AM;
- The bicycle/pedestrian path was kept open at all times during the duration of construction. The paths were protected and kept safely isolated from construction activities;
- Reductions to the channel widths below the bridges were minimized through coordination with the United States Coast Guard (USCG) and the local mariners. All channel reductions were scheduled during cold weather months, when the channels were less utilized, and channel closures were scheduled during overnight hours;
- A full-time Community Liaison is on staff to respond to inquiries from the local communities, issue public notifications and publish project newsletters.

Figure 16.   Pre-construction conditions at Floyd Bennett Field (2009).

Figure 17.   Tidal wetland mitigation plantings at Floyd Bennett Field (2015).

Table 2.   Projected/Actual costs Belt Parkway Bridges.

Bridge	Projected/Actual construction cost
Gerritsen Inlet	$104.2 M (Projected)
Mill Basin	$263.7 M (Projected)
Paerdegat Basin	$204.5 M (Actual)
Rockaway Parkway	$58.0 M (Actual)
Fresh Creek Basin	$121.8 M (Actual)
Total	$752.2 M

5   PROJECT COST

Although the final costs will be not be known until the project is completed, the overall program is on schedule to be completed early and under the original budgets, as shown in Table 2.

6   CONCLUSIONS

After 75 years of service to the Borough of Brooklyn, the replacement of the five Belt Parkway bridges between Knapp Street (Exit 9) and Pennsylvania Avenue (Exit 14) is nearing completion. As the five mile stretch of parkway will be in full compliance with all federal and state standards, the structures on the new Belt section in Brooklyn are expected to outlast the original structures.

The success of the project can be attributed to a range of factors, starting with the intense planning and coordination of individual contracts during the advanced design stage. NYC-DOT gathered input from a vast array of industry experts in order to package the contracts in the most coherent and efficient manner possible. As the project evolved and individual contracts entered construction, issues encountered in earlier contracts were addressed in the subsequent design packages. Throughout the progression of design and construction, the high level of cooperation, ingenuity and professionalism exhibited by the team of agency, consultant and contractor personnel has been exemplary.

Although the final costs will be not be known until the project is completed, the overall program is on schedule to be completed early and under the original budgets.

REFERENCES

New York City Department of Parks and Recreation. 1937. New Parkways in New York City.
New York City Department of Parks and Recreation. 1940. The Belt Parkway.
New York City Department of Transportation, Contract Number HBK 1023. 2015. Contract Specifications and Drawings—Replacement of the Belt Parkway Bridge Over Mill Basin, Borough of Brooklyn.
New York City Department of Transportation, Contract Number HBK 1024. 2009. Contract Specifications and Drawings—Replacement of the Belt Parkway Bridge Over Paerdegat Basin, Borough of Brooklyn.
New York City Department of Transportation, Contract Number HBK 1072. 2009. Contract Specifications and Drawings—Replacement of the Belt Parkway Bridge Over Fresh Creek Basin, Borough of Brooklyn.
New York City Department of Transportation, Contract Number HBK 1072TW. 2011. Contract Specifications and Drawings—Tidal Wetland Mitigation, Borough of Brooklyn.
New York City Department of Transportation, Contract Number HBK 1091. 2009. Contract Specifications and Drawings—Replacement of the Belt Parkway Bridge Over Rockaway Parkway, Borough of Brooklyn.
New York City Department of Transportation, Contract Number HBK 643. 2012. Contract Specifications and Drawings—Replacement of the Belt Parkway Bridge Over Gerritsen Inlet, Borough of Brooklyn.
New York City Department of Transportation. 2016. NYCDOT 2015 Bridges and Tunnels Annual Condition Report.
The New York Times. 4/25/1971. Brooklyn Truckway Proposal Draws Wide Opposition.
The New York Times. 5/21/1971. Rockefeller Agrees To Drop Plan to Construct Truck Lanes.
The New York Times. 6/29/1940. Belt Road To Open to Traffic Today.
www.nycroads.com/roads/belt.

Chapter 10

Accelerated bridge construction project and research databases

D. Garber
Florida International University, Miami, Florida

M. Ralls
Ralls Newman LLC, Austin, Texas

ABSTRACT: The use of accelerated bridge construction (ABC) is becoming more critical, as many of the nation's bridges approach the end of their service life and require replacement while maintaining traffic flow. Most states have built at least one bridge with some aspect of ABC. Many states, however, remain unsure of the best ways to utilize these streamlined technologies and would like to know how other states have achieved successful ABC implementation. To assist the states, the ABC University Transportation Center (ABC-UTC) at Florida International University has imported the Federal Highway Administration's ABC project database and enhanced it to provide additional functionality and capacity including the incorporation of ABC research projects. This paper describes the ABC-UTC's project and research databases and what they offer bridge professionals working to implement successful ABC projects.

1 INTRODUCTION

1.1 *Background on need*

In spite of significant work to upgrade substandard bridges, the American Society of Civil Engineers' 2017 Infrastructure Report Card continues to give bridges a grade of C+ ("mediocre, requires attention"). As detailed in the Report Card (ASCE, 2017), "The U.S. has 614,387 bridges, almost four in 10 of which are 50 years or older. 56,007 – 9.1% – of the nation's bridges were structurally deficient in 2016, and on average there were 188 million trips across a structurally deficient bridge each day. While the number of bridges that are in such poor condition as to be considered structurally deficient is decreasing, the average age of America's bridges keeps going up and many of the nation's bridges are approaching the end of their design life. The most recent estimate puts the nation's backlog of bridge rehabilitation needs at $123 billion".

The Report Card states that, as of 2016, the average bridge in the US is 43 years old, just seven years short of the typical 50-yr design life of most of those bridges. With millions of trips daily across deficient bridges, the need to rapidly replace these substandard bridges while maintaining traffic flow is becoming more critical. To assist in this need the ABC-UTC Project and Research Databases provide details on projects constructed using ABC and on ABC innovations coming from research.

1.2 *Background on databases*

The 2017 Infrastructure Report Card also discusses the need for innovation to address the challenges of upgrading the nation's substandard bridge inventory. Upgrading bridges more quickly to last longer is possible with the use of ABC technologies and high-performance materials. ABC technologies include prefabricated bridge elements and systems (PBES) that allow bridge components to be built offsite or adjacent to the site and then installed in a matter of hours or days. High-performance materials include ultra-high performance concrete (UHPC), for example, in deck closure joints to provide improved durability. These ABC technologies and materials are being used in various states; information needs to get out about this use.

The original ABC project database was developed by the Federal Highway Administration (FHWA) and included 100 completed projects that incorporate ABC technologies. To facilitate broader use, in 2015 the data were transferred to the Accelerated Bridge Construction University Transportation Center (ABC-UTC) at Florida International University in Miami, Florida. At the ABC-UTC a research project (Garber, 2016a) enhanced the project database.

The ABC research database was developed in an ABC-UTC research project (Garber, 2016b) to provide a repository for completed, ongoing, and proposed research related to ABC. The initial contents for the research database were provided by the Transportation Research Board's ABC Joint Subcommittee.

As discussed later in this paper, both databases include a submission process to allow any user to submit a project or research for proposed inclusion in the databases. It is anticipated that, with time, the databases will include hundreds of ABC projects and research.

This paper describes the databases, how to access them, their search capabilities, and their submission processes.

2   ABC PROJECT DATABASE

2.1   *Description of project database*

The ABC Project Database provides compiled information on existing completed ABC projects. The information is available in a user-friendly format that allows easy access to project details and reference documents.

Project details typically include the bridge project name, location description, state, latitude and longitude, owner, year ABC component of the bridge was built, local and national bridge identification numbers, owner contact information, mobility impact of construction relative to conventional methods, drivers for ABC use, bridge dimensions and materials, construction description, costs, and funding including any incentive programs. Searchable database keywords are included for the ABC technologies used in the project, with breakdown for planning, geotechnical solutions, and structural solutions. Downloadable documents including contract plans, specifications, bid tabs, construction schedule, photos, and other related information are also available for many of the projects. This project information is available as approved by the bridge owner for posting in the database.

2.2   *Accessing the project database*

The ABC Project Database is available from the left vertical menu on the ABC-UTC home page (www.abc-utc.fiu.edu) by clicking "Project and Research Databases." This takes the user to a page that provides options to access the project database user guide, to view the keywords used for database searches, and to go directly to the database. This page also explains the two-step process to submit a project proposed for the database.

The user accesses the database by clicking the blue box titled "ABC Project Database." This takes the user to the database located on a separate server, as shown in Figure 1. A vertical menu on the right side of the page provides additional options, including training videos that explain how to use the database.

2.3   *Project database search capabilities*

Three methods to search the database are available to users of the project database—an interactive map search, faceted navigation, and a keyword search. Each of these methods is described below.

Figure 1.   ABC-UTC project database home page.

2.3.1   *Interactive map search*

The interactive map, shown in Figure 1, identifies the locations of all the ABC projects contained in the database. By clicking an icon on the map and then clicking the bridge name that appears above the icon, the user can access details about the specific ABC project represented by that icon.

2.3.2   *Faceted navigation search*

The faceted navigation search option allows users to reduce the number of ABC projects to show only those of interest. This is achieved by selecting the desired search filters in various categories on the faceted navigation pane, shown in Figure 2. The categories in the faceted navigation pane are state, beam material, location, spans, construction equipment, traffic impact category, maximum span length, total bridge length, year ABC was built, average daily traffic count, structural solutions, geotechnical solutions, project planning, and funding source. The search functions as an "OR" criterion within categories and "AND" criterion between categories. For example, if within "Beam Material" both "Concrete" and "Steel" are selected, bridges will be included that have either concrete or steel beams. Across categories, if "Concrete" beam material and "Rural" location are selected, only bridges with concrete beams in a rural location will be included.

An example of a completed faceted navigation search is shown in Figure 3 for a one-span concrete-beam bridge installed using self-propelled modular transporters (SPMTs). Currently two bridges with those criteria are in the database. Note that thumbnail information is provided on the two bridges.

Significant additional information on a project can be accessed by clicking the bridge names shown in Figure 3. For example, Figure 4 shows a partial view of the project summary sheet for the first bridge thumbnail shown in Figure 3. The extensive information includes detailed descriptions of the existing and replacement bridges and construction methods, specific ABC technologies used in the project, and stakeholder contact information in addition to downloadable documents such as contract plans and specifications.

Faceted navigation allows only a broad search for projects that include ABC technologies under the "Structural Solutions," "Geotechnical Solutions," and "Project Planning" categories.

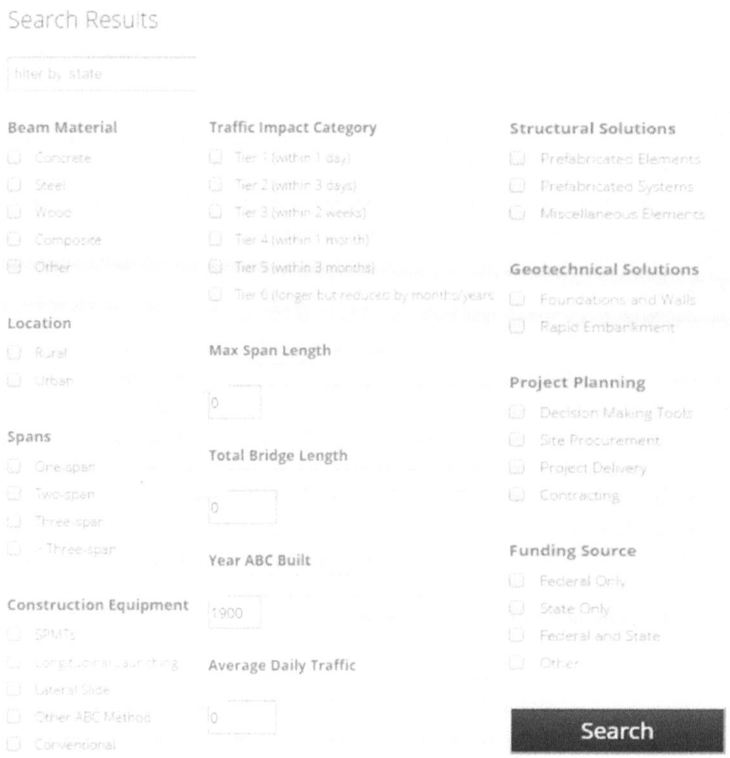

Figure 2.   Faceted navigation pane in ABC project database.

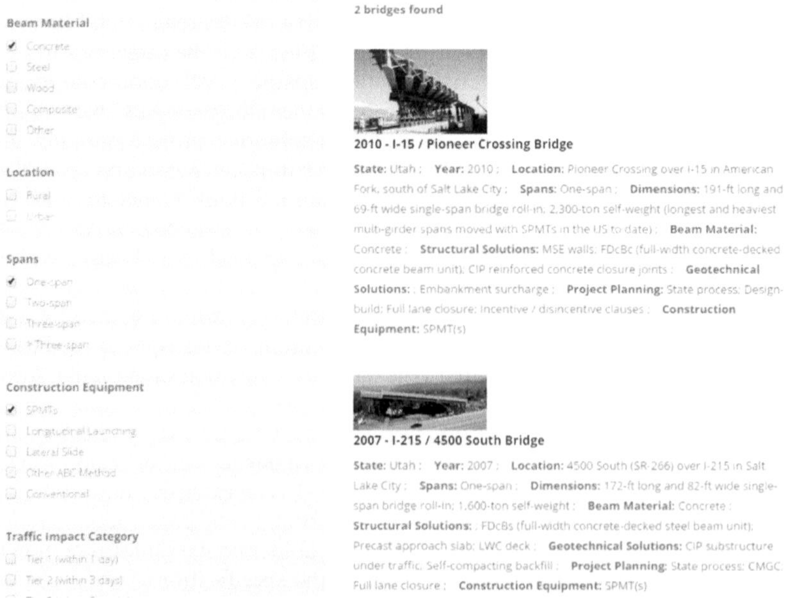

Figure 3.   Example of a faceted navigation search.

Figure 4.   Partial view of example project summary sheet.

To conduct more detailed searches, faceted navigation can be combined with the keyword search option described below, using the listing of specific keywords provided in the database.

2.3.3   *Keyword search*

The keyword search allows users to search the database for the specific ABC technologies of interest. This feature adds search flexibility and can supplement the faceted navigation feature. For ABC technologies listed under the categories of "Structural Solutions," "Geotechnical Solutions," and "Project Planning," users are encouraged to copy and paste the specific terminology for the ABC technologies of interest from the "Keywords" page to ensure all projects with those technologies are accessed in the search. For example, if the user is interested in accessing all bridges built with Modular Decked Beams having steel beams and concrete deck, in the keyword search the user would input "MDcBs" from the Keywords listing in the database, shown under "Deck Beam Elements" in Figure 5.

As previously mentioned, the keyword search can supplement the faceted navigation feature to further define the projects of interest. When a keyword(s) is entered into the keyword search bar in combination with selected faceted navigation categories, the keyword search will act as an additional category with an "AND" criterion. For example, if the user wants to access only one-span concrete-beam bridge projects that were installed with SPMTs and that used the Construction Manager/General Contractor project delivery method, the user would keep the input on the navigation pane shown in Figure 3 and add "CMGC" in the keyword search bar; see Figure 6.

2.4   *Project submission process*

The Project Database has a two-step submission and review process in which any user can propose a completed ABC project, but only projects approved by the bridge owner and consistent with the current definition of ABC are posted in the database. This submission process is shown in Figure 7 and described below.

2.4.1   *Step one of project submission process*

The first step in the submission process is simple and can be accessed via the "Submit Project" option. The first-step submission, shown in Figure 8, can be quickly and easily completed as it only requires information on the submitter and project owner, and basic information on the project including a brief statement of why it should be considered for the database.

Deck Beam Elements

- Adjacent Deck Bulb T Beam
- Adjacent T Beam
- Adjacent Inverted T Beam
- Adjacent Box Beam
- Adjacent Slab Beam
- Adjacent Slab Beam w/Backwall
- MDcBc {Modular concrete-Decked concrete Beam}
- MDcBs {Modular concrete-Decked steel Beam}
- PT Concrete Through-Girder
- Other Deck Beam Element

Full-Width Beam Elements

- Truss Span w/o Deck
- Arch Span w/o Deck
- Precast Segmental
- Steel Segmental
- Other Full-Width Beam Element

Pier Elements

- Precast Pile Cap
- Precast Cap Shell
- Precast Cap & Column(s)
- Precast Column Cap
- Precast Column(s)
- Precast Footing Shell
- Precast Footing(s)
- Precast Caisson Cap

- Steel Pile Cap
- Steel Column Cap
- Steel Column(s)
- Steel Cap & Column(s)
- Other Pier Element

Figure 5.   Example of keyword terminology to copy and paste for keyword search.

Figure 6.   Example of a faceted navigation search (from Figure 3) combined with keyword search.

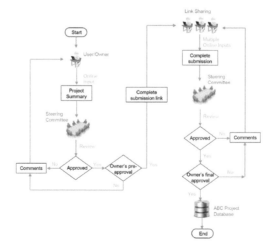

Figure 7.   Flowchart of submission and review process for ABC project database.

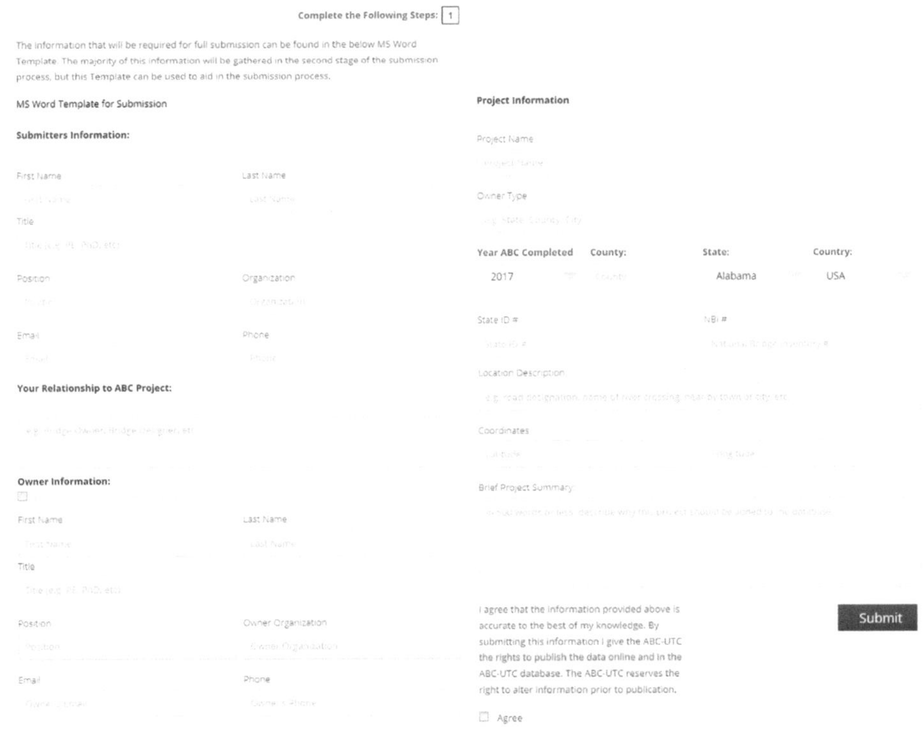

Figure 8.   Complete view of step-one submission page.

Upon completion of step one, the submitter sees a confirmation page. The chair of the ABC Project Database Steering Committee reviews the submission and discusses it with the committee as needed. If the project is deemed appropriate by the committee and the owner gives approval for the project to be included in the database, then a submission link for the step-two submission is sent to the submitter. If it is decided that the project is not an ABC project or the owner does not give approval, then the submitter is informed accordingly and no further action is taken.

2.4.2   *Step two of project submission process*

The second step of the submission process is more detailed. The goal of step two is to collect all remaining data on the project needed for the project database. Data include general project and bridge information, ABC benefits, ABC technologies used in the project, cost and funding information, and contact information for the primary stakeholders. Documents to be uploaded may include contract plans, specifications, bid tabs, construction schedule, photos, and other documents related to the project.

To complete step two after step-one approval, the submitter receives a URL link that is connected to the project. This unique link allows multiple users to work on the submission by link sharing. Information on each page can be saved in a temporary database by clicking the "Next" button at the bottom of the page. Saved data reappear when the page is reopened. The project submission is completed by pressing the "Submit" button on the last page of the step-two submission.

After the project is submitted, the submitter sees a confirmation page similar to step one. The Steering Committee chair then reviews the final submission and contacts the submitter as needed to clarify or update data. The chair discusses with the committee any issues related to posting the project in the database prior to asking the bridge owner for final review and approval to post. This process facilitates the addition of projects to the database while ensuring database integrity is maintained.

3 ABC RESEARCH DATABASE

3.1 *Description of research database*

The ABC Research Database provides compiled information on completed (published and unpublished), ongoing, and planned ABC-related research projects. The information is available in a user-friendly format that allows easy access to research details. The ABC Research Database mirrors and links with the ABC Project Database.

Research details typically include the research project name, research status, topic, abstract, keywords, specific ABC aspect, budget and timeline, primary sponsor, primary performing organization, and downloadable documents such as the final report.

3.2 *Accessing the research database*

Similar to the ABC Project Database, the ABC Research Database is available from the left vertical menu on the ABC-UTC home page (www.abc-utc.fiu.edu) by clicking "Project and Research Databases." This takes the user to the page that provides various options including a link to go directly to the research database. The page also explains the one-step process that can be used to submit research for the database.

The user accesses the database by clicking the blue box titled "ABC Research Database." This takes the user to the research database located on a separate server, as shown in Figure 9.

3.3 *Research database search capabilities*

Two methods to search are available to users of the research database—faceted navigation and the keyword search. Each of these methods is described below.

3.3.1 *Faceted navigation search*

The faceted navigation search option allows users to reduce the number of research projects to show only those of interest. This is achieved by selecting the desired search filters in various categories on the faceted navigation pane, shown in Figure 9 on the left side of the page. The categories in the faceted navigation pane are status, topics, states, and project start date. The search functions as an "AND" criterion within and between categories.

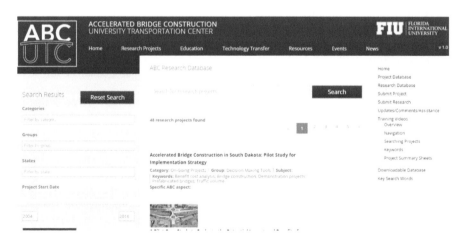

Figure 9.   ABC-UTC research database home page.

3.3.2 *Keyword search*

The keyword search allows users to search the database for the specific ABC research of interest. This feature adds search flexibility and can supplement the faceted navigation feature.

3.4 *Research submission process*

The Research Database has a one-step submission and review process in which any user can propose research but, similar to the Project Database, only research consistent with the current definition of ABC is approved for posting in the database. The submission process can be completed relatively quickly as it only requires information on the submitter, primary research sponsor, primary performing organization, and basic research details including an abstract.

Upon completion of the submission, the submitter sees a confirmation page. The FIU Research Database contact reviews the submission. If the project is deemed appropriate, it is posted. This process facilitates the addition of research to the database while ensuring database integrity.

4 SUMMARY

The first use of an innovation such as an ABC technology can increase the risk of project delay, cost overrun, and other challenges. These challenges can be lessened with knowledge of similar details in projects successfully completed by others. The ABC Project Database houses completed bridge construction projects that can be viewed by others developing similar projects, to increase their understanding of the ABC technologies used in those projects. The database includes significant project details as well as reference documents for many of the projects.

Similarly, the ABC Research Database is a repository of completed, ongoing, and proposed research on ABC-related topics that can be searched by users interested in determining the latest innovations in ABC.

These databases allow bridge professionals to easily search and view information on construction projects and research related to ABC and to submit information on additional ABC projects and research. The search interface is accomplished through faceted navigation and the keyword search for both databases, and additionally through an interactive map for the project database. The submission process allows any user to submit a project or research to be considered for the databases, while ensuring that only projects and research consistent with the current definition of ABC are included in the databases.

It is anticipated that the submission of additional construction projects and research will result in robust databases that will facilitate the implementation of successful ABC projects and fill ABC-related research gaps in addition to expanding the number and enhancing the quality of ABC products from research.

REFERENCES

ABC Joint Subcommittee. Transportation Research Board. Website: http://www.trbaff103.com/ [March 10, 2017].
ABC-UTC. Accelerated Bridge Construction University Transportation Center—home. Website: http://abc-utc.fiu.edu/ [March 10, 2017].
American Society of Civil Engineers (ASCE). 2017. Infrastructure report card. Website: http://www.infrastructurereportcard.org [March 10, 2017].
Garber, D. 2016a. "Compilation of accelerated bridge construction bridges." Research final report. Florida International University, Miami, FL. Website: http://abc-utc.fiu.edu/research-projects/compilation-of-abc-solutions/ [March 10, 2017].
Garber, D. 2016b. "International database of ABC research." Research project. Florida International University, Miami, FL. Website: http://abc-utc.fiu.edu/research-projects/international-database-of-abc-research/ [March 10, 2017].

Chapter 11

Analysis and design of the South Road double composite steel tub girder bridge

M. Loureiro & M. Ingram
Jacobs, Chicago, IL, USA

M. Loizias
Jacobs, Morristown, NJ, USA

ABSTRACT: The South Road Bridge is part of the $620 million Darlington Upgrade design/build project which provides for the improvement of approximately 3.2 kilometers of one of the most important transit corridors in Adelaide, Australia. The bridge is 180 meters long and consists of three-span continuous twin curved steel tub-girders carrying a multi-use path as well as vehicular traffic over a major expressway. The design of the bridge is innovative in its use of double composite concrete construction, where the girder section over the piers acts compositely with the deck as well with the concrete poured in the bottom of the steel tub section to resist negative bending moment demands. To minimize closure times on the expressway, the contractor chose to build the bridge in its entirety at a nearby site and move the bridge into place using self-propelled modular transporters (SPMTs).

1 PROJECT BACKGROUND

1.1 Project motivation

The North-South Corridor is one of Adelaide's most important transportation corridors. It serves as a major route for northbound and southbound traffic including freight vehicles running between Old Noarlunga suburbs on the south to the town of Gawler on the north, a distance of approximately 80 kilometers. The development of a non-stop North-South corridor for Adelaide is part of a strategic initiative to reduce Adelaide's urban traffic congestion and to stimulate economic growth by providing an efficient transportation link to strategic ports and interstate destinations in the region.

The Darlington Upgrade project is an important section in the delivery of Adelaide's North-South corridor as it will provide full reconstruction of approximately 3.2 kilometers of the existing Main South Road. This section of South Road is frequently congested for northbound traffic in the morning peak period and southbound traffic in the afternoon peak period. The Darlington precinct allows vehicular flow from the south via the Main South Road and the Southern Expressway which often comes to a halt as these two main arteries come to a main intersection slightly south of Flinders Drive—a popular destination.

In order to alleviate the congestion at the intersection of Main South Road and the Southern Expressway a three-span overpass bridge structure was proposed to carry the Main South Road northbound traffic over the Southern Expressway as shown by Figure 1.

1.2 Bridge geometry

The bridge span arrangement comprises of three span continuous twin composite steel box girders of variable depth. The span lengths consist of 71.6 m, 66.8 m, and 40.1 m long and

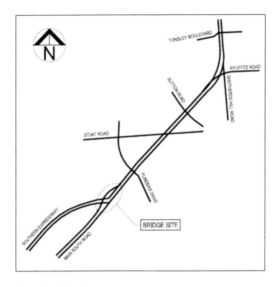

Figure 1.   Bridge site and project location.

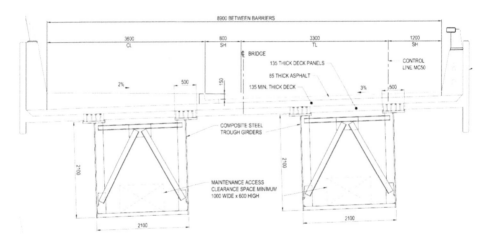

Figure 2.   Bridge cross section.

the depth of the box girders vary between 2.1 m at the bridge abutments to 3.8 m at the piers. The bridge horizontal alignment is set on a 75.3 m radius at a superelevation of 3% grade. The bridge width is 8.9 m between traffic barriers. Due to the project overall traffic staging the bridge will be temporarily open to two lanes of traffic and in its final condition it will accommodate one lane of traffic and a pedestrian carriageway. Medium performance traffic barriers are provided along the edges of the deck. The barriers are a combination of 0.9 m high single slope concrete barrier, and a steel post and rail system attached at the top of the barrier. An overall cross section of the proposed bridge is shown in Figure 2.

1.3   *Substructure*

The substructure comprises of two reinforced concrete piled abutments and two reinforced concrete piled piers. All piles are continuous flight auger (CFA) piles (drilled shafts) founded

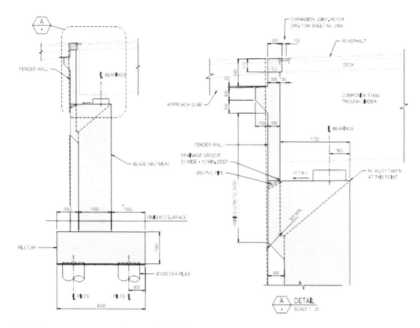

Figure 3.   Abutment cross-section and details.

on sand/clay/gravel type soil layers which sit atop a siltstone substrate. The abutments varied in height from 10 m at the bridge south end to 6 m at the north end, and consisted of a 1.6 m thick blade wall up to the underside of the box girder and a 0.3 m thick backwall. The abutment wall retains soil backfill and it is supported on two rows of three 1050 mm diameter CFA piles as shown by Figure 3. Mechanically stabilized earth (MSE) walls are used to retain the side approaches to the abutments.

The bridge superstructure is supported by two piers. Pier 1 comprises of a single 2.0 m diameter column connected to a 1.5 m deep pile cap that is supported by six 1050 mm diameter CFA piles. Pier 2 comprises of a pair of 1.5 m diameter columns connected to a 1.5 m deep pile cap that is supported by six 1050 mm diameter CFA piles as shown by Figure 4.

The abutments and the piers were designed to sustain a collision load of 3,000 kN which corresponds to a design speed of 80 km/hr. Additionally, the substructure was designed to resist lateral loading from earthquake and to meet the minimum lateral provisions of the Australian Standard AS 5100.2.

1.4   *Superstructure*

The bridge superstructure design incorporated several requests from the Contractor that primarily emphasized accelerated bridge construction. The superstructure construction is carried out off-line in an assembly yard with the intent of reducing the construction time and to minimize traffic impact to the Southern Expressway.

The tub girders are pre-fabricated in six segments and transported to an assembly yard in close proximity to the bridge site. The tubs consisted of 350 MPa grade steel and measured 2.1 m wide with straight vertical webs. The tub top flanges varied from 600 mm wide at the piers to 535 wide at the midspans. The bottom flange thicknesses varied from a minimum of 25 mm to a maximum of 40 mm plates. The tub segments will be supported on temporary support towers in the assembly yard. All tub segments contain internal bracing and top flange lateral bracing. Additionally, full height bearing stiffeners are provided at temporary support locations and at supports points for the SPMT lift.

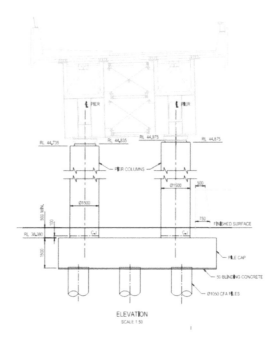

Figure 4.   Pier 2 elevation view.

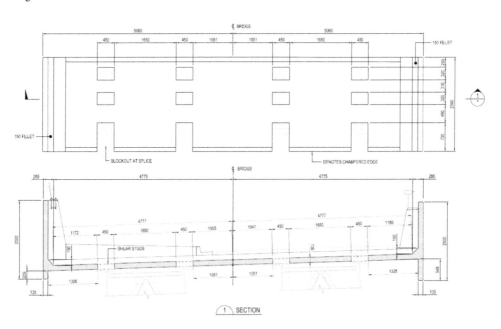

Figure 5.   3.0 meters long precast deck panels and topping slab.

The bridge deck utilizes precast deck panels 135 mm thick with pockets for shear stud placement. The deck concrete strength was 50 MPa for the precast deck panel and 40 MPa for the deck cast-in-place portion. The outer shell of the traffic barrier is made integral with the precast deck panel to expedite construction. The precast panels span the entire width of the bridge and are approximately 3.0 m in length as shown by Figure 5. After the deck panels are

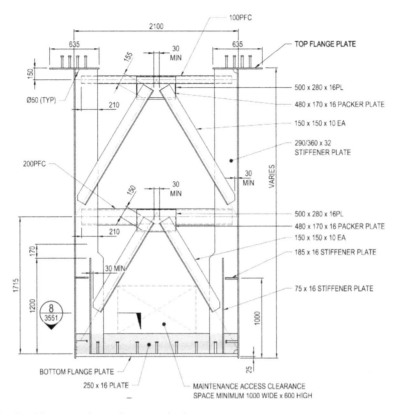

Figure 6. Double-composite section over pier 1.

placed on the tub girders, a 135 mm cast in place concrete slab is poured over the panels and the shear pockets to obtain composite action. The topping slab has a single layer of longitudinal reinforcing bars acting as tensile reinforcement in the negative bending moment regions and as general crack control reinforcement for the positive moment regions. The topping slab also includes transverse bars that lap and connect with the traffic barrier starter bars.

The superstructure design used a double-composite section over Pier 1 in order to increase the load carrying capacity of the tub section for negative bending moment demand. Due to the unbalanced span arrangement the lower negative moment demand at Pier 2 did not necessitate a double-composite section. The double-composite section was achieved by providing a 250 mm thick stiffening slab at the bottom of the tub section as shown by Figure 6. Transverse stiffeners across the bottom flange were provided to prevent the bulging out of the tub bottom flange during pouring of the bottom slab. Shear studs were provided at the bottom flange to obtain composite action and minimum reinforcing steel was provided to prevent concrete cracking.

2 DESIGN APPROACH

2.1 Design standard and live loading

The design standard used was the Australian Standard AS 5100 (AS5100 2004) and the project specific structural design requirements set by the Owner. In addition, the bridge superstructure was also designed according to the AASHTO LRFD Bridge Design Specifications (AASHTO LRFD 2014). In general, both design standards, the AS 5100 and the AASHTO LRFD, were in close agreement. Minor discrepancies were found between both design codes that yielded different sets of results. In these instances, the more conservative design approach was implemented in the final design.

The applied live loading used for design was strictly based on the AS 5100.2 design standard (AS5100 2004) which provides for a broader spectrum of live loading than the AASHTO LRFD (AASHTO LRFD 2014). The applied live loads consisted of the following:

- W80 wheel load: this live load models an individual heavy wheel load. The load consists of an 80 kN load uniformly applied on a contact area 400 mm × 250 mm. This wheel load is applied anywhere on the roadway surface.
- A160 axle load: this live load models an individual heavy axle. It consists of an 160 kN axle load with wheels spaced at 2 m with a surface contact area of 400 mm × 250 mm placed on a 3.2 m wide standard design lane.
- M1600: this live load models moving traffic load and it is shown in Figure 7. The M1600 load is positioned laterally within a 3.2 m standard design lane. The uniformly distributed load component of the M1600 load can be applied in a continuous or discontinuous manner and of any length to produce the most adverse effects.
- S1600: this live load models a stationary queue of traffic. The application of this live load is similar to the M1600. The distinct difference of the S1600 loading is that its single triaxial group weighs 240 kN compared to the 360 kN for the M1600. However, the S1600 loading has a considerably higher uniform distributed load of 24 kN/m compared to the 6 kN/m for the M1600.

Final live loading results indicated that the S16 stationary load controlled over the negative moment regions, while the M1600 moving load controlled over the positive moment regions.

2.2   *Bridge modeling*

The bridge staged construction on temporary supports combined with the different pour sequences for the bottom and top slab and the SPMT bridge move necessitated comprehensive structural modeling to capture the locked-in force effects from the different stages. Additionally, the bridge was analyzed for different scenarios of differential displacements including bridge twist conditions that may occur during the SPMT move and that may negatively affect the bridge superstructure.

A full 3D finite element model of the bridge substructure and superstructure was developed with Midas Civil as shown by Figure 8. The full construction sequence including construction on temporary supports to the final SPMT move was modeled in the Construction Stage Analysis module of Midas. The effects of creep and shrinkage were also accounted in the final model. The post-processing which included the assembly of all load combinations (service and factored load combinations) was internally created within Midas.

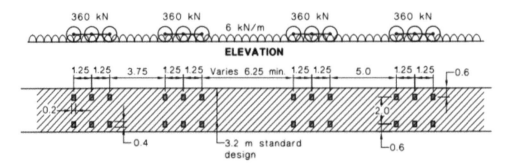

Figure 7.   AS 5100.2 M1600 moving traffic load configuration.

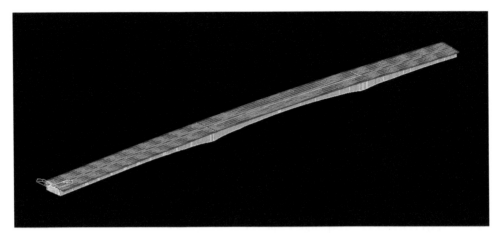

Figure 8.   Bridge FEM model (all construction stages were combined into one single model).

Figure 9.   FEA model (containing tub girders internal bracing and internal stiffeners).

The design also took into consideration the analysis of local force effects to specific structural members such as internal bracing members, top flange lateral bracing, and points of local support for the SPMT move. Additionally, the transverse self-weight distribution of the precast panels over the tub top flanges was analyzed. Due to the presence of the integral barrier at the end of the panels the exterior tub flanges carried a higher portion of the precast panel self-weight than the interior flanges. The unbalanced loads on the tub flanges led to an increase in the top flange lateral bracing which was accounted for in the design. The analysis for local force effects was performed via finite element analysis with a model that contained the true representation of the superstructure with its internal bracing, transverse and longitudinal web stiffeners and the deck on its non-composite and composite state as shown by Figure 9.

2.3   *Double-composite design*

Double composite design is not a recent development in the bridge industry. Its application has been primarily implemented in Europe with its known first application on the Ciérvana Bridge built in Spain in 1978. Since then bridges with double-composite action have grown in application and that include the Arroyo de Las Piedras Bridge in Spain, and the A5 Highway Viaduct Bridge in Portugal. Its application in the United States has been nearly non-existent and most interest has focused on research studies.

The concept of double-composite has been predominantly used for short and medium span bridges with rectangular or trapezoidal box sections. Despite the growing application of double-composite design there is limited literature and design guidance in standard bridge design codes.

The overall design for the double-composite section adopted herein utilized the findings and recommendations by the research performed by the University of South Florida on the evaluation of double-composite action (Sen and Stroh, 2010), the AASHTO LRFD standards (AASHTO LRFD 2014), and a full 3D finite element model. The design approach adopted the following guidelines, but not limited to:

- The maximum compressive stress in the bottom slab at the strength limit state was limited to a maximum of 0.6fc'
- The section was evaluated as a non-compact section
- The section satisfies AASHTO LRFD ductility requirement as per Section 6.10.7.3
- Reinforcing steel is provided at the bottom slab to prevent cracking due to shrinkage effects
- Transverse stiffeners are placed along the tub bottom flange to limit bulging out of the flange due to bottom slab wet concrete weight

2.4   *SPMT move*

Considerable design effort was devoted for the analysis of the SPMT move. The continuous span arrangement combined with a reasonably flexible superstructure makes the structure susceptible to a large combination of undesirable differential displacements during the bridge move. In addition to the analysis of the overall structural integrity of the bridge structure during the SPMT move, the analysis of deck stresses is of utmost concern in order to prevent widespread and undesirable cracking of the deck.

The proposed sequence of the SPMT move added an additional layer of complexity as the required bridge placement to its final position necessitated a series of translational and rotational movements as shown in Figure 10. This complex series of movements creates numerous combinations of potential differential settlements between SPMT supports that may include horizontal and vertical bending, and twist of the superstructure.

The analysis for the SPMT move included the development of numerous differential settlements scenarios that could potentially occur during the bridge transport. Each scenario was inputted to the bridge 3D FEM model by means of a support settlement command and the superstructure forces and the deck stresses are analyzed as shown by Figure 11. A limiting value of differential settlement was obtained for each scenario. The limiting value ensured that the deck stresses remained below the modulus of rupture of the concrete in order to prevent cracking of the deck.

After the controlling cases were obtained along with the limiting values a monitoring plan was developed in conjunction with the contractor responsible for the bridge move. The monitoring system uses laser beams aligned along the traffic barriers at one end of the bridge and reflecting mirrors at the opposite end. Readings are obtained on a real-time basis and monitored so not to exceed the established limiting values.

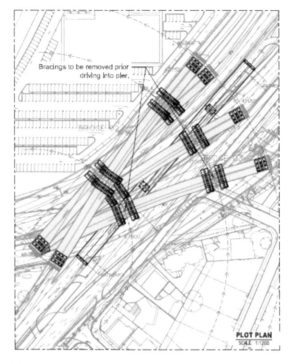

Figure 10.   SPMT bridge moves.

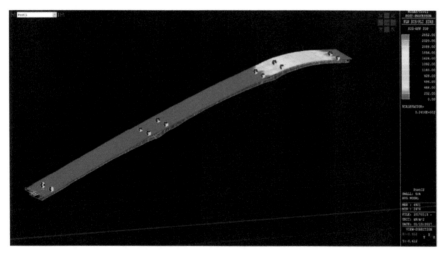

Figure 11.   3D FEM model showing SPMT bridge move differential support displacement and deck stress contours.

3   CONCLUSIONS

The South Road Bridge is part of the $620 million Darlington Upgrade design/build project which will provide for the improvement of approximately 3.2 kilometers of one of the most important transit corridors in Adelaide, Australia. The South Road Bridge will be a three-span overpass bridge structure that will carry Main South Road northbound traffic over the Southern Expressway.

The new bridge structure is designed to take advantage of several concepts of accelerated bridge construction that include the use of precast deck panels with integral barriers and shear stud pockets, off-site construction and the use of SPMTs. The use of accelerated bridge construction techniques in this project will allow for safe and cost-effective construction time with minimum traffic impact in this busy traffic corridor. The design also takes advantage of the double-composite action that can be created over the negative moment regions by placing a concrete slab at the girder bottom flange.

The concept of double-composite girder bridges is not new, but its application has not been widely implemented across the United States or in other countries. The Darlington Upgrade project is unique on itself as several bridges along this corridor adopt the double-composite design.

REFERENCES

AASHTO LRFD Bridge Design Specifications 7th Edition, 2014.
AS 5100 Bridge Design Standard Australian Code, 2004.
Sen, R. and Stroh, S., 2010. Design and evaluation of steel bridges with double composite action. University of South Florida.

Chapter 12

Repairs to 13 movable bridges in New York City after Hurricane Sandy

E. Kelly
HNTB Corporation, NY, USA

B. Gusani
New York City Department of Transportation, New York, USA

ABSTRACT: Hurricane Sandy hit New York City with a devastating blow in October 2012. Every portion of the City's infrastructure was damaged and needed repair. This included New York City Department of Transportation (NYCDOT) roads and bridges. Particularly hit hard were NYCDOT's movable bridges, which are located in low lying areas. In November 2012, through an Engineering Services Agreement, NYCDOT retained the services of HNTB Corporation to assist NYCDOT with the inspection, assessment and recommendations of their movable bridges to determine the repairs necessary to place the bridges back to pre-Sandy condition. This paper will discuss the impact that Sandy had on the bridges, the measures and procedures that NYCDOT had to follow to ensure permitting and funding was obtained as well as the hurdles that needed to be crossed to develop biddable repair contracts in a short time and in compliance with Federal guidelines.

1 INTRODUCTION

1.1 *Description of NYCDOT bridges*

NYCDOT oversees 789 bridges and tunnels within the 5 Boroughs of New York City. Within that inventory are 25 diversified movable bridges include two retractile bridges, seven swing bridges, four lift bridges, and twelve bascule (drawbridge) bridges of both single and double leaf types (NYCDOT 2017). Figure 1 shows the location of NYCDOT movable bridges.

These bridges vary in size, age and condition, that are required to operate on demand for marine traffic except where the United States Coast Guard has given a grant to relax the operation requirement to a specific schedule. The bridges are generally clustered over specific bodies of water. There are eight bridges that span across the Harlem River between Manhattan and the Bronx, six that span the Newtown Creek and its tributaries in Queens, five that span across the Gowanus Canal in Brooklyn, four that span across the Bronx River, Eastchester and Westchester Creeks in the Bronx, one that spans the East River to Roosevelt Island and one over Mill Basin in Brooklyn.

Figure 2 shows Metropolitan Avenue Bridge over English Kills in Brooklyn, while Figure 3 shows Madison Avenue Bridge over Harlem River in Manhattan.

1.2 *Description of Superstorm (Hurricane) Sandy*

In late October 2012, a tropical wave in the Caribbean developed on October 19. It quickly developed into a tropical storm within 6 hours and upgraded to a hurricane on October 24, 2012. Hurricane Sandy made landfall in the United States (US) about 8 p.m. EDT on October 29, striking near Atlantic City, NJ with winds of 80 mph See (Figure 4). A full

147

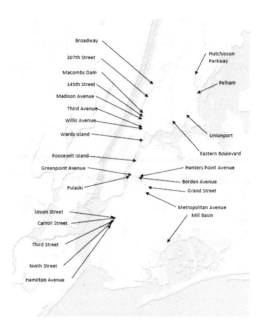

Figure 1.   Map showing location of NYCDOT movable bridges.

Figure 2.   Metropolitan Avenue Bridge over English Kills, Brooklyn—typical view of low-lying bridge. Note proximity of machinery to water level.

moon occurring at that time made high tides 20 percent higher than normal and amplified Hurricane Sandy's storm surge. Sandy, the 10th hurricane of the 2012 Atlantic hurricane season was the second largest Atlantic tropical cyclone on record according to the National Hurricane Center and had the lowest measured central pressure at 940 millibars (27.76 inches) to make landfall north of Cape Hatteras, NC. When ranking a hurricane by strength, the choice would be to compare wind speed, but since measurements of most extreme winds is difficult, hurricanes are compared by their lowest central pressure. The lower the pressure in a hurricane, the stronger its winds. The previous record holder was the 1938 "Long Island Express" Hurricane, which dropped as low as 946 millibars in 1938 (CNN 2016).

Sandy's strength and angle of approach combined to produce a record storm surge of water into New York City. The surge level at Battery Park topped 13.88 feet at 9:24 pm Monday

Figure 3.   Madison Avenue Bridge over Harlem River, Manhattan—typical view for larger structure with significant height above water level.

Figure 4.   Image of Hurricane Sandy after making landfall on the eastern coast of the US (NOAA 2012).

surpassing the 10.02 feet record set by Hurricane Donna in 1960. New York Harbor's surf also reached a record level when a buoy measured a 32.5 foot wave on Monday. That was 6.5 feet higher than a 25 foot wave churned up by Hurricane Irene in 2011 (CNN 2016).

2   RESULTS OF THE STORM DAMAGE

The results of the storm were felt not only along the east coast of the US, but as far in as the Great Lakes. The damage was devastating not only due to loss of life, but also the destruction of property and the impact of the sustained damage that is still being repaired to this date. News reports tried to explain and show the impacts the storm had on the infrastructure of New York City both during and after the storm, but the breadth and magnitude were difficult to describe. Some of the statistics follow:

- 43 deaths
- 6,500 patients evacuated from hospitals and nursing homes
- Nearly 90,000 buildings in the inundation zone
- 110 homes burned to the ground in the Breezy Point neighborhood of Queens million New York City children unable to attend school for a week
- Close to 2 million people without power
- 11 million travelers affected daily
- New York Stock Exchange was closed for 2 days
- Subway service suspended for 3 days
- Airports closed for 2 days, reopen with limited service (NYC 2013)

3   ASSESSMENT OF CONDITIONS

NYCDOT was affected as well with 2 of their 3 tunnels flooded and 500 miles of roads suffered significant damage. Major bridges reopened after winds dissipated and some portions of the transportation network, not directly flooded, experienced minor damage. The conditions of the movable bridges had to be assessed.

NYCDOT wanted to perform the repairs in accordance with the Emergency Relief Program and to do this needed to perform an assessment of all their assets, including the condition of their movable bridges. This was to determine the extent of the damage as well as obtain an estimated cost to perform the repairs. The first step was for NYCDOT to examine their facilities to see what damage, if any, occurred. This was immediately performed after the storm by NYCDOT maintenance forces. All bridges were visually examined and from this initial assessment, it was determined that 20 of the 25 movable bridges had been affected in some form or another by the storm.

Due to the magnitude of the damage and NYCDOT having to perform other vital functions, NYCDOT looked to retain the services of an engineering firm familiar with the inspection and design of movable bridges to perform an inspection of the 20 bridges identified by NYCDOT. The goal of this was to assess and document the conditions caused by the storm. The urgency of this assignment prevented standard procurement procedures from being followed so NYCDOT utilized their Engineering Services Agreement (ESA) to retain HNTB Corporation.

HNTB was informed of the scope of work for the assignment on November 20, 2012. HNTB developed a preliminary fee estimate and staffing for the assignment and was given a letter of intent (conditional NTP) by November 27, 2012. HNTB immediately developed a schedule for collecting available documents on the bridges, becoming familiar with the structures, and scheduling the inspection. The inspection would be performed on the 20 bridges from December 4 through December 13 with an average of 3 bridges being inspected per day. Reports on the findings, conditions, recommendations and costs were completed by December 25.

3.1   *Damage inspection and collection of data*

Upon arrival to a structure, the first task was to determine if the structure sustained wind damage. This was performed by visually examining the exterior of the structures as well as the traffic control components exposed to the outdoors such as the gates and signals. Maintenance personnel who were first on site were interviewed to see if repairs had been performed to wind damaged components that needed to be fixed for safety reasons such as a hanging traffic signal or a broken gate arm.

Final elevation of maximum water level needed to be determined to ascertain the level of possible damage due to flooding. Since the bridges cross different bodies of water in different locations in New York City, the surge tide varied from bridge to bridge. Once the level was obtained all areas of the bridge were examined for water intrusion. This was done by first visually examining mechanical and electrical equipment for water marks, collection of debris as well as rust staining or corrosion due to brackish salt water intrusion. The team of mechanical and electrical inspectors would then systematically document areas of damage and extent of water inundation. At the same time, personnel back in HNTB's office were reviewing existing plans of the bridges to confirm the field observed water level to the theoretical level based upon NOAA and FEMA flood maps that were being revised and becoming available at the time of inspection.

Conditions were generally typical for the major components. Any component that was below the surge level was immersed in salt water and the corrosive effects were noticeable (See Figure 5). Conduits and electrical boxes were flooded and in some instances still contained water even after a month (See Figure 6). Mechanical equipment including gearboxes and bearings, which have passages for lubrication and breathers, were also filled with salt water. Energized electrical equipment at the time of the storm such as transformers shorted and

Figure 5.   Electrical box with salt/debris build up and corrosion to conduit.

Figure 6.   Electrical box in tail lock area with water still flowing out after 1 month.

failed. Control panels for sump pumps became immersed and failed to operate, preventing discharge after the event subsided.

The following description of one bridge is typical of the condition observed. The Metropolitan Avenue Bridge over the English Kills is a twin double leaf trunnion bascule carrying four lanes of traffic between the Boroughs of Queens and Brooklyn. Top of roadway is approximately 16.66 feet above mean high water (MHW). Due to the low height of the structure the bascule spans are situated in a pit pier whose base is approximately 11 feet below MHW. There is a 3 story control house. The lower level (below roadway) houses the electrical equipment, the second (roadway) level is where the Assistant Bridge Operator resides and upper level is the control room. The electrical equipment including the transformer, motor drives and Motor Control Center are located in the lower level. The operating machinery including the tail locks and span machinery are located below the roadway under the bridge.

A summary of the findings excerpted from the report for the Metropolitan Avenue Bridge are as follows:

– It was determined that the Electrical Room was flooded to a depth of approximately 26–28 inches (varies as floor is not completely level), affecting much of the 480/277 Volt electrical distribution equipment. The bridge is presently operating on utility power, as much of the 208/120 volt distribution system was unaffected by the flooding due to its location above the flood line. Therefore, certain parts of the bridge lighting and heating systems are operational, but many circuits were non-functioning, disconnected, or tripped due to the variety of damage and temporary repairs.
– The depth of water at the bascule pier entrances on exterior to the bascule girders was approximately 58 to 60 inches deep, and the water reached to approximately 126 inches above the level of the floor at the front pier walls, next to the channel. The top of the flood water therefore reached a level just a few inches beneath the trunnion shafts, fully immersing the operating machinery and the tail lock machinery, including backup hydraulic systems, and the majority of electrical equipment in the machinery rooms. Navigation lights defining the edges of the navigation channel on the fenders were also submerged, as were channel flood lights.
– With regards to other mechanical and electrical equipment, it was reported that no damage occurred to either the span locks at the center of the bascule span or to the traffic gates and barrier gates on the roadway.

As can be seen in the above assessment, the water had free reign to penetrate any portion of the structure that was below the flood level and the water entered and resided within the equipment for a period of time.

On the larger structures that are higher above the water level, lesser but more specific damage was experienced. One such example is the Madison Avenue Bridge over the Harlem River.

The Madison Avenue Bridge is a 300 ft long rim bearing swing span that carries four lanes of traffic and 2 sidewalks between Madison and Fifth Avenues and East 138th Street in Manhattan and East 138th Street and Grand Concourse in the Bronx over the Harlem River. The swing span is driven by operating machinery mounted beneath the roadway and above the center pier, with two pinions engaging a circular rack just above the rollers.

The depth of water at the fenders submerged the navigation lights and a portion of the electrical enclosure mounted on the fender that services these lights. The operating machinery and the end wedge machinery, except for the rollers, lower track, racks, and pinions were not affected. No visible flood line was found on the piers.

A flood line was found 12 inches above the finished floor in the transformer room located near FDR Drive that immersed the 300 kva dry type transformer and submerged the 800 Amp main service circuit breaker interrupting the Manhattan electrical service to the bridge.

A similar example is the 145th Street Bridge. At this facility, the only observed damage occurred to the Submarine Cable Terminal Cabinet Platform and Cabinets on the center

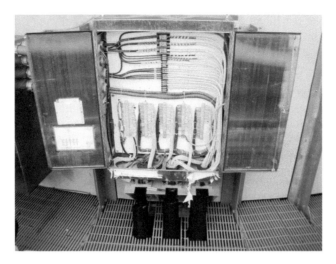

Figure 7.   Submarine cable terminal box partially submerged—note clear line showing water elevation.

pivot pier. These cabinets were flooded. The height of water level was 63 inches above the top of platform which is approximately 22 inches from base of cabinets, causing the bottom 1/3 of the terminal cabinets to be immersed (See Figure 7). This allowed the salt water to enter the ends of the submarine cable conductors jeopardizing the long term reliability of the conductors within the submarine cable.

3.2   *Damage assessment*

From the field assessment, the inspectors went back to the office to determine the repairs that would be needed to assure future long term reliable service of the mechanical and electrical systems. This was done by examining the conditions and investigating the corrosive effects salt water has on the components. Based on this assessment the safe and conservative approach was to replace components that were damaged from the salt water. This included electrical equipment, the wiring whose terminations were below the surge level as well as any conduit below the surge level. For the mechanical equipment, any lubricated surface that could be disassembled and cleaned could be reused, but components that utilized smaller high precision components such as roller bearings had to be replaced due to the corrosive effects of salt pitting the contact surfaces.

As part of this assessment, preventive measures needed to be considered to try and prevent future damage to any portion the bridge system. These mitigation measures could be in any form and would be considered so long as it was a cost-effective solution whose investment would minimize the damage if the situation were to occur again.

3.3   *Documentation submission*

Cost for repairs to the structures could be covered under the Emergency Relief Program by the Federal Highway Administration (FHWA) since Hurricane Sandy was a natural disaster.

All repair work at first was thought to be repaired as emergency repairs. Emergency Repairs are those repairs during and immediately following a disaster to restore essential traffic, to minimize the extent of damage, or to protect the remaining facilities. The cost of these repairs is covered 100% by the FHWA. These repairs need to be performed within 180 days of the occurrence of the disaster.

With continued discussions with FHWA, it was determined that since vehicular traffic could travel on the structures and the extent of repairs were too great to be performed within

180 days, that the rehabilitation would fall under the second category in the Emergency Relief Manual which is Permanent Repairs. Permanent Repairs are those repairs undertaken (usually after emergency repairs have been completed) to restore the highway to its pre-disaster condition. Permanent repairs must have prior FHWA approval and authorization unless done as part of the emergency repairs. The cost for these repairs are shared between FHWA and the owner (NYCDOT) with 90% Federal share for Interstates and 80% Federal share for all other Federally-aided highways. These repairs must be initiated before the end of the second Fiscal Year following the year in which the disaster occurred. Also, these repairs are to restore the structure to its pre-disaster condition. The exceptions to this are if betterments are introduced to prevent future damage or to bring a portion of the structure up to current code requirements.

It was determined that the repairs listed in the inspection report were eligible and the reconstruction would be Federally funded and supervised through New York State Department of Transportation (NYSDOT). NYSDOT indicated that their procedures needed to be followed for the funding to be used. Submitting the previous Inspection and Assessment report was the first step in the process. The findings of the reports were summarized in Detailed Damage Inspection Reports (DDIR) following FHWA requirements.

The next step in the documentation process was to develop the Design Approval Document (DAD) following NYSDOT guidelines. Since these emergency repairs were following in-kind replacement, the documentation would not have to be as detailed as a reconstruction project, but still had to be developed to ensure the permitting requirements are being followed.

Once the DAD was approved by NYSDOT (and FHWA), the development of repair plans could be performed. The contract documents (Plans and Specifications) had to be developed to ensure the repairs were clearly shown in a set of biddable documents.

Notification of the various agencies was required for the repairs to occur. Since it was an in-kind replacement, most of the Agencies found concurrence with the repairs and their approval was straight forward. This was true for all except the Nationwide Permit for the Army Corps of Engineers to allow the trenching for the replacement of damage submarine cables. This permitting process took longer since most of the waterways which these structures (Borden Avenue, Grand Street, 207th Street and Union Street Bridges) cross has contaminated material on the river bottom and suitable disposal measures for the trench material had to be included in the contracts.

4   REPAIR CONTRACT DEVELOPMENT

4.1   *Schedule*

The rehabilitation of the thirteen bridges was configured into three separate contracts. The first was to repair the NYCDOT's busiest bridge which is the Metropolitan Avenue Bridge. The second contract would encompass eight bridges that did not require permitting for any portion of the work and the third contract would encompass the four bridges that required permits for the submarine cable replacement. These were the Borden Avenue, Grand Street, 207th Street and the Union Street Bridges. This ensured that the permitting process would not delay the repairs to the other bridges.

The development of the contract plans had to be completed to allow bid opening by the end of October 2014. This made for the development of the plans and specifications on a very tight schedule and HNTB looked for ways to streamline the process since the development of repair plans started in April 2013.

4.2   *Contract development*

Replacement in-kind was the objective to meet Federal requirements. It was determined that any piece of equipment, either mechanical or electrical, that had sensitive components, which had been immersed, would be replaced. This included electrical equipment and any wiring whose terminations became submerged as well as conduits. This was difficult in many instances since the equipment on some of these structures had been in place for many years and were no longer manufactured. Similar newer equipment was available, but its adequacy, compatibility and ability to fit in the same location had to be researched.

For mechanical equipment, if the component had a greased surface, replacement was not warranted, rather the component would be disassembled, cleaned, examined, reassembled and reinstalled. If examination revealed excessive damage, then replacement would be considered on a case by case basis. Mechanical equipment that had roller bearings instead of plain bronze bearings were to have the roller elements replaced. These determinations came from investigation during the development phase where the roller bearings needed to be replaced was the result of comparing three separate sources that supports a roller element subject to salt exposure would pit and result in premature failure. This was confirmed by a bearing manufacturer, a bearing user and a client whose equipment had been previously flood damaged. The need to confirm this requirement was necessary because the gear reducers on several of these bridges had roller bearings and to properly replace the bearing required the reducers to be removed and disassembled. This increased the complexity and cost of the project.

The repairs to the mechanical and electrical systems would disable the bridge for an extended period, disrupting marine and vehicular traffic. Some of the bridges had to be operational as soon as possible and to allow that, temporary drives were included into the contracts. These were included to allow the contractor to perform the work on the mechanical and electrical systems without a tight critical deadline being imposed to complete the assignment as quickly as possible. It was felt that the added cost for the temporary drive was needed to ensure time would not be a factor to the contractor and have the contractor work overtime, have larger crew sizes and allow adequate time to perform the work properly both in the field and at the fabrication shops.

The temporary drives had to be compact and independent from the existing mechanical and electrical systems to allow the contractor to work on the existing systems while they were in place. The use of hydraulics and winches were determined to be the best method for the temporary drives (See Figure 8).

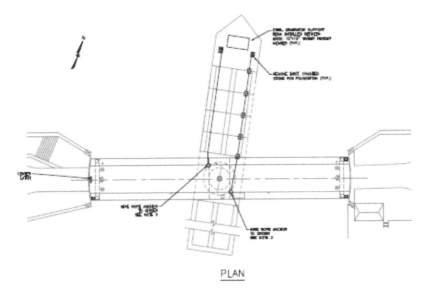

PLAN

Figure 8. Example of temporary drive for swing bridge through use of winches and cables.

The development of the drawings, predominantly the electrical drawings, were a production issue due to the amount of information that had to be depicted in such a short amount of time. To expedite the process, HNTB determined the best way was to use the current available plans as much as possible and point to the work that needed to be performed. This was done by taking the existing plans as a pdf and inserting them into a border. Once the base image was developed, call outs and dimensions were inserted by CADD. For the electrical drawings, the components and wiring to components that needed replacement would be darkened to define what was to be replaced and what was to remain.

4.3 *Mitigation measures*

As part of the design, the investigation into mitigation measures was conducted to see if any could be incorporated into the repairs. These mitigations could be any type of repair that would prevent the situation from occurring should a catastrophic event occur again. These measures could be large or small in breadth and scope. The understanding is the Federal Government will invest cost at the time of construction to minimize the damage in the future as well as prevent incidents from reoccurring.

Due to working within the confines of the existing movable bridge, the opportunities to develop mitigative measures were limited. HNTB looked primarily at two approaches; restrict flood water from entering and elevate equipment where possible.

At the Metropolitan Avenue Bridge, the electrical room had all the openings sealed, flood windows and doors installed while conduit openings and the conduits themselves were plugged using a special sealant.

Electrical equipment at other locations had pre-manufactured flood barrier walls installed to protect from a future event. The installation required foundations to be built around the perimeter of the facility having the equipment. This foundation served to support the barrier posts, provide a smooth level surface for sealing as well as a barrier to slow ground water penetration into the area. If an event were to occur, the posts would be put in place and flood barrier panels stacked to construct the wall. These removable posts and panels are stored when not in use.

At other locations, it was determined the most cost effective method would be to elevate the equipment above the anticipated flood level. The decision was to make sure the elevation is above the 0.2% BFE (500 yr year flood). This was done for the submarine cable terminal cabinets at 145th Street, Macombs Dam and University Heights Bridges.

Elevating of the equipment was also performed at the Carroll Street Bridge. The Carroll Street Bridge over the Gowanus Canal is a historical retractile bridge which utilizes wire ropes, sheaves and a winch drum to open and close the span (See Figure 9). The structure is

Figure 9. Carroll Street Bridge machinery outdoors and historic control house were inundated by Hurricane Sandy.

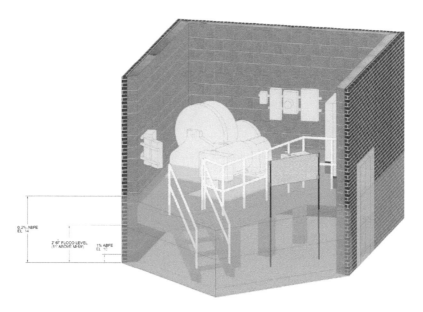

Figure 10.   Carroll street bridge equipment elevated in control house.

low lying over the Gowanus Canal and was overwhelmed when Sandy hit. Water level was measured to be 30 inches above the floor of the control house. Alternate methods of operating the bridge that would clear the flood level were investigated, but would alter the historical appearance as well as not being cost effective. It was determined to leave the equipment outside (sheaves, trucks and ropes) as is. The building could not be sealed due to the large opening in the floor that allows the ropes to go to the movable span. Also, the building construction is a CMU with red brick façade and it would not be able to take the water pressure without structural reinforcing.

Since all of the machinery and most of the electrical equipment within the house were affected by the flood, all the equipment was planned to be removed. The building, fortunately, has ceilings over 12 feet above the floor. This allowed a structural support system to be installed to elevate the machinery and electrical equipment above the 500 year flood level. Positioning of the equipment and detailing of the platform were critical to be able to fit into the confines of the building and allow the machinery to utilize the existing openings in the floor while still providing access through the existing door. The platform raised the machinery over 36 inches while still providing over 7 ft -9 in of head clearance on the platform. Operation of the bridge is from a small control desk in this room and the operator utilizes the windows to view the operation. Raising the control desk would have forced the operator to have to crouch down to look through the window which was not acceptable. The determination was to place the control desk on vertical guide rail system and utilize an electric hoist to raise and lower the desk. The desk would be stored in the raised position and when a bridge opening is required, the desk would be lowered into position to allow the operator to perform the opening. Once the opening is complete, the desk is again raised up and out of the way (See Figure 10).

5   CONCLUSIONS

Hurricane Sandy wreaked havoc to the communities and infrastructure of New York City. The advantage of a movable bridge, being a low-lying structure that can accommodate a large vertical clearance by opening, results in these structures being susceptible to inundation by water during storm events. Although it is not practical to design the structure for an extreme event, consideration to placement of critical and costly equipment such as the motor control center or the submarine cable terminal cabinets should be considered when designing a rehabilitation of an existing movable structure or a new movable structure.

REFERENCES

City of New York, A Stronger, More Resilient New York 2013.
CNN.com, Hurricane Sandy Fast Facts November 2016.
NOAA/NASA Geostationary Satellite Server 2012.
NYCDOT, http://www.nyc.gov/html/dot/html/infrastructure/movable-bridges.shtml 2017.

Bridge analysis & design

Chapter 13

Fatigue damage assessment of stay cables for the light rail transit bridges

J. Jiang
WSP Canada, Vancouver, British Columbia, Canada

R. Coughlin
City of Vancouver, Vancouver, British Columbia, Canada

ABSTRACT: This paper presents a recent study completed on assessing fatigue induced damage to stay cables for an in-service 616 m long light rail transit bridge with a main span of 340 m. In addition to numerical investigations, the study carried out field investigations of bridge deck ambient free-vibration and selected stay cable vibration frequency measurements to establish fundamental dynamic characteristics of the in-service bridge and permanent tension forces in the stay cables. The field results were used to calibrate the numerical model to facilitate the fatigue damage assessment of the stay cables. The fatigue assessment required realistic train loading information including vehicle types and train configurations; historic and future train passage data; and passenger traffic counts during peak and off-peak hours to develop a realistic train histogram to calculate the fatigue demands. As the positioning of the simultaneous train loading on the bridge deck is critical to the assessment, a train crossing study was completed to determine the timing and positioning of two bound trains crossing over the bridge deck simultaneously. The study identified a number of stay cables that are highly vulnerable to pre-mature fatigue failure under the current loading conditions. However, the study also demonstrates that fatigue life of the critical stay cables can be extended by optimizing departure time of trains in opposing directions to avoid simultaneous train passage over the bridge deck at the most critical locations. It can be concluded from the study that the fatigue assessment of stay cables is more complicated and imperative for light rail transit bridges than for highway bridges and that stay cables can be vulnerable to pre-mature fatigue failure even if the original design was based on modern bridge design standards.

1 INTRODUCTION

The use of cable stayed structures for long span bridges gained popularity in the late 20th century with the development of advanced construction materials and equipment, and advancement of sophisticated structural design and modelling capabilities. There are many Light Rail Transit (LRT) cable-stayed bridges in use today. Due to repetitive cyclic loading from passage of similar trains, the potential risk of fatigue-induced damage to stay cables is significantly higher for LRT bridges than for roadway bridges. As LRT bridges carry heavier trains and provide increased level of services, questions about the fatigue induced damage to stay cables and its impact on the remaining service life has been frequently raised by transit authorities.

For any type of bridge structure the intended use and actual vehicular loading changes over time. During the design process assumptions are made as to the future use and expected growth in demand. Particularly with cable-stayed LRT bridges, as train technology advances and passenger demands grow, heavier train loading and more frequent services result in increased fatigue-induced stresses, which may exceed initial assumptions made by the designer. As part of on-going bridge asset management programs it is important that transit authorities review and assess the impact of any significant changes to train capacities and service frequency on the fatigue life of the stay cables. Additionally, transit authorities should identify potential timelines for intervention, if necessary, including reprograming of train operating schedules to reduce demands or replacement of the stay cables that have reached their anticipated fatigue life.

A recent study completed by the authors to assess fatigue-induced damage to stay cables was completed for an in-service 616 m long LRT Bridge, built between 1987 and 1989, carrying two tracks with a main span of 340 m and a post-tensioned precast concrete deck supported by semi-parallel 7 mm galvanized wire bundle stay cables. The fatigue assessment required realistic train loading information including vehicle types and train configurations, historic and future train passage data, and passenger traffic counts during peak and off-peak hours to develop a train loading histogram to calculate the fatigue demands.

In addition to the numerical analysis, the study carried out field investigations of bridge deck ambient free-vibration and selected stay cable vibration frequency measurements to establish fundamental dynamic characteristics of the in-service bridge structure and permanent tension forces in the stay cables. The field results were used to calibrate the numerical model to facilitate the fatigue damage assessment of the stay cables.

The study also investigated the effects of simultaneous train passage over the bridge and their impact on the fatigue life of the stay cables. As the positioning of the simultaneous train loading on the bridge deck is critical to determine the fatigue demands, train passage data from the transit authority was analyzed to determine the timing and positioning of two bound trains crossing over the bridge deck simultaneously.

In general the study was carried out in three phases:

1. Field investigations to determine in-situ response of the bridge structure and its stay cables.
2. Structural analysis of the cable-stayed bridge and calibration with field measured data.
3. Fatigue damage assessment of the bridge stay cables.

2   FIELD INVESTIGATIONS

In order to assess the fundamental dynamic response of the LRT Bridge and to adequately calibrate the three-dimensional finite element model for the fatigue assessment, field-testing was conducted to determine the in-situ dynamic response of the bridge structure. Past and current field investigations utilized for the fatigue assessment included bridge deck surveys, deck ambient free-vibration testing and modal analysis, stay cable vibration frequency measurements under both structural dead load and static empty train loads, and an automated photographic timing record of train passages over the mid-span of the LRT Bridge for assessment of simultaneous train loading.

2.1   *Bridge deck surveys*

Development of an accurate structural model and evaluation of creep and shrinkage effects is heavily dependent on the availability of as-built bridge deck geometric information at completion of construction and subsequently over the service life of the bridge. As the as-built survey of the bridge deck at completion is not available a bridge deck survey was conducted, after 16 years in revenue services, to serve as a baseline followed by additional deck surveys in subsequent years to evaluate the change in deck profile and assess the creep and shrinkage effects.

2.2   *Cable free-vibration tests*

In 2010, a series of free-vibration testing was conducted for all stay cables on the LRT Bridge. Under the dead load condition, the permanent tension forces in all cables were calculated based on the measured natural frequencies, mode shapes, and estimated cable free length.

The results indicated that the tension forces in the cables due to dead load were generally higher than those specified on the design drawings. Further to the 2010 stay cable free vibration testing, additional free-vibration testing on nine selected stay cables was conducted in 2015 to further confirm the cable vibration frequency measurements under dead load alone and to determine if there is any noticeable changes to the stay cable free-vibration characteristics under static empty train loading.

2.3 Deck ambient free-vibration tests

To facilitate better calibration of the structural model and to ensure that overall dynamic response of the bridge is captured by the analysis, the bridge deck ambient free-vibration testing was conducted in 2015. The deck ambient free-vibration testing provided actual data, which was used to determine predominant free-vibration frequencies and associated mode shapes of the bridge structure. Testing included ground measurements, three-dimensional measurements on the deck, and measurements of the towers at deck level.

3 STRUCTURAL MODEL CALIBRATION

Fatigue damage assessment of the LRT Bridge stay cables included development of a three-dimensional finite element model utilizing nonlinear cable elements to capture the catenary behavior of the stay cables. Deck survey data and free-vibration measurements collected from field investigations were utilized to verify the structural model to ensure it captures the physical conditions and behavior of the bridge, e.g. cable permanent tension forces, deck free-vibration characteristics (e.g. natural frequencies and mode shapes).

3.1 Calibration to deck ambient free-vibration measurements

The analysis from the deck ambient free-vibration measurements indicated a number of well-defined modes which were used to calibrate the structural model. Table 1 provides a summary of the selected free-vibration modes used for comparison with the structural model.

   The bridge deck free-vibration frequencies from the field measurements and the structural model were found to have a good correlation, and differences are generally small, providing verification that the model was able to capture the in-situ response of the bridge structure.

3.2 Calibration to cable free-vibration measurements

As part of the field investigations, two static empty train loading cases were considered for the cable free-vibration frequency measurements, e.g. one or two empty four-car trains placed at mid-span of the bridge, respectively. To facilitate calibration of the model, the same loading conditions were simulated by the model. All total cable tension forces from the model were found to be within ten percent of the forces obtained from cable free-vibration measurements.

Table 1. Frequency comparison—field measured and calculated.

Mode description	Field measured (Hz)	Calculated (Hz)
1st Transverse	0.205	0.166
1st Vertical	0.282	0.270
1st Torsional	0.350	0.410
2nd Vertical	0.566	0.563
3rd Vertical	0.714	0.696

As the fatigue assessment is based on the stress ranges experienced by the stay cables, a further investigation of the incremental change of the cable forces due to static empty train loading was carried out. The incremental cable forces from the model were verified by comparing the increase in cable tension force to the increase in cable frequency as measured in free-vibration testing under the static empty train loading. As the relationship between cable tension force and the cable frequency is nonlinear, the equation from the Taut String Theory was used to convert the cable tension force, T, from the model to a corresponding frequency:

$$T = \frac{4wL^2 f_n^2}{n^2 g} \tag{1}$$

where w = cable unit weight; L = cable free length; f_n = n-th natural frequency of vibration; and g is the gravitational constant.

Schmieder et al. (2012) discussed the use of the Taut String Theory with natural frequency-based methods to calculate the cable tension force for a known cable length and unit weight. Using the cable properties of the respectively selected cables for free-vibration measurements, the cable tension forces from the model were converted to frequency by Equation 1.

In general, the increase in cable tension forces due to static empty train loading was found to be higher for the structural model as compared to the field measurements. The difference in cable frequency between the field measurements and the structural model is generally within two percent. A close correlation of the results further validates that the structural model is capable of capturing cable behavior of the LRT Bridge.

4   FATIGUE DAMAGE ASSESSMENT

Fatigue damage assessment of the LRT Bridge stay cables consisted of establishing train transit histograms and passenger peak and off-peak hour loading scenarios for all vehicle types as per historic and future projected data; developing loading scenarios for simultaneous train crossings on the bridge deck; and fatigue damage assessment using historic and projected train transits to determine the remaining fatigue service life of the stay cables.

4.1   *Train transit and loading information*

It is imperative for the fatigue damage assessment that historic train data is available and future train traffic is adequately projected by the transit authority so that an accurate train transit histogram can be developed. Three main factors were considered in development of the train transit histogram for use in the fatigue damage assessment: vehicle live loading for each type of vehicles, annual train transit volume of each type of vehicles, and passenger traffic counts at peak and off-peak hours during typical working days and weekends/holidays.

4.1.1   *Maximum vehicle live loading for fatigue limit state*
The vehicle self-weights and maximum passenger capacities are defined by the train manufacturer for all vehicle types including MK I, MK II, and MK III vehicles. The operating trains are comprised of two-, four-, five-, and six-car trains of different vehicle types. For the general design purposes, a fully crush loaded vehicle is based on eight passengers per square meter of train usable floor area. Such high design live loads are typically used for ultimate limit states but do not represent realistic loading conditions for the fatigue limit state. Through observations of local passengers' tolerance for crowdedness of LRT trains, the transit authority determined that the maximum passenger loading for a fully occupied train is six passengers per square meter, which was utilized to calculate the maximum train live loading for the fatigue limit state.

4.1.2 *Annual train transit volume*

The transit authority provided the historic train passage data over the bridge deck from the opening of the bridge to the present day. The data provides the number of train transits over the bridge in each year including a distribution by vehicle types. Similar annual data has to be projected for future transit volume over the remaining service life of the bridge. The transit authority estimated the future demands based on assumed annual population growth in the region, planned future expansion of the LRT system, and increased service frequency.

4.1.3 *Passenger traffic survey*

In addition to the maximum vehicle live loading and annual train transit data, the fatigue damage assessment also requires passenger traffic surveys for typical peak and off-peak hours on typical working days and weekends or holidays. A three-day passenger traffic survey was conducted at two adjacent stations of the bridge to develop an hourly account of passengers riding on trains in each direction throughout a typical service day.

For the passenger traffic survey, four different passenger loading categories were defined: 0–225 passengers, 225–450 passengers, 450–675 passengers, and 675+ passengers per train. For trains used for this study, the 675+ category represents passenger volumes ranging from 675 passengers to the fully loaded capacity of six passengers per square meter for each train.

As different type of vehicle has a different passenger capacity, when conducting the passenger traffic survey, it is important to record the exact number of passengers per train rather than recording passenger volumes by a percentage of a fully loaded vehicle because the later is dependent on the vehicle type. Since the passenger capacity varies for different vehicle types, a vehicle recorded as 50% of fully loaded would represent a higher live loading for a large capacity vehicle than a small capacity vehicle. In reality the number of passengers on a train at a given time is not dependent on the vehicle type unless the vehicle capacity is exceeded. Therefore, the passenger loading based on a percentage of fully loaded vehicle may not be representative of the actual number of passengers carried. The passenger traffic survey recorded the number of passengers for each train so that the passenger loading can be scaled accordingly to the future ridership projections and easily distributed to the various vehicle types that form a particular train configuration for any given year of interest.

The live load demands for the fatigue assessment are constructed based on the three-day passenger traffic survey performed in 2015, which served as the baseline year for all projected future live load demand calculations as shown in Figure 1. Over time as the ridership increases the required train capacities are expected to increase by deploying larger vehicles and/or longer

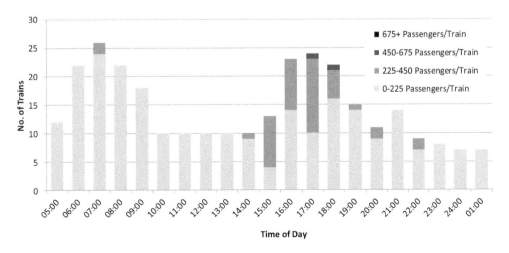

Figure 1.   Baseline passenger demands for 2015—Southbound.

train configurations. Based on the annual ridership and trip data of all train configurations provided by the transit authority for a given year, the passenger demands for the baseline year are scaled to determine the distribution of passengers on the trains for each year. The analysis accounts for the vehicle types and train configurations in the year of interest such that if the numbers of passengers for a given trip exceed the train capacity the passengers in excess of the capacity will be distributed to the next train. With the calculated passenger distribution for every year considered in the analysis, the number of train passages for each loading category can be estimated for each year over the entire service life of the bridge.

Over time as the capacity of the vehicles increases, the distribution of more heavily loaded vehicles also increases. The full distribution of loaded trains for each year allows for the calculation of an equivalent stress range representative of the actual train live load demands.

4.2   *Cable fatigue demands*

4.2.1   *Axial tensile stresses*

Using the calibrated three-dimensional finite element model, the different vehicle loading configurations were simulated over the entire bridge length to determine the baseline demands for all types of empty vehicles. For the fatigue damage assessment, the demands were scaled accordingly from an empty vehicle to a fully loaded vehicle based on the passenger loading categories discussed in Section 4.1.3 in order to represent realistic train live loading on the bridge through a typical service day.

For the fatigue damage assessment, the live load demand was computed using influence lines by the structural model to determine the range of stresses for the passage of a single vehicle over the entire length of the bridge. The range of stresses resulting from the passage of a vehicle allowed for the number and magnitude of stress cycles in each cable to be determined for each vehicle type using rainflow cycle counting (Rychlik 1987). Rainflow cycle counting reduced the spectrum of varying stresses into a set of simple stress reversals to facilitate determination of the number of fatigue loading cycles. An example of the stress reversals for empty vehicles is shown in Figure 2 for a backstay cable.

The results of cycle counting are represented as an equivalent stress range using Miner's Summation cumulative damage model (Maddox 2003).

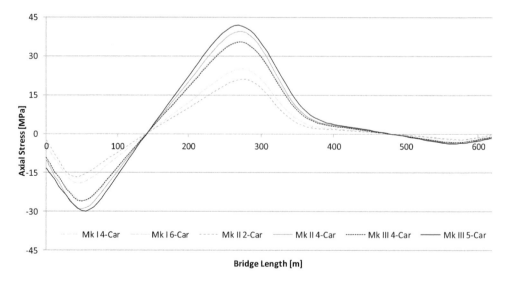

Figure 2.   Backstay cable stress reversal.

Based on the live loading for each train configuration, a stress range was calculated and assigned to each case for the analysis. Using the estimated annual train trips over the bridge and distribution of vehicle type and passengers for a given year, the annual fatigue damage was calculated for each cable element. As specified by the transit authority, a dynamic load allowance of ten percent of the live load demand was considered to account for an increase in load amplitude due to the dynamic effects of train passage over the bridge.

4.2.2   *Secondary bending stresses*

The fatigue stress in the cables is caused by axial tensile forces and secondary bending at the anchorages. The cable sag varies as the train loading passes over the bridge, resulting in a rotation at the cable anchorage. The change in angle at the cable anchorage induces secondary bending stresses into the cable which contribute to additional fatigue damage. Gimsing & Georgakis (2012) derived the following expression to represent the local bending stress, σ_b, in the cable due to an angular change at the cable anchorage:

$$\sigma_b = 2\Delta\varphi\sqrt{\frac{ET}{A_{cb}}} \tag{2}$$

where $\Delta\varphi$ = angular change at a stay cable anchorage; E = modulus of elasticity of the cable material; T = axial tension force in the cable; and A_{cb} = cross-sectional area of the cable.

From the variation in cable axial stresses caused by the passage of a train over the bridge, the above expression can be used to approximately calculate the local bending effects in the cable. The maximum and minimum stress in each cable determined from the live load analysis for each vehicle passage is used to calculate the angular change at the anchorage which is then used to obtain the secondary bending stresses in the cable.

The analysis considered the elastomeric dampers at each cable anchorage, which, based on evidence provided by the original designer, effectively reduce the secondary bending effects by one half.

4.2.3   *Simultaneous train loading*

The critical loading case that induces the largest amount of fatigue damage in the cables is the event where two trains are on the bridge deck at the same time. For all live loading events consisting of a single train passage the maximum stress range is experienced by all cables, each occurring at a different loading position. This is not the case when two trains are on the bridge deck as they may not always be positioned in such a way that they produce the maximum effects. Analysis of the train data provided by the transit authority indicates that the position of two trains crossing each other on the bridge deck is quite variable and random, as shown in Figure 3.

For the simultaneous train loading occurrences the stress reversals from a single train passage of a northbound and southbound train, respectively, are superimposed to determine the resultant stress reversal for the two train loading condition. As the speed and crossing time for northbound and southbound trains differ, the variation in stress was related to the crossing time of each train. The stress reversals for each simultaneous loading case are positioned such that the northbound and southbound trains would cross at the "critical locations" on the bridge deck to cause the worst loading effect for the specific cable of interest. Based on the influence lines for each cable, the stress reversal can be calculated for any location of the cables where two trains cross each other on the bridge deck.

Given that the simultaneous train loading data analyzed was not conclusive based on the variability in the data, two cases were considered: (1) all simultaneous train crossings on the bridge deck occur at the "critical locations" for each respective cable and (2) simultaneous train crossings are taken as the distribution as observed from the train data of Figure 3.

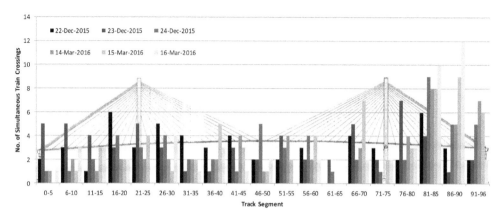

Figure 3.   Simultaneous train passage survey data.

The results of second case are provided for reference only and are not considered reliable to represent the realistic loading condition because the sample size of data used to develop the distribution is very small, thus there is low confidence in the result. Therefore, the first case was used to determine the cumulative fatigue damage from simultaneous train loading scenario. Although this is conservative, it is justifiable given the low confidence in the data for the second case.

5   FATIGUE ASSESSMENT RESULTS

5.1   *Fatigue damage under current projections*

The cumulative fatigue damage index was calculated for all 120 stay cables of the LRT Bridge using the criteria outlined in the previous sections. The fatigue damage was calculated for the two assumed simultaneous loading cases as discussed in section 4.2.3. The second case is for reference purposes only.

The assessment found that the fatigue life of 12 stay cables would be exhausted prior to reaching its 100 years of service life in the original design. The first backstay cables are found to reach the fatigue limit between approximately years 60 and 85, followed by the second backstay cables between approximately years 75 and 100, and the third backstays between approximately year 95 and beyond 100.

As shown in Figure 4, when the fatigue damage index reaches 1.0 it represents the end of the fatigue life of the stay cable. The lighter gray areas represent the contribution to fatigue damage from simultaneous train crossing and single train crossing events, respectively. The fatigue damage caused by simultaneous train crossings is critical and has a significant impact on the fatigue life of the stay cables.

A summary of the cumulative fatigue damage for all stay cables is shown in Figure 5.

5.2   *Fatigue damage with modified train timetable*

The fatigue damage index for the critical stay cables as shown in Figure 4 clearly demonstrates the impact that the simultaneous train crossing events have on the fatigue life of the stay cables. A further investigation was carried out to determine the feasibility of extending the fatigue life of the stay cables by modifying the train operating timetable such that no simultaneous train passages would occur on the bridge. In this case there are four (4) stay cables that are found to reach their fatigue life limit before the 100 years of service life.

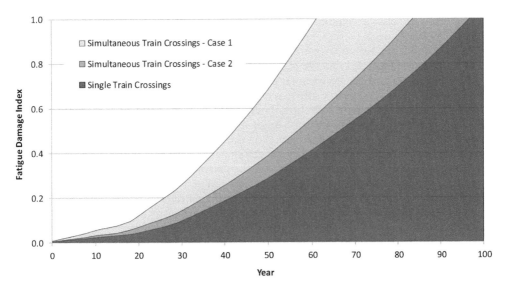

Figure 4.   Fatigue damage index for backstay cable.

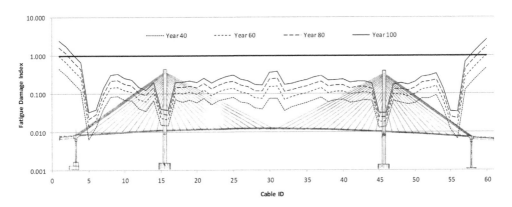

Figure 5.   Cumulative fatigue damage summary.

The first backstay cable is found to reach the fatigue life limit between approximately years 85 and 95.

The investigation assumes that as of year 2017 the train timetables are revised such that only a single train is permitted on the bridge at a time. Thus, the cumulative fatigue damage due to simultaneous train crossings remains constant beyond year 2017 and only further fatigue damage comes from single train crossings.

6   CONCLUSIONS

The results of the stay cable fatigue damage assessment show that the train transit histogram used and assumptions made for simultaneous train crossings are critical to the estimated fatigue life of the stay cables. Given that variation in the assumed train transit histograms and the simultaneous train loading positions has a significant impact on the fatigue damage assessment of the stay cables, it is paramount that the transit authority shall have accurate data of the train passage volume, time record, and passenger loading profile over the service life of the LRT Bridge.

As projections for future train transits introduce additional uncertainty for the fatigue analysis, it is recommended that the transit authority conduct additional fatigue assessment in the future using more accurate train transit information and updated projections for future train transits and ridership to confirm assumptions and fatigue damage assessments from previous studies. Future data collection should focus on using available technology to track annual changes in the train passenger loading and to monitor and document frequency and positioning of simultaneous train crossings on the LRT Bridges.

The results of the analysis show that the simultaneous train crossings contribute to a large portion of the fatigue damage in the stay cables. A scenario was investigated where the train timetable can be modified such that only a single train crosses the bridge at one time, thus eliminating simultaneous train loading conditions and extending the fatigue life of the stay cables considerably. It is recommended that the transit authority consider modifications to train timetables to reduce simultaneous train loading effects or eliminate them completely. If modifications are made to the train timetable, the changes should be well documented for future fatigue assessment of the stay cables.

It can be concluded from the study that the fatigue assessment of stay cables is more complicated and imperative for light rail transit bridges than for highway bridges, and the stay cables can be vulnerable to pre-mature fatigue failure even if the original design was based on the modern bridge design standards. It is important for transit authorities to be aware of such risks and carry out fatigue damage assessment if such assessment has not be performed, especially for simultaneous train loading conditions and if they are planning for deployment of larger vehicles or longer train configurations.

REFERENCES

Gimsing N.J. & Georgakis C.T. 2012. *Cable Supported Bridges: Concept and Design. Third Edition.* United Kingdom: John Wiley & Sons.

Maddox S.J. 2003. Review of fatigue assessment procedure for welded aluminum structures. *International Journal of Fatigue* 25: 1359–1378.

Rychlik I. 1987. A new definition of the rainflow cycle counting method. *International Journal of Fatigue* 9: 119–121.

Schmieder M., Taylor-Noonan A. & Heere R. 2012. Rapid non-contact tension force measurements on stay cables. In Fabio Biondini & Dan M. Frangono; (eds), Bridge maintenance, safety, management, resilience and sustainability; *Proceedings of the Sixth International IABMAS Conference, Italy, 8-12 July, 2012.* Stresa: CRC Press.

Chapter 14

The design and construction of Kemaliye Bridge, Turkey

Ş. Caculi & E. Namlı
Emay International Engineering and Consultancy Inc., Istanbul, Turkey

ABSTRACT: The design of Kemaliye Bridge, with a total length of 290.0 m and width of 17.0 m, has been carried out by Emay International Engineering and Consultancy Inc. Within the scope of the Project entitled "Design and Engineering Services for Kemaliye and Kozlupınar Bridges located on Kemaliye-Dutluca Road", implemented by the Client Republic of Turkey, Ministry of Transport, Maritime Affairs and Communications, General Directorate of Highways. The Kemaliye Bridge has 3 spans with 140.0 m. central span length and side spans each of 75.0 m length. As measured from the top level of the foundations, the pier heights are 60.0 m each. The bridge deck has been designed as prestressed post-tensioned concrete by cantilever method. The segment lengths are 3.0 m, 4.0 m, and 5.0 m. The depth of the box girder deck varies between 8.50 m on piers and 3.50 m at midspan.

During the design stage of the Kemaliye Bridge, which will be constructed by the balanced cantilever method, deflections and calculations involving time dependent factors such as creep and shrinkage have been carried out following the completion of each segment. Furthermore, losses in cable prestressing forces due to various factors (creep, shrinkage, elastic loss, relaxation, friction, anchor set) and also stresses and deformations due to uniform temperature differences (thermal gradient) have been taken into account in the calculations.

The environmental effects of the construction of the bridge have been minimized by virtue of the balanced cantilever method of construction, and following the completion of the construction work, an esthetic and elegant structure will emerge which will be harmonious with the environmental scenery and attractions.

1   INTRODUCTION

From the past up to the present, bridges have been important engineering structures which span over difficult geographical obstacles and topographical conditions such as rivers and deep valleys and thus they ensure a continuous transportation. Nowadays, by virtue of the technological advances and improvements in the structural quality, using prestressed post-tensioned concrete, the number of bridges built by balanced cantilever method of construction which are suitable for spanning over difficult terrain such as rivers and deep valleys in particular, has increased considerably. The Kemaliye Bridge, planned to be constructed over the Euphrates River is one example of such bridges which are to be built by using the balanced cantilever method. Upon the completion of the project, Iliç and Kemaliye districts of Erzincan province will be connected directly to Arapgir district of Malatya province as seen in (Figure 1). Due to the nature of construction method, the environmental effects of bridge construction by balanced-cantilever method will be a minimum; following the construction period the bridge will appear to be an aesthetic and slender structure, being harmonious with the natural surroundings blending well with the natural scenery.

The Kemaliye Bridge, which is to be built within the scope of the "Design and Engineering Services for Kemaliye and Kozlupınar Bridges located on Kemaliye-Dutluca Road" implemented by the Client, "Republic of Turkey, Ministry of Transport Maritime Affairs and Communications, General Directorate of Highways", has a total length of 290 m and width of 17 m. It has been designed by Emay International Engineering and Consultancy Inc. In this paper, the design criteria, the methods of analysis including the construction stage and

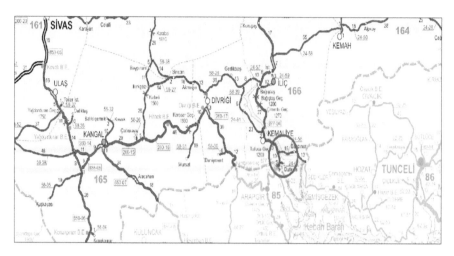

Figure 1.   Location of Kemaliye Bridge.

the results obtained concerning the Kemaliye Bridge to be constructed by the balanced—cantilever method, has been evaluated.

2   BALANCED CANTILEVER METHOD

The balanced cantilever method of construction is preferred for bridges spanning over deep valleys where large span lengths are required. From the consideration of cost, precast segments and cast in situ box section concrete segments are suitable for span length ranges of 60–110 m, and 110–200 m respectively; the segments are too heavy in the latter range for precast construction. Depending on the weights, the segment lengths vary between 3–6 m. The box girder deck depth is variable.

In this method, the initial segments are constructed over the completed piers. Following the completion of the pier foundation, the pier column is constructed by climbing form, starting from the foundation top up to the underside of the deck structure. Upon the completion of the pier (column) the assembly of the joint segment which is monolithic with the pier concrete is started. The concrete is cast at three stages, namely, the bottom, web and top sections. Following the geometrical checks the reinforcing steel is placed. After placing the reinforcing steel at the required positions, the ducts for the tendons are located within the formwork, in accordance with the number of tendons required as shown in the drawings. Afterwards, starting from the central pier segment, one extra segment is assembled at each side of the pier segment, applying the same procedure, maintaining the "balance". After the completion of a pair of segment, another pair is added in a similar fashion until the advancing segments physically join at the centre of span. At each segment, when the concrete compressive strength reaches the required value, the tendons which are located inside the previously installed ducts at their required position and tensioned at each segment end. The concrete at the final segments at abutments is cast by diaphragm formwork system. Finally, the bridge box girder deck construction is completed by completing the key segments at midspan.

3   TECHNICAL INFORMATION CONCERNING THE BRIDGE

The Kemaliye Bridge has three spans with a main span length of 140 m, and each side span of 75 m, and thus the total length is 290 m (Figures 2 and 3). The bridge is

Figure 2. General plan view of the bridge.

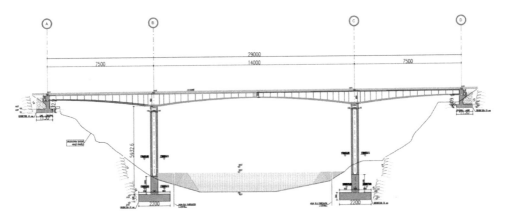

Figure 3. Longitudinal section of the bridge.

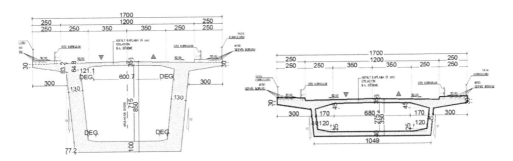

Figure 4. Typical bridge cross section (at pier and at midspan).

designed to be built by the balanced cantilever method. The bridge deck has a box section type of cross-section, the deck material being prestressed post-tensioned cast in
place concrete.

The depth of the box girder deck was necessarily selected to be variable, taken as 8.5 m at
pier section and 3.5 m at mispan (Figure 4). The pier segment which is monolithic with the
pier top, 8.0 m long. Apart from the key segments at mid-main span and abutment segments,
there are 14 intermediate segments at each side of a pier, namely $2 \times 1 \times 3$ m + $2 \times 3 \times 4$ m +
$2 \times 10 \times 5$ in order, from the pier towards the midspan (main span) and/or abutment. The
abutment segment and the midspan key segment has lengths of 7.5 m and 2 m respectively.
The initial 3 m (at the abutment side) of the 7.5 m long abutment segment has a solid section. The pot bearings have been designed as guided sliding type, allowing movements in

the bridge longitudinal direction and fixed in transverse direction of the bridge. There are four pot bearings at each abutment. The vertical and horizantal load carrying capacities of each pot bearing is 19917 kN and 1950 kN respectively. At each abutment two shear keys were designed so as to counteract the horizontal forces. The pier cross section is solid with dimensions of 9.488 m × 7 m for a height of 12 m as measured from the foundation top, and the rest of the pier being hollow (box section) measuring 9.488 × 7 m, with wall thickness of 1.3 m. The bridge piers and abutment foundations were determined to be a shallow foundation type in accordance with the Geotechnical Report. The pier footing measures 26 m × 28 m with a 5 m thickness and abutment footing measures 11.3 m × 18 m with 2.5 m thickness.

4 MATERIALS

The quality of the materials used in the load bearing deck and substructure members is given in the table below (Table 1).

5 DESIGN CRITERIA

The dimensioning, load assumptions, load combinations, construction stages and design checks pertaining to balanced cantilever bridge members have been performed in accordance with AASHTO LRFD Bridge Design Specification (2012).

The Seismic design spectrum used in the calculation of the earthquake forces acting on the bridge has been determined by using KGM (2014) Table K.E.K.7.1.1.

Design Guide Prestressed Concrete Bridges Using the Cantilever Method. SETRA (2003) has been made use of during the bridge design stage.

Truck loadings indicated in AASHTO LRFD Bridge Design Specification (2012) have been adopted for live load calculations on bridge deck.

The stresses occurring at bridge deck concrete and post-tensioning tendons have been checked in accordance with stress limits indicated in AASHTO LRFD Bridge Design Specification (2012), both for construction stage and for service stage.

Table 1. Summary of the quality of the materials.

Concrete:	
Prestressed post-tensioned box girder deck	C50 (f_{ck} = 50 MPa)
Pier columns	C40 (f_{ck} = 40 MPa)
Other reinforced concrete members	C35 (f_{ck} = 35 MPa)
Levelling and infill concrete	C16 (f_{ck} = 16 MPa)
Reinforcement:	
All substructure pier columns and deck	B500C (fyk = 500 MPa)
Structural steel at shear key	S355JR (fy = 355 Mpa)
Posttensioning strands:	
Type of strand: Low relaxation ASTM A416/A416M-2 Grade 1860 [270]	
Tensile strength (f_{pu})	1860 MPa
Yield strength (f_y)	(0.9 f_{pu}) = 1674 MPa
Cross section area	150 mm²/strand
Post-tensioning tendons	19C15 (Freysinnet-C Range)

6 LOADS

The loads used in the calculations are presented herewith:

Dead loads:
Reinforced concrete and prestressed concrete structural members: $\sigma_b = 25$ kN/m³
Asphaltic coating, levelling: $\sigma_k = 23$ kN/m³
Sidewalk (pavement) weight (25 cm thickness): 31.25 kN/m
Railing weight (guard rail+pedestrian railing): 3.00 kN/m
Asphalt layer weight (6 cm thickness): 23.46 kN/m

Live loads:
HL-93 Design Truck Load, Design Lane Load and Design Tandem Vehicle Load as shown in AASHTO LRFD (2012) Section 3.6.1.2 has been used as live load on the bridge. Live loads are defined on particular lanes in Midas Civil analysis model, and thereby the most unfavourable forces and moments encountered in the system have been taken into consideration.

Wind loads:
Wind loads acting on the structure have been calculated in accordance with AASHTO LRFD (2012) Section 3.8; in the calculation of wind load acting on the structure itself and on live loading a wind speed of 180.246 km/h has been taken AASHTO LRFD (2012) Section 3.8.1.1-1.
    Wind loads on structures (horizontal): 3.0 kN/m²
    Wind loads on structures (vertical): 0.96 kN/m
    Wind loads on vehicles: 1.45 kN/m²

Temperature effects:
The temperature effects on the structure has been considered by assuming Uniform Temperature and Temperature Gradient in accordance with AASHTO LRFD (2012) Section 3.12.2 and 3.12.3. Uniform temperature effect has been taken into account at Midas Civil Model as between +20°C and −25°C according to cold climate category. As for temperature gradient effect, the map of radiation regions in Turkiye, temperature table T_i in the KGM (2014) Section K.1.17 has been used. As Kemaliye district is located in Zone 1, temperature values were obtained as $T_1 = 28$°C and $T_2 = 6$°C.

Effects of Creep and Shrinkage (SR, SH):
The effects of creep and shrinkage has been taken into consideration in accordance with the clauses of CEB (2001) by defining material properties in the model as a result of which such effects have been calculated automatically by the program. As the deck cross sectional area is variable, there is no fixed section ratio dependent on creep. Creep coefficients were derived from the relevant equations in accordance with the specification. Approximately creep and shrinkage values corresponding to a 30 year period were taken into account in calculations.

Earthquake Loading (EQ):
The seismic loads acting on the bridge has been taken into account by using multi-mode spectral analysis method in accordance with AASHTO LRFD (2012) Table 4.7.4.3.1-1 "seismic design". The design seismic spectrum has been derived from KGM (2014) Table K.EK.7.1.1. In the Guidelines, values of $S_s = 0.9686$ and $S_1 = 0.7203$ were given for Kemaliye Bridge. The soil category has been taken as class B for bridge soil. In order to establish the design acceleration spectrum, the necessary values of T_0 and T_s were found to be $T_0 = 0.149$ s and $T_s = 0.744$ s respectively for local soil class B.

7   STRUCTURAL ANALYSIS AND DESIGN

During the structural modelling stage, the general purpose structural analysis programs such as Midas Civil and SAP2000 computer programs has been used. The bridge abutments and prior foundation were modelled in SAP2000 program. Other members and structural units in the system were modelled in Midas Civil Program. In the Midas Civil Program, a 2-dimensional structural model was established and an analysis was performed, taken into account the construction stage, creep and shrinkage effects in concrete and relaxation of post-tensioning tendons. The cross-section of the post-tensioning tendons extending in the bridge longitudinal direction has been determined by taking dead loads and live loads into account. In the preliminary tendon calculation, the procedure was as follows: for cantilever tendons the maximum cantilever case of construction stage, and for tendons at section bottom service loading (all dead and live loads) together with a continuous structural system condition was considered.

The typical cross-sections showing the positions of post-tensioning tendons in the section, is shown in (Figure 5).

The post-tensioning tendons so determined were defined in the model by placing them in suitable positions (Figure 6) and as a result of the analysis, concrete stresses and deformations in the deck sections were checked. Transverse post-tensioning tendons had to be used since the width of the bridge is 17 m. These tendons also were defined in Midas Civil Program.

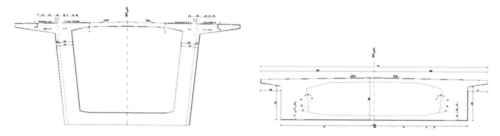

Figure 5.   The layout of post-tensioning tendons (cross section).

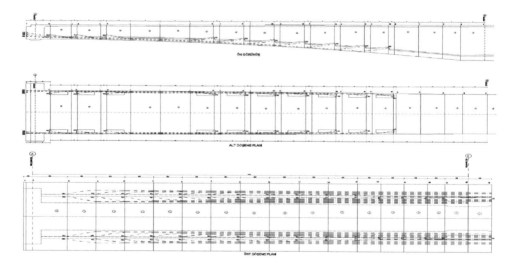

Figure 6.   The layout of post-tensioning tendons.

8   CONSTRUCTION STAGE ANALYSIS

The construction stage analysis of the bridge was done by using the Midas Civil Program, which enabled the checking of stresses and displacements derived from sections and which took into consideration the time dependent effects. The construction stages defined in the calculation model were as follows:

1. Firstly, the column pier and the pier segment is constructed.
2. Subsequently, both adjacent segments are constructed simultaneously (symmetrical cantilever construction).
    It was assumed that a period of 12 days would be required for all the construction work in each segment, which was defined in the program;
    1. Day: Removal of the traveller forms and their assembly for the next segment
    2. Day: Positioning and placing the steel reinforcement and tendon ducts
    3. Day: Concrete casting
    4.~6. Days: Curing of concrete
    7. Day: Tensioning the post-tensioning tendons
3. Abutment connection work, following the completion of all segments.
4. Continuity tendons at side spans are tensioned.
5. Construction of key segment at main span.
6. Continuity tendons at main spans are tensioned.
7. After the completion of the construction stage analysis, the structure is aged to 30 years and 100 years in order to take into account the time dependent effects of material such as creep and shrinkage.
    The construction of Kemaliye and Kozlupınar Bridges has not yet started, some information concerning the construction stages and details was essential for the completion of the design calculations and drawings.

9   ANALYSIS RESULTS

The prestressed concrete box girder deck stresses and the post-tensioning tendon stresses derived from the results of service limit state analysis and construction stage analysis performed in relation to computer model, have been compared with the stress limits stated in AASHTO LRFD (2012). Furthermore, the reinforced concrete strength design method of calculation was carried out for the deck, pier columns, abutments and foundations according to the dynamic analysis results. However, only the stresses and displacements has been shown in this paper.

9.1   *Design checks for the deck*

The stresses at prestressed concrete box girder deck due to both construction and service limit state were checked. It was observed that the compressive and tensile stresses at the top and bottom concrete fiber were well within the limiting values (Figure 7). A summary of maximum compressive and tensile stresses are presented in the table below (Table 2).

Furthermore, the deck displacement curves for each construction stage have been evaluated and the maximum displacement following a 100 years period was found to be 9.75 cm (Figure 8).

Finally, box girder deck stress values and displacement curves were observed to be within the limiting values stated in AASHTO LRFD (2012).

9.2   *Post-tensioning tendon checks*

It was observed that the amount of post-tensioning tendons derived from the analysis performed was sufficient and satisfies the stress limitations. There are 52 tendons at each pier section, 10 tendons under the section at midspan of main span and 7 tendons under the

Compressive stress at bottom fiber under construction condition

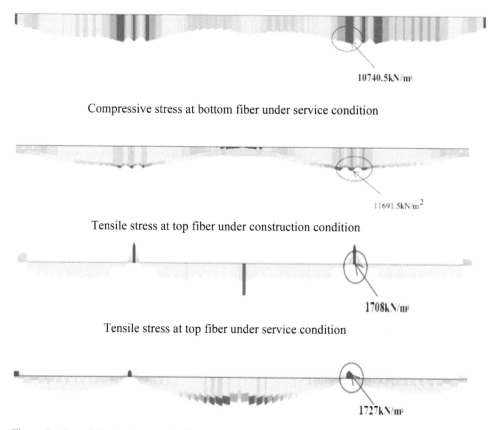

10740.5kN/m²

Compressive stress at bottom fiber under service condition

11691.5kN/m²

Tensile stress at top fiber under construction condition

1708kN/m²

Tensile stress at top fiber under service condition

1727kN/m²

Figure 7.   Box girder deck stress checks.

Table 2.   Summary of maximum stresses at the deck.

	Compressive stress		
	Allowable stress	Maximum stress	Element no
Construction stage	22500 kN/m²	10740 kN/m²	50
Service limit state	30000 kN/m²	11691.5 kN/m²	53
	Tension stress		
	Allowable stress	Maximum stress	Element no
Construction stage	3520 kN/m²	1708.88 kN/m²	52–53
Service limit state	3520 kN/m²	1727 kN/m²	52–53

section at midspan of each side span. The number of transverse tendons, placed at 70 cm spacing, varies from segment to segment.

The tendon stresses obtained from the calculation model analysis were checked for stress limitting values stated in AASHTO (2012) Table 5.9.3.1 the resulting tendon stresses were found to be lower than the limits.

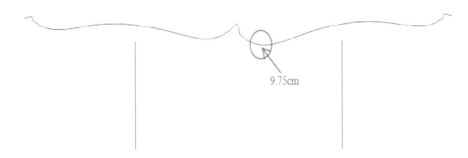

9.75cm

Figure 8. Maximum displacement curve for deck.
Construction stage no. 20, displacement curve (36000 days-100 years).

10 CONCLUSIONS

The Kemaliye Bridge will be built within the scope of the "Design and Engineering Services for Kemaliye and Kozlupınar Bridges located on Kemaliye-Dutluca Road" implemented by the client, "Republic of Turkey, Ministry of Transport Maritime Affairs and Communications, General Directorate of Highways". The bridge has been designed by Emay International Engineering and Consultancy Inc. In this paper, the design criteria, the methods of analysis including the construction stage and the results obtained concerning the Kemaliye Bridge to be constructed by the balanced-cantilever method, has been evaluated. The environmental effects of the construction of bridge has been minimized by virtue of the balanced cantilever method of construction, and following the completion of the construction work an aesthetic and elegant structure will emerge which will be harmonious with the environmental scenery and attractions.

Some experience has been acquired concerning the design for large span bridges built by the balanced cantilever method of construction through similar previous design work and relevant technical information; although the construction of Kemaliye and Kozlupınar Bridges has not yet started, some information concerning the construction stages and details was essential for the completion of the design calculations and drawings, which contributed to the relevant experience. No doubt much more experience will be gained during the construction stage.

REFERENCES

AASHTO American Association of State Highway and Transportation Officials (2012). LRFD Bridge Design Specification.
CEB (2001). CEB-FIP 1990 Model Code.
KGM General Directorate of Highways (2014). Technical Guide lines for the Developments in Bridge Engineering Design and Construction in Türkiye.
SETRA Ministry of Transport and Infrastructure of France (2003). Design Guide Prestressed Concrete Bridges Using the Cantilever Method.

Chapter 15

Design of a short span suspension footbridge: Detailing for success in rural Kenya

J. Smith & M. Bowser
WSP, Oakville, Canada

K. Severns
WSP/PB, Washington, DC, USA

ABSTRACT: The design for the Peace Bridge, a 45 m (148 ft) span footbridge, was detailed for fabrication and construction in very remote parts of Sub-Saharan Africa. The design was developed with two objectives: to utilize materials that are readily available in rural Kenya and to provide structural details that are tailored for fabrication with the tools (or lack of tools) available in the area. For example, the towers were detailed to be erected by hand with the aid of only a wire rope winch. A simplified method for the structural analysis of the bridge is presented, as well as details used for the anchors, suspenders, saddles, and tower foundations. The Peace Bridge is the latest bridge to be completed by Bridging the Gap Africa, a non-profit that enables remote communities to build footbridges that will improve the quality of their lives and help prevent river-crossing fatalities.

1 INTRODUCTION

1.1 Background

In rural Kenya rivers create barriers that restrict access to education, health care, and commerce. Bridging the Gap Africa (BtGA) is a non-profit that enables communities to build footbridges across these dangerous rivers. To date, the organization has facilitated the construction of 60 footbridges and it is estimated that hundreds of bridges are still needed throughout Kenya. BtGA believes that their approach of enabling local communities to build these bridges will result in a sustainable bridge program that is capable of reaching hundreds of communities throughout Kenya.

The latest bridge to be completed by BtGA is the Peace Bridge.

1.2 Bridge site

The Peace Bridge crosses the Nzoia River in western Kenya at a site approximately 40 km (25 miles) south of Kitale or 370 km (230 miles) northwest of Nairobi (GPS Coordinates 0.7521°N; 34.9880°E).

Two crossing locations were reviewed for the proposed bridge site. The design team determined that a 45 m (148 ft) span suspension bridge was feasible at either the upstream or downstream location but chose the downstream location primarily because the abutment foundations and anchors could be dowelled into granite bedrock whereas the upstream location would have required a spread footing for the tower foundation and a reinforced concrete passive earth beam anchor for the cables. Drilling into the granite bedrock at the downstream location enabled smaller foundations and anchors, while significantly reducing the risk for undermining of the tower foundations and the cable anchors.

2   DESCRIPTION OF STRUCTURE

The Peace Bridge is a short span suspension footbridge with a span of 45 m (148 ft). Earth filled ramps provide access to the bridge deck at each approach. A photo of the bridge is shown in Figure 1. A brief description of the key structural components of the bridge is provided in the following sections.

2.1   *Steel towers*

The towers for the bridge consist of two steel rectangular hollow sections (RHS), each 6.0 m (19'8") in height. Steel angles provide cross bracing between the RHS columns in the upper halves of the towers and the steel towers are pinned to the reinforced concrete tower foundations.

   Several structural details for the tower saddles and bottom hinge connections were reviewed during preliminary design and the detailing chosen for these components uses readily available steel sections consisting of steel angle, steel plate, and steel pipe. Since torches for cutting steel are not commonly available in rural Kenya, details that require 'round' cuts in steel were avoided. All cuts for the structural steel are straight cuts, often cut by hand using a hack saw.

2.2   *Tower foundations and anchors*

The reinforced concrete tower foundations consist of square columns connected by a ballast wall that is placed monolithic with the columns. The tower columns and ballast wall sit on a reinforced concrete pad that is founded on, and dowelled into, bedrock. The steel towers are connected to reinforced concrete bearing pedestals using embedded steel pipes and concrete filled steel pipe hinge pins.

   The anchors for the main cables on each side of the river consist of reinforced concrete anchor blocks that are doweled into bedrock. A hammer drill with a rock bit was used to drill

Figure 1.   Profile of the Peace Bridge.

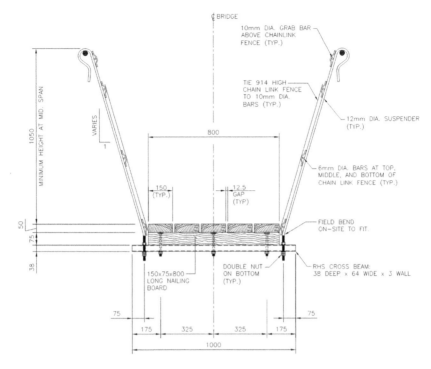

Figure 2.   Deck cross-section.

Figure 3.   View of underside of deck showing crossbeams.

into the granite substrate and the steel dowels were grouted in. The connection between the anchor and the main cables of the bridge is facilitated above ground by using RHS sections that extend from the reinforced concrete anchor.

2.3   *Cables, suspenders, and bridge deck*

This suspension bridge is supported by two 32 mm (1¼") diameter wire rope cables. The suspenders are fabricated using steel round bar with a 'shepherd's hook' bend at the top of each suspender to facilitate the connection to the main cables while the bottom of each suspender is threaded to facilitate a bolted connection to the deck cross beams. The deck cross beams consist of small RHS sections with a timber nailing strip that is bolted to the RHS cross beam to facilitate nailing of the timber deck. The deck surface consists of five longitudinal timber deck boards. A cross-section of the deck, showing these details, is presented in Figure 2. Along the entire length of the deck, lateral bracing is provided by fastening diagonal timbers in a Z pattern on the underside of the deck between each cross beam. Wire fencing is attached to the inside of the suspenders on each side of the deck along the full length of the bridge. A view of the bridge showing the diagonal timbers and fencing is presented in Figure 3.

3   STRUCTURAL DESIGN

The primary reference used for the design of the Peace Bridge was the CAN/CSA S6-06 Canadian Highway Bridge Design Code (CHBDC). The following sections present the design methodology and some of the structural systems within this bridge.

The design loads for the Peace Bridge consist of live (pedestrian) loading, wind loads, and dead loads due to the self-weight of the bridge. The unfactored pedestrian live load for the design of the Peace Bridge is 3.5 kPa (73 psf), calculated in accordance with Clause 3.8.9 of the CHBDC. The bridge is not designed specifically for transportation of cattle or motorbikes, but it is noted that demands to the bridge main cables, towers, and anchors resulting from cattle or motorbikes are comparable or less than the demands that result from full pedestrian loading.

The communities using the bridge have been instructed that using the bridge to transport cattle will result in premature wear of the deck boards and that there is also a risk of hooves punching through the bridge deck, which would result in damage to the bridge and could injure the livestock. If the bridge is used to transport motorbikes, the rider and any passengers are to dismount and walk the motorbike across the bridge.

Since specific wind data was not available for this site, the design team used a lateral wind load equal to 0.40 kN/m (27.4 lb/ft) along the main span and applied in the transverse direction to the bridge deck. This lateral wind load is recommended by the Swiss organization Helvetas and it has been used successfully for the design of hundreds of short span suspension bridges constructed in Nepal.

3.1   *Analysis*

The design of the Peace Bridge is based on a static linear structural analysis. The method used to calculate the unfactored demands on the main cables and towers is as follows.

Cable tension at midspan and cable tension in the back span are calculated by making two 'cuts' within the structural system of the bridge to isolate the main cable within half of the bridge span as a free body diagram. The first 'cut' is made to the wire rope at midspan. At this location the wire rope is horizontal so there is no vertical tension component in the cable. The second 'cut' is made through the main cable in the back span. The third 'cut' is made through the steel tower. The resulting free body diagram of the isolated half-span of the bridge is shown in Figure 4. This system is statically determinate and the midspan cable tension of

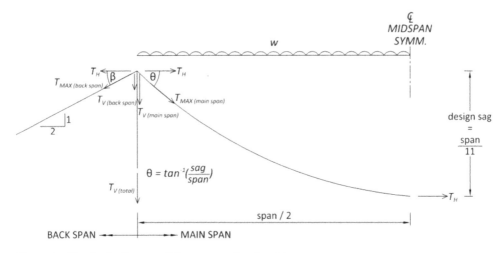

Figure 4.   Free body diagram of the tower and main cable.

the bridge can be found by summing moments at the top of the saddle. The next step in the analysis is to sum horizontal forces to prove that tension in the main cable at the midspan of the bridge is equal to the horizontal component of the tension in the back span cables. Finally, the horizontal component of force in the back span cables is used to determine the tension in the back span cables based on the inclination of the cables.

To find the vertical forces that are applied to the steel towers, a new free body diagram is isolated by making a 'cut' through the saddle at the top of the towers and through the cables on either side of the saddle. The horizontal component of force in the cables within the main span at the saddle is determined by summing horizontal forces. Using the inclination of the main cable, the vertical component and the tension in the main cable are calculated. The compression applied to the top of the towers is equal to the sum of the vertical components of the tension in the back span and main span cables.

3.2 *Design of towers*

The structural system for the steel towers consists of two RHS columns that are connected by horizontal angle braces at the mid-height and top of the columns along with steel angle 'X' braces.

The structural capacity of the RHS columns is governed by Euler buckling. The strong axis of the RHS section is orientated in the longitudinal direction of the bridge and has a K value of 1.0 corresponding with an unbraced length equal to the entire height of the column (measured from top of saddle to centre of hingepin). The weak axis of the RHS section is orientated in the transverse direction of the bridge and has a K value of 1.0 corresponding with an unbraced length equal to half of the column height. The saddles and hinge connection for the RHS columns are detailed so that the compressive force is applied at the centroid of the column; however, the design allows for the bending in the RHS columns that would result from an eccentrically applied compressive load to account for fabrication and construction tolerances. A detail of the saddle is shown in Figure 5. In addition to the bending forces that result from an eccentric load there are also bending forces in the weak axis of the RHS columns that result from transverse wind loads. The design of the towers conservatively assumes that the entire transverse wind load is transferred from the deck, into the main cables, and applied to the saddles located at the tops of the towers.

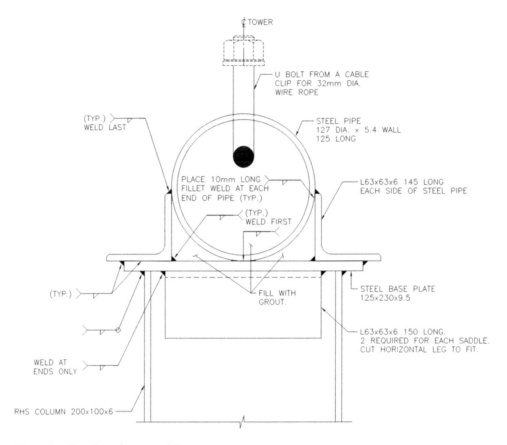

Figure 5.   Elevation of tower saddle.

The towers are designed for compression, bending, and combined axial compression and bending, in accordance with the relevant clauses of the CHBDC.

3.3   *Design of foundations and anchors*

The reinforced concrete bearing pedestals are designed for the compressive (gravity) loads transferred from the steel towers to the tower foundations. The bearing pedestals were also checked for the shear forces in the longitudinal direction of the bridge that occur when erecting the steel towers. The reinforced concrete towers are sized based on site geometry and the reinforcing steel was detailed based on minimum requirements for reinforced concrete components in accordance with the CHBDC. The tower foundations are supported on a reinforced concrete footing that is founded on granite bedrock. Steel reinforcing dowels anchor the tower foundation to a granite substrate.

Lateral earth pressures acting on the back side of the tower foundation produce an overturning moment. Given that the steel dowels fix the tower foundations to the bedrock, an at-rest lateral earth pressure was used for the analysis. Righting moments result from the self-weight of the tower foundation, dead load transferred from the steel towers to the bearing pedestals, and from the reaction of the steel dowels that anchor the foundation to bedrock. These steel dowels were checked for yield and pull-out capacity. A failure resulting in pull-out of a block of granite was also checked.

3.4 *Wire rope and setting sag*

BtGA currently has an inventory of repurposed wire rope that is used for the main cables of their footbridges that was donated by a crane company.

Wire rope conforming to ASTM A603 is permitted for use as cables in bridges according to CHBDC Clause 10.4.8.3. The reduction factor for wire rope in tension is defined in Clause 10.5.7(f) as 0.55. Other applicable sections of CHBDC include Clause 10.17.2.8.1, which states that cables and hangers used in suspension bridge construction need not be designed for fatigue unless special fatigue provisions are required in the judgement of the Engineer.

The authors do not anticipate any significant reduction in capacity of the repurposed wire rope in comparison to new wire rope; however, to account for potential wear in the repurposed wire rope, the resistance factor listed in the CHBDC was reduced to 0.50.

The method used for determining the hoisting sag for the main cables was completed using the following general process:

1. The tension in the cables that corresponds with the unfactored dead and live loads was calculated based on the design sag, which is equal to span / 11.
2. The unfactored tension and cable length were used to determine the total stretch in the cable (from anchor to anchor) using the method and equation recommended in the Wire Rope User's Manual that is published by the American Iron and Steel Institute.
3. The stretch in the cable was subtracted from the original length of the cable within the mainspan to provide a new (shortened) length for the cable within the mainspan.
4. The hoisting sag was calculated by using a parabolic equation to solve for the sag that corresponds with the shortened length of the cable.

The setting sag could be refined further through an iterative process by repeating the sequence described above; however, the authors found that further iteration resulted in minimal refinement in this instance. The authors note that a parabolic equation was used to simplify the calculations and provides an acceptable approximation of the actual cable profile.

3.5 *Crossbeams and decking*

The steel RHS cross beams are designed for both moment and shear. The timber nailing strip that is bolted to the RHS cross beams is conservatively considered to be a non-structural member; however, it is acknowledged that the wood nailing strip does stiffen the cross beam and provides some additional capacity.

The wood deck panels are designed for both moment and shear. The deck panels are designed for two separate load cases: a uniform distributed live load and a concentrated load resulting from someone stepping directly in the midspan of a deck board.

Diagonal timbers are fastened to the underside of the deck to provide lateral bracing that stiffens the overall deck system. These diagonal timbers are not included in the design, that is, they are not sized for any specific load case, as they are considered to be a non-essential component of the bridge.

The deck is fastened to each ballast wall at the end of the bridge using a steel angle that is bolted to the reinforced concrete ballast wall. The fixed connection resists moments at the ends of the deck in the lateral transverse direction.

3.6 *Tower erection*

The connection at the base of the towers is designed to provide a pinned connection for the RHS columns and to facilitate tower erection. The Peace Bridge design is based on the following erection process:

1. Connect the base of the tower to the bearing pedestal using the hinge pin with the tower in a horizontal position leaning back towards the anchor.

2. Use man power to lift the tower so that the elevation of the saddle is 1.0 m (3'3") (minimum) above the elevation of the hinge.
3. Set the tower down on temporary timber supports.
4. Prefabricate the knee braces that will be used to provide temporary support to the towers during construction.
5. Tie a heavy nylon tow strap to the top of the tower back to the bridge anchor (or to an approved alternate 'deadman'). This strap will serve two purposes: it will prevent the tower from falling toward the river in the event that the towers are erected past their vertical position and it will enable the use of a winch to set the final position of the tower.
6. String a steel cable that connects the top of the tower to the tower foundation on the opposite side of the river.
7. Use a wire rope winch (Tirfor machine) to lift the tower. Minimize the slack in the nylon tow strap during erection.
8. Using the winch, set the towers to a true vertical position.
9. Install the knee braces to provide temporary support to the RHS columns. The knee braces remain in place until construction of the entire bridge is complete.
10. Tighten the cables clamps located in the saddles at the tops of the towers then remove the temporary knee braces.

The maximum tension in the cable can be reduced by tilting up the towers higher prior to using the winch. The temporary forces in the bearing pedestals were reviewed during detail design and these forces are well below the capacity of the reinforced concrete pedestals.

Temporary restraint needs to be provided to tie the towers back to the anchors during erection because when the towers start approaching their vertical position (but are still tilted back towards the anchor) the weight of the winch cable spanning across the river can easily (and suddenly) pull the towers past vertical resulting in them falling down towards the river.

The towers are designed to be raised to a true vertical position. Consideration was given to tilting the towers back towards that anchors by a small amount so that under full live load the towers would reach a true vertical position due to their slight rotation towards the river; however, under full live load it was found that the tops of the towers would rotate less than 20 mm (3/4") toward the river. This rotation is so small that it did not seem reasonable to compensate for this movement by rotating the towers back towards the anchors. It is important to note that if passive earth beam anchors were used (rather than anchors dowelled into bedrock), then it would have been advisable to design the towers to be rotated back towards the anchors as a passive earth anchor would be expected to move towards the towers under its design load, which may result in noticeable movement of the towers.

4   CONCLUSIONS

4.1   *Why footbridges in Africa?*

BtGA is dedicated to saving lives and improving the quality of life for rural African families and communities by enabling them to construct pedestrian footbridges. The need for these bridges can be seen when considering that an estimated 900 million rural people in developing countries do not have reliable, year-round access to road networks and that every third person is without access to a motorized vehicle (Lebo & Schelling 2001). Rural developing communities essentially live in a walking world.

Aid dollars and infrastructure improvements often focus on paved highways and major vehicular bridges because a country's ability to maximize its economic potential is closely linked to the efficiency of its transportation system (Haynes et al. 2003). These investments still leave 300 million rural citizens without access to the most basic services and opportunities, though. Investment in rural infrastructure and transportation improvements helps to alleviate poverty by improving access to markets, medical clinics, and opportunities for education.

BtGA works in remote areas throughout Kenya including Laikipia, Trans Nzoia, Trans Mara, Tsavo East, and West Pokot. To date, BtGA has enabled 60 pedestrian footbridges over many different rivers and ravines. These areas are isolated, rugged, and mountainous and are prone to drought and flash flooding. These bridges are saving an estimated 600 lives every year by eliminating hippo and crocodile attacks and drownings for those living in a world where walking is the main mode of transportation.

When walking is still the main, and often the only, means of transportation, rivers are often barriers that prevent easy movement from one place to another. Many studies have been conducted to determine what challenges members of these walking communities face. A study by Mannock Consultants (1997) that looked at communities in Zimbabwe found that:

- A typical rural household spends on average 70 hours travelling per week.
- The majority of the trips (approximately 86%) undertaken are on foot.
- The travel time is excessively long despite the short distances travelled.
- Women carry a disproportionate amount of the travel and transport burden and predominantly through carrying loads on their heads (e.g. 95% of water transportation is carried by women and girls.)
- The average amount carried by a household for subsistence needs by all modes equals to 60 tonne-kilometres (37 ton-miles) per year out of which 48 tonne-kilometres (29 ton-miles) was the responsibility of women.

Animal attacks are also a constant risk for many in African communities. In 2010, in one small village, six people were killed by crocodiles in just under a year (New York Daily News 2010). Although hippos are not carnivorous, and therefore not hunters, they aggressively protect their territory and often attack boats or people who encroach on their space. Despite their harmless appearance, hippos kill nearly 100 people per year just in Tanzania (Fitzpatrick 1999).

Rivers, streams, and annual floods isolate and inhibit many rural communities. Many of the world's poorest people are faced with the disadvantage of having no direct access to basic amenities or an adequate infrastructure system necessary to reach them. River crossings can be located miles downriver and reaching a school or market may take hours or even a full day. Flooding and animal attacks can often prove to be life threatening. Lack of access reinforces the cycle of poverty for nearly 50% of the world's people living in rural isolation (United Nations 2005). Providing pedestrian bridge crossings results in low-cost solutions for access to resources for those with the greatest need in the most remote locations. These structures are relatively quick and simple to construct. The fact that they are constructed by crews of two or three technically competent workers alongside two dozen mostly unskilled workers for approximately US$50,000 per bridge is testament to this.

4.2   *Concluding remarks*

Opening day of the Peace Bridge is shown below in Figure 6. The short span suspension bridge design presented in this paper not only resulted in the construction of a new pedestrian bridge, it has also initiated a new prototype for short span suspension bridges facilitated by BtGA in rural Kenya. A second short span suspension bridge is currently under construction and there are several bridge requests from communities across Kenya that require funding. BtGA has the privilege of enabling footbridges that save lives and bring social change to a walking world.

Figure 6.   Opening day of the Peace Bridge.

REFERENCES

Fitzpatrick, M. 1999. *Lonely Planet: Tanzania.* Lonely Plant Publications.

Haynes, R., Lovett, A. & Sunnengerg, G. 2003. Potential Accessibility, Travel Time and Consumer Choice: Geographical Variations in General Medical Practice Registrations in Eastern England. *Environment and Planning* 35: 1733–1750.

Lebo, J. & Schelling, D. 2001. Design and the World Bank. *World Bank Technical Paper No. 496.*

Mannock Management Consultants 1997. *Rural Tranport Study in Three Districts of Zimbabwe: Main report presenting the findings from the survey [in Zaka, Rushinga, and Chipinge].* International Labour Organisation, Zimbabwe Ministry of Transport and Energy.

New York Daily News (online), *Crocodiles terrorize Kenya village, gobbling up human prey.* May 24, 2010. http://articles.nydailynews.com/2010-05-309 24/news/27065244_1_crocodiles-villagers-tana-river last accessed September 2011).

United Nations 2005. *Department of Economic and Social Affairs, Population Division, World Population.*

Chapter 16

Innovative cable system designs

T. Klein
WireCo WorldGroup, St. Joseph, MO, USA

ABSTRACT: Cables produced with high strength fibers have gained acceptance in many applications due to their enhanced mechanical properties and resistance to the elements. This technical paper describes the design requirements and advancements of high performance fiber cable systems with a focus on structural applications. Innovative structures have been designed and built around the world incorporating advanced non-traditional synthetic cable designs successfully without sacrificing quality and lifecycle. Cables produced from synthetic fibers provide strengths higher than that of conventional steel wire with a corresponding weight seven times lower. These products are currently deployed in high cycle fatigue environments where other materials cannot provide long term performance. The mechanical characteristics of these cable structures are beneficial to the owner-operator of the structure for cost comparison analysis in regards to maintenance and life cycle. These cables may also provide the designer the ability to reduce the cable diameter or quantity of the structural members without compromising the strength of the cable. This will save considerable funds and reduce projects costs while increasing the functioning value. A thorough explanation of the mechanical properties and benefits of these cable constructions verses those currently specified for use in structural applications will be detailed as well as the innovative nature of these products will be outlined.

1 INTRODUCTION

1.1 *Synthetic fiber rope systems*

The synthetic fiber rope market is advancing at an unparalleled rate due to the increased demand for safety and quality in the products used. Fiber rope systems are rapidly becoming standard tension members in the global market as they provide a unique set of properties that improves on the weaknesses that hinder traditional steel products. Synthetic fiber ropes come with their own rules and limitations but the long term performance of the products naturally lend themselves for use in structural applications.

1.2 *Fiber types*

The paper focuses on two high performance fibers that are commonly used in lifting applications. These fibers are aramids and high modulus polyethylene (HMPE) types. These are generic fiber types as there are several grades and types with various coatings associated with numerous producers of these items located around the world. Fibers are constituted by very long shaped molecules and atoms chains, known as macromolecules or polymers, which may be organic or inorganic in nature. The molecules aggregate strongly with each other forming crystalline regions. The degree of orientation in these regions is an important determinant of the usefulness of the fiber for specific application (Davies, 2011). Each fiber provides unique advantages and shows the performance value when compared to traditional steel. The individual fibers are very small and are bundled into yarns. This bundled yarn can be

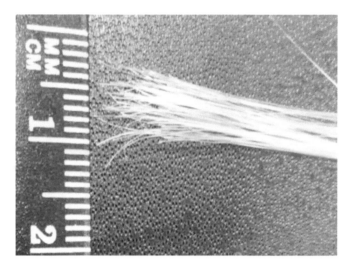

Figure 1.   Fibers separated from an individual yarn.

compared to the wires in steel ropes. The yarns are on average one millimeter or less and one end may contain up to several hundred tiny filaments that are 10 to 50 um diameter. In any fiber rope structure there will be several hundred thousand and often millions of fibers. The overall characteristics of the rope structure depends on the sum of the fibers of which it is composed. Figure 1 below provides a very descriptive view of fiber from an individual yarn.

HMPE fibers are composed of exceedingly long polyethylene molecules increasing the molecular weight of the product. A differentiating polymer processing method is needed to make available the high potential mechanical properties of the molecule, since conventional polymer processing techniques are not possible. All of the physical and chemical properties of a standard polyethylene material remain in the fibers. The resulting mechanical properties from the high chain extension (stretching), the high molecular orientation and the high crystallinity. The initial stiffness of a new HMPE fiber rope increases with load level in a bedding-in process resulting from both molecular alignment and construction reorientation.

Aramid is the generic name for polyparaphenylene-terepthalamide. Aramid fibers have long chain molecules highly oriented in the axis of the fiber. Because of the stability of the aromatic rings and the added strength of the amide linkages, aramids exhibit high tensile strength and thermal resistance up to relatively elevated temperatures near 250 degrees Celcius. The high impact resistance of the p-aramids makes them popular for "bullet-proof" body armor and rope fibers that need to resist external abrasion sources. Additional benefits are low elongation characteristics over time and a high stiffness from first loading. Proper prestretching and measuring practices of the Aramid fiber rope in the assembly fabrication process limits the permanent elongation characteristics of the product in service. From a fatigue perspective the bending over sheave tests indicate longer lifetime cycles for the aramid fibers in comparison to the HMPE due simply to the abrasion resistance of the fiber.

1.3   *Fiber versus steel comparison*

For years the traditional standards required a polypropylene or natural fiber core be used in the wire rope system for movable vertical lift structures. This is still prevalent today for many of the wire ropes used in movable structures. This requirement was based on historical success of the product design and the ideas behind using a fiber in combination with steel outer strands. Due to significant advancements in steel processes and the design capabilities available for the rope engineer fiber core ropes are almost extinct except on special order projects. Not only can the high performance core fibers now be used as tension members but the

Table 1.   Cable minimum breaking force and weight comparison.

Cable diameter MM	Synthetic rope		Steel rope		Difference	
	MBF KN	Weight kg/m	MBF KN	Weight kg/m	MBF	Weight
12	158	0.097	118	0.685	134.9%	707%
19	282	0.174	262	1.55	107.6%	809%
26	489	0.310	460	2.75	106.3%	887%
38	979	0.680	1014	6.19	96.5%	910%
50	1699	1.24	1761	11.00	96.5%	887%

ability to produce fully synthetic lifting components is becoming more prevalent as standards are developed around the synthetic fibers. High strength synthetic fibers are utilized across most industries due to their unique properties and performance over typical steel limitations. These fibers are used to produce high performance ropes and cables in demanding applications, with proven reliability and safe usage. In static and dynamic product applications, the synthetic fiber rope is an easy selection for the designer.

In comparison to steel the fibers are on average 1/8 the weight while providing a marginal increase in tensile strength. Table 1 below lists the average Minimum Breaking Force and Weight Comparison of synthetic and steel rope along with the weight per foot of each product (WRTB, 2005). The differences in the two products are also compared showing the drastic differences in weight.

In static structural applications synthetic fibers have shown to improve the vibration dampening characteristics of the structure. Although a larger diameter rope may be required to match the stiffness of steel rope the synthetic fibers will provide superior resistance to tension-tension fatigue and negligible creep effects. The Modulus of Elasticity for synthetic products is very comparable to steel wire rope if the products are properly fabricated. Synthetic products can provide approximate Modulus of Elasticity values of 127.7 mpa for standard rope constructions.

2   ROPE CONSTRUCTIONS

2.1   *Structure*

Synthetic fiber ropes are fabricated very similar to the practices used for producing wire rope. During the production of fiber ropes the individual fibers are assembled into yarns, strands and then closed into the final rope constructions. Braided jackets are very common as these offer superior protection to the individual yarns by increasing the density and compaction of the rope structure. The fine filaments that are the basic component of synthetic fiber ropes are susceptible to damage from prolonged exposure to ultraviolet light. With this knowledge the construction of the rope should be designed with a jacket blocking the penetration from affecting the superficial layers.

One of the largest concerns surrounding the use of a fiber products is the elongation of the product in service. In all structural applications the static tensile member must have low elongation and creep and the rope structure cannot change shape, either temporarily or permanently, while under load. The stiffness characteristics of wire rope are essentially linear in normal operating conditions. The fiber ropes are produced of material with viscoelastic properties. Thus their stiffness characteristics vary with design, load duration, and magnitude. Aligning the fibers is critical in the rope design to increase the stiffness of the final product. Figures 2, 3, and 4 below shows the common constructions of fiber rope applicable to the discussion. Each of these constructions provide a benefit to the product and thus can be matched the intended application.

As with steel wire products the alignment of the tensile members increases the strength of the rope as well as increase the stiffness of the product. The parallel members in the figures

Figure 2.   Twelve strand braided rope with braided jacket.

Figure 3.   Parallel sub-core ropes with braided jacket.

Figure 4.   Parallel yarns with braided jacket.

above provide an efficient design that provides a torque neutral product with minimal efficiency loss in the design. The ability to design a product with a high efficiency allows smaller component diameters as there is minimal loss in tensile strength of the fibers and yarns. These products are regularly used for mooring structures offshore because of their strength and resistance to corrosion. These types of ropes are now regularly used as tower guys for large antennas and pendants for boom applications where the optimal characteristics of the fiber rope benefit the owner significantly with less down time.

3   PRODUCT COMPARISON

3.1   *Forward looking*

The advantage of the discussed materials lies in their low maintenance and resistance to the effects of corrosion and fatigue. The life cycles of fiber rope systems can easily double what is currently seen with steel products and these systems are optimized for long term dimensional stability. Thus, the fiber rope systems may become prime contenders for static bridge tendons, marine structures, and movable structures where the prime removal factor is deterioration due to corrosion. When evaluated on a life cycle cost basis, or compared against a high cost corrosion protection system, the higher initial cost of fiber ropes may be significantly mitigated. In all cable system applications, the long-term costs associated with maintenance or possible assembly change outs should be strongly considered as the fiber system will have significantly reduced yearly maintenance and extended fatigue life cycle. These are benefits that have not previously been available to the structural designer.

3.2   *Cyclical bending over sheave tests*

The materials and construction naturally lend high performance fiber ropes to be resistant to cyclical bending fatigue operations over sheaves. Numerous fatigue life test cycles have shown

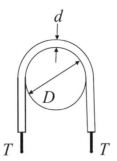

Figure 5.   Cyclical bend over sheave testing schematic.

a high performance fiber rope will provide a fatigue life cycle 4 times longer than that of steel rope. The product limitation comes in the form of internal heat generation that can lead to premature failures of the fiber. Heating of the fiber rope due to internal and external friction should not be a concern due to the limited use of the products in vertical lift applications. The fiber rope products also allow significant reduction in the diameter of the sheave diameters as optimum size range is 20 times the nominal rope diameter. Figure 5 below depicts the typical simple bending test schematic performed on rope samples.

3.3   *Static rope systems*

Additional benefits of the fiber rope systems include low weight factors and heat stability (Faria, 2014). The braided jacket is standard to protect the cables from damaging ultraviolet and abrasion however other benefits can be enhanced into the design. In high wind applications aerodynamic spoilers and cable separators can be applied into the jacket covering for cable stability. Flame resistant polyurethane coating can be applied to the exterior of the jacket system provided protection from any external flame or heat sources.

Fiber rope systems can be used in suspension bridges and cable stayed bridges. As the span length of a new suspension bridges increases, the weight of the cables is required to increase in relation to the total weight of the suspended structure. This is applicable to both the main cable and the suspension cables. A higher percentage of the cable stress is therefore related to the self-weight of the cables themselves. The use of lightweight cables will greatly reduce the stress in the overall system dynamics allowing for lighter structures that will be capable of longer spans.

Isolated structural systems that benefit from less maintenance due to accessibility are top candidates for fiber systems. Figure 6 below shows fiber ropes installed on a suspended roof that require minimal maintenance and were light weight to lessen the stress on other components in the system.

3.4   *Installation*

Carbon fiber tendons have been successfully used as prestressing tendons for reinforcing concrete and in stay cable applications on vehicular bridges in Europe and the United States. The restrictions on these were the brittle resin that was required to encase the structural fibers and the associated costs of purchasing, handling, and installing. Technology has advanced the spinning and drawing processes allowing yarn and stranding fabrication that results in very similar products when compared to traditional steel structural members. The assembly fabrication and equipment must also be tailored to account for the initial elongation of the fiber rope. Precision dimensional tolerances are possible with the assemblies.

The initial cost of the fibers will be approximately 3–4 times that of steel. The owners must balance this cost with the safety and assurance that comes with the fiber rope product.

Figure 6.   Fiber assemblies installed on roof structure.

With corrosion no longer a factor in the annual maintenance cycle of the tensile members there is a removed burden from the owner. The majority of rehabilitations required in the industry are due to corrosion and deterioration of the steel tension members. The life extension of the structural members on bridges would far exceed the other stress components. Cable life could be extended well past 100 years with the fiber technology available. Although inspection of the fiber rope system will still be required on an annual basis the visual inspection process for any damages is quick and easy. Smart technology has also been advanced to allow a smart system to be incorporated into the system that alerts the owner of issues making inspection and maintenance easy.

4   CONCLUSIONS

The fiber systems discussed in this paper are still considered to be new and innovative in most of the markets served. This paper should by no means be considered a complete analysis of these fiber rope constructions or designs. This paper was written to educate the bridge engineering community on the developments in the field of structural fiber cables. The current ASCE specifications do not include these types of cable systems in the text or discussion. Innovative materials are available that will significantly impact the life cycle of cable structures.

REFERENCES

Davies, P. & Reaud, Y. 2011. Mechanical Behavior of HMPE and Aramid Fibers for Deep Sea Handling, Ocean Engineering.
Faria, R. 2014. Synthetic Fiber Rope: Theory and Properties. Lankhorst Euronete.
WRTB. 2005. Wire Rope User's Manual 3rd Edition.

Chapter 17

Bending stresses in parallel wire cables of suspension bridges

A. Gjelsvik
Columbia University, New York, USA

B. Yanev
Department of Transportation, New York, USA

ABSTRACT: The bending stresses induced in parallel wire suspension bridge cables at cable bands are correlated to the forces in the suspenders. The ratio between the bending stress and the total stress in the cable wires is presented as a function of the ratio between the forces in the suspenders and the tension in the main cables. The forces induced by dead and live loads in the suspenders and in the main cables of suspension bridges are critically important design and performance parameters, modeled analytically and, to the extent possible, measured in the field. Using these established parameters, the present method estimates the magnitudes of secondary stresses at cable bands for existing structures. The extreme cases of fully restrained and free wire slippage are considered. Numerical examples reveal consistent trends.

1   INTRODUCTION

Mayrbaurl and Camo (2004) reported 52 suspension bridges in North America. The cables of 29 of them are aerially spun, 21 have helical strands and 2 have Prefabricated Wire Strands (PWS). Analysis typically assumes that the high-strength parallel cable wires are subjected to uniform uniaxial tension. The dominant nature of the primary design load justifies that assumption, however significant "secondary" bending stresses are also present. Bending a cylindrical wire with radius r to (or from) a curvature with radius R induces in it the following bending moment M and corresponding maximum stress σ_b^n:

$$M = \frac{EI}{R} \tag{1}$$

$$\sigma_b^n = \frac{M}{S} = E\frac{r}{R} \tag{2}$$

where $I = \pi r^4/4$ is the moment of inertia of the wire section; $S = I/r$ is the wire section modulus; and the superscript "n" denotes bending due to "natural" wire curvature.

Figure 1 illustrates the evolution of high-strength suspension cable wires over the course of the 20th century. The air-spinning method of construction was dominant during most of this period. It subjects parallel wires to bending in three primary ways. The highest bending stresses, combined with compression occur at the anchorages where wires wrap around shoes with a very small radius. Gimsing (1997) reports a radius of 400 mm at the Great Belt Bridge. Older strand shoes are smaller, as shown in Figure 2. The wires bent around them have clearly developed plastic deformation because they retain it when removed, for example during re-anchoring of strands.

Bending to a larger radius and compression are combined at the tower saddles as well. For a radius $R = 7000$ mm, $2r_{wire} = 5$ mm and $E = 205$ GPa, Gimsing (1997) obtains bending stress $\sigma_b^n = 73$ MPa (11 ksi), or roughly 10% of the cable's working stress. Cold-drawn cable wires however, are not straight. During the air spinning their pre-existing curvature matches that of the saddle, thus reducing the bending stress. After attaining their final cross-section during the drawing process, the wires cool off to a naturally curved shape with a diameter that could be as small as 1500 mm. Based on X-ray diffraction tests, Mayrbaurl and Camo (2004) estimated bending stresses of up to ± 250 MPa (36 ksi) due to straightening. Bowden and Seely (1933) report that the wires on the George Washington Bridge were spun from a reel with $2R$ = 1830 mm (6 ft). Straightening a wire with $2r$ = 4.98 mm (0.196 in) and a

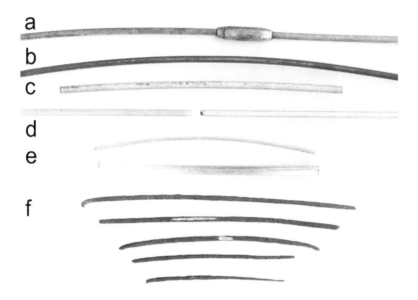

Figure 1.   Suspension cable wires: a) Traditional galvanized wire with splicing ferule; b) Non-galvanized "black" wire; c) The type of wire used in the Great Belt Bridge cables; d) Pre-straightened wire, used in PWS strands with a ductile break; e) Z-shaped wrapping wire; f) Non-galvanized wires from the cables of the Williamsburg Bridge.

Figure 2.   Anchorage of suspension cable built by the air-spinning method.

stress-free curvature with radius $R = 915$ mm would induce a maximum stress $\sigma_b = 558$ MPa (79 ksi), surpassing the uniform working stress for which older suspension cables have been designed. Wires invariably crack on the side where straightening has produced tension. As shown in Figure 1-f, cable wires retain some of their original curvature after 100 years of service, demonstrating that they were never stressed to yield.

Recent suspension bridge cables are built of pre-straightened wires (as shown in Figure 1-d), bundled in Prefabricated Wire Strands (PWS), and socketed at the anchorages. Saddles are profiled and the compacted hexagonal cables are rotated as shown in Figures 3 and 4 ($X'OY'$). Thus local wire bending is reduced or eliminated. A fourth source of "secondary" bending are the cable bands where suspenders transfer the bridge loads to the main cables, as illustrated in Figures 3 and 5.

Figure 3.   Profiled tower saddle, cable and cable band on the Akashi-Kaikyo Bridge.

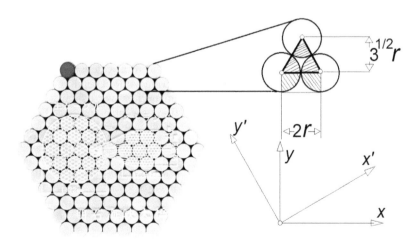

Figure 4.   Perfectly compacted 127 PW Strand of Akashi-Kaikyo Bridge.

Figure 5.   Cable, bands, suspenders and mid-span clamp at the Great Belt Bridge.

Gimsing (1997) modeled the secondary bending as a function of the in-plane rotation of the cable bands. Lee and Kim (2016) expanded the model to consider variable wire friction between cable bands and correlated their findings with test results. The present model adopts the same general assumptions, appropriate for the behavior of a string with localized bending stiffness at the application point of a transverse force. Rather than assume a rotation at the cable band, it models the equilibrium of the cable, subjected to the transverse force of a suspender. The suspender forces are critical to both bridge design and service, and hence, their estimates and measurements are relatively available. Consequently, the model can estimate the secondary stresses at cable bands for a number of existing structures.

2   GENERAL ASSUMPTIONS

2.1   *Geometry and load*

According to Irvine (1981), "by the late 17th century the Bernoulli brothers (Jacob [1654–1705] and Johann [1667–1748]), Leibnitz [1646–1716] and Huygens [1629–1695], more or less jointly discovered the catenary". Leonard Euler (1707–1783) defined the catenary or funicular curve as the optimal shape minimizing potential energy of a string without bending stiffness under a load uniformly distributed along its length. Rough estimates however, continue to be made for uniform loads distributed along the horizontal projection of a cable, suspended from two end points at the same elevation, as shown in Figure 6. The following notation is adopted:

F = cable sag; p = uniform vertical load distributed along the cable horizontal projection; L = horizontal distance between the supports; T = tension in the cable, variable along its length; T_{max} = tension at the supports; H, V = horizontal and vertical components of T_{max}.

The resulting shape ω satisfies the following relationships:

$$\omega = 4f\left(\frac{x}{L}\right)^2 ; f = \frac{pL^2}{8H} ; V = \frac{pL}{2}$$

$$T_{min} = H = \frac{pL^2}{8f} \quad \text{at } x = 0$$

(3)

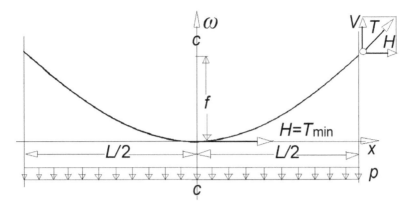

Figure 6.   String subjected to uniformly distributed load along the horizontal chord.

$$T_{max} = \sqrt{\left(\frac{pL^2}{8f}\right)^2 + \left(\frac{pL}{2}\right)^2} = \frac{pL}{2}\sqrt{\left(\frac{L}{4f}\right)^2 + 1} \tag{4}$$

For the typical $L/f = 10$:

$$T_{min} = 1.25\,pL$$
$$T_{max} \approx 1.35L$$

The uniform load p would induce in N uniformly spaced suspenders tension P equal to:

$$P = \frac{pL}{N} \tag{5}$$

Consequently, mid-span:

$$\frac{P}{T_{min}} = \frac{8f}{LN} \tag{6}$$

$$\frac{P}{T_{min}} = \frac{1}{N} \text{ for } \frac{L}{f} = 8$$

2.2   Wire behavior

Gimsing (1997) and Lee and Kim (2016) demonstrate that the secondary bending induced at cable bands propagates along the cable, depending on the level of friction between the wires. Cable wire friction is never fully known, but inferred from the degree of wire compaction. Compaction, in turn is quantified by comparing the cable circumference to the net area of the wires. It is also correlated to the tension in the wrapping wire and in the clamping bolts of the cable bands. It serves two main purposes. For one, it reduces the voids where humidity can condense into water and cause corrosion. Second, friction should restore the working stress state of broken wires over the so-called "clamping distance". Estimating the clamping length became critically important at the Williamsburg Bridge, where many cable wires were lost to corrosion (Figure 1-f). Unlimited clamping length would have implied that the continuous wires were stressed to unacceptably high levels. Steinman Boynton Gronquist & Birdsall (1988) obtained average initial retractions of 0.449 in (11.31 mm) and 0.723 in (18.4 mm) for wires cut in the main and side spans, respectively. Maximum retractions observed in the field were less than 1 in (25.2 mm). It was concluded that "undisturbed cable bands retain over 80% of the wire tensions after cutting. Thus two bands will provide complete recovery for a wire break, without counting any effect of the wire wrapping, so that the total clamping length across the unloaded length of a wire break would be three panel lengths [...] or 60 ft (18.30 m)". This motivated the decision of the Williamsburg Bridge Technical Advisory Committee (1988) to rehabilitate, rather than replace the bridge.

Gjelsvik (1991) showed that under perfect compaction the clamping distance would be relatively short, due to the Poisson effect, causing a broken wire to expand at the stress-released end. Raoof and Huang (1992) presented a more practical estimate of the recovery length of a fractured wire, which took into account the normal stiffness between the individual wires in line contact, and the transition from the full-slip to no-slip inter-wire friction as one moves away from the fractured end of the wire.

Under perfect compaction each wire is in contact with the 6 surrounding ones, as shown in Figure 4. The equilateral triangle defined by the centers of three identical tangent circles is the simplest repetitive unit of the resulting hexagon and it obtains the maximum "void ratio" V_r between the solid and the total area of the cross section, as shown in Equation 7.

$$V_r = \frac{\pi r^2}{2r^2\sqrt{3}} = 0.907 \tag{7}$$

In the outer half-layer the void ratio is $\pi/4 = 0.785$. The number of external wires is $N_1 = 6N_L$ in an ideally compacted hexagonal cable with N_L wire layers, and $N_1 \approx \pi(N_{wires}/V_r)^{1/2}$ in a cylindrical parallel wire cable with total number of wires N_{wires}. The number of wires in air-spun cables and their measured circumference suggest a typical void ratio $V_r \approx 80\%$. Compaction is higher for PWS cables.

Wires are carefully compacted during construction, spiral wrapping wire is tensioned, and cable bands are tightened against slippage. Nevertheless, Lee and Kim (2016) report that "the "fixing" cable band on the Yi Sun-sin Bridge was tightened only after the PWS cables had adjusted their position to dead load in order to reduce secondary bending stresses. Gimsing (1998) describes the stress analysis of a cable band at the Great Belt Bridge as follows: "A 2-D FEM model was established to investigate the very complex stress distribution in the clamp and the interconnection between the cable and the clamp. The model comprised elements of the cable, the upper cable clamp part, and the lower cable clamp part, including the eye plate. The upper and lower parts were interconnected by elements representing the screwed rods. Between the cable elements and the clamp elements special surface elements allowing sliding were applied. The Young's modulus for the cables in the cross-section analysis was determined so that the diameter will be reduced by 5 mm when the bolt tensioning load is applied."

As a result of optimizing the various conflicted construction and service constrains, the friction in the wires is maximized under the cable bands, and remains somewhat indeterminate between them. Gimsing (1997) assumes that wires are fully restrained within the cable bands and free to slip between them. Then both conditions are relaxed over an "effective length".

3   THE DEFLECTED STRING

Figure 7 illustrates the wire deflection under the extreme cases of fixed and free sliding wires. The simplified models assuming cable band effects ranging from local to global are illustrated in Figure 8.

Assuming that slippage is constrained by the cable band, but reduced to a negligible level within half the distance to the next one, would allow treating the case shown in Figure 8 b) as sufficiently similar to the ideal case of Figure 8 a), whereas case 8 c) would not apply. The deflected shape $\omega(x)$ of Figure 8 a) is described by Equation 8:

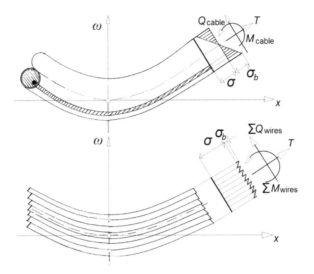

Figure 7.   Deflection and stress distribution in fixed and sliding parallel wires.

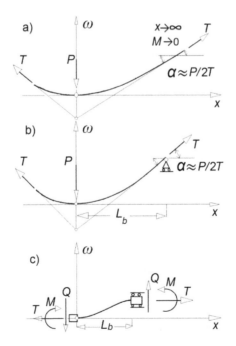

Figure 8.   Models of string deflection: a) Local bending effect, decaying as $x \to \infty$; b) Local bending effect, sufficiently decaying as $x \to L_b/2$; c) Bending effect extending over the distance L_b between cable bands.

$$\frac{d^4\omega}{dx^4} - \lambda^2 \frac{d^2\omega}{dx^2} = 0 \qquad (8)$$

where $\lambda = (T/EI)^{1/2}$ [1/length]; $T =$ string tension force; and $E =$ Young's modulus. The cross section is a circle with radius r, area $A = \pi r^2$, moment of inertia $I = Ar^2/4$, and section modulus $S = I/r$.

The solution of Equation 8 has the following form and boundary conditions:

$$\omega = Ax + B + C\sinh \lambda x + D\cosh \lambda x$$
$$x = 0 : \omega = 0; \frac{d\omega}{dx} = 0 \tag{9}$$
$$x \to \infty : \frac{d^2\omega}{dx^2} = 0; \frac{d\omega}{dx} = \alpha$$

Applying the first three B.C. obtains the following form:

$$\omega = C\left(1 - \lambda x + \sinh \lambda x - \cosh \lambda x\right) \tag{10}$$

$$\frac{d\omega}{dx} = \lambda C\left(\cosh \lambda x - \sinh \lambda x - 1\right) \tag{11}$$

For small α, Equation 11 obtains:

$$C = -\frac{\alpha}{\lambda}; \alpha = \frac{P}{2T}; C = -\frac{P}{2\lambda T}$$

Hence:

$$\omega = \frac{P}{2\lambda T}\left(\lambda x - 1 + \cosh \lambda x - \sinh \lambda x\right) \tag{12}$$

Equation 13 is obtained after the following substitutions in Equation 12:

$$\cosh \lambda x = \frac{e^{\lambda x} + e^{-\lambda x}}{2}; \quad \sinh \lambda x = \frac{e^{\lambda x} - e^{-\lambda x}}{2}$$
$$\omega = \frac{P}{2\lambda T}\left(\lambda x - 1 + e^{-\lambda x}\right) \tag{13}$$

The slope, moment M and shear Q are obtained by derivation in Equations 14, 15, and 16.

$$\frac{d\omega}{dx} = \frac{P}{2T}\left(1 - e^{-\lambda x}\right) \tag{14}$$

$$M = EI\frac{d^2\omega}{dx^2} = \frac{P}{2\lambda}e^{-\lambda x} \tag{15}$$

$$Q = -\frac{dM}{dx} = \frac{P}{2}e^{-\lambda x} \tag{16}$$

The results are illustrated in Figure 9. Setting $e^{-\lambda x} = 0.5$ obtains the "half-length of influence" over which M and Q decay by 50% of their maximum values at $x = 0$. The half-length of influence is related to the radius r of the string in Equations 17, 18, and 19:

$$x = \frac{0.69}{\lambda} = \frac{0.69}{\sqrt{T/EI}} \tag{17}$$

$$\frac{x}{r} = \frac{0.69}{\lambda r} \tag{18}$$

$$\lambda r = r\sqrt{T/EI} = 2r\sqrt{\frac{T}{EAr^2}} = 2\sqrt{\sigma/E} \tag{19}$$

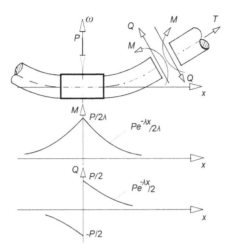

Figure 9.   Decay of M and Q with respect to x.

where $\sigma = T/A$ is the tension stress in the string.

The respective half-lengths of influence x_c and x_w for a "solid" cable with n wires and radius R_{cable}, and for n independent wires with radii r_{wire} subjected to the same uniformly distributed tension are compared in Equation 20. It is noted that $\lambda_{cable}/\lambda_{wire} = r_{wire}/R_{cable}$.

$$x_c/x_w = R_{cable}/r_{wire} \tag{20}$$

At the Williamsburg Bridge, $r_{wire} = 0.096$ in (2.95 mm), $n = 7696$. The "net steel" cable radius should be approximately $R_{cable} = 8.4$ in (0.214 m). Assume $\sigma = 44$ ksi (304 MPa). Substituting these values in Equations 17 through 20 obtains:

$$x_c = 8.86R_{cable} = 74.4 \text{ in (1.9 m)}; \ x_w = 8.86r_{wire} = 0.85 \text{ in (21.6 mm)}; \ x_c/x_w = 87.5$$

Table 1 contains values of $\lambda r = 2(\sigma/E)^{1/2}$ and $x/r = 0.69/(\lambda r)$ for a practical range of stresses σ. Young's modulus $E = 29,000$ ksi (205 MPa). Figure 10 illustrates the reduction of the influence length x with the increase of the stress σ.

From Equation 15 it follows that the maximum bending moment M_{max} occurring at $x = 0$ is equal to:

$$M_{max} = \frac{P}{2\lambda} \tag{21}$$

The maximum bending stress σ_b can be expressed in one of the forms in Equations 22 and 23:

$$\sigma_b = \frac{M}{S} = \frac{rP}{2I\sqrt{T/EI}} = P\sqrt{E/TA} \tag{22}$$

$$\sigma_b = \frac{P}{T}\sqrt{E\sigma} = 2\alpha\sqrt{E\sigma} \tag{23}$$

The last expression in Equation 23 matches Equation 2.82 in Gimsing (1997). As noted therein, σ_b depends on the tension stress σ in the string (be that cable or wire), and not on its diameter. The ratio σ_b/σ between the bending and the axial stress is as follows:

Table 1.　$\lambda r = 2(\sigma/E)^{1/2}$ and x/r with respect to the string tension σ.

σ				
ksi	MPa	λr	x/r	$(E/\sigma)^{1/2}$
25	173	0.059	11.95	34.05
50	346	0.083	8.37	24.08
75	518	0.102	6.80	19.66
100	690	0.117	5.90	17.03
125	863	0.131	5.35	15.23

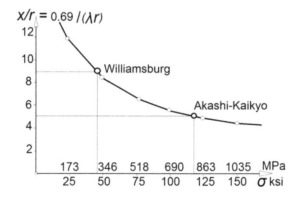

Figure 10.　Relationship of λr, x/r and σ.

$$\frac{\sigma_b}{\sigma} = P\frac{\sqrt{E/TA}}{T/A} = \frac{P}{T}\sqrt{E/\sigma} \qquad (24)$$

4　EXAMPLES

The relationships derived thus far are applied to the bridges enumerated in Tables 2 through 5. The iconic Brooklyn Bridge is omitted, because it is a hybrid combining suspenders and stays. The following notation is used:

　P_{cable} = the cable weight distributed uniformly along the span L; p = specified maximum vertical load uniformly distributed along the span L; T_{max} = maximum cable tension at the supports induced by p; T_{min} = minimum cable tension mid-span, induced by p; $\sigma_{max} = T_{max}/A$ is the maximum design stress of the cable wires; $\sigma_{min} = T_{min}/A$; $\sigma_{tensile}$ = specified wire strength; N_{wires} = wires per cable, N = suspenders per cable.

　The following sources are quoted in Table 2:

Williamsburg (4 cables):

Steinman Boynton Gronquist & Birdsall (1988) measured forces at the anchorage eye-bars, averaging at roughly 9000 kips per cable. Design had assumed $\sigma_{tensile}$ = 198 ksi (1367 MPa), however tests (ibid.) obtained 220 ksi (1518 MPa).

Manhattan (4 cables):

During the recent replacement the suspender forces were set at P = 130 kips (580 KN). In this case the vertical load transmitted to the cable by the suspenders p_b and the total load on the cable p are:

Table 2.   Design parameters of representative bridges.

Bridge	L			d_{wire}			A_{cable}		$\sigma_{tensile}$		
	ft	m	L/f	in	mm	N_{wires}	in²	m²	ksi	MPa	N
Williamsburg	1600	488	9.09	0.192	4.88	7,696	222	0.144	220	1518	77
Manhattan	1472	449	9.09	0.192	4.88	9,462	280	0.180	220	1518	81
George Washington	3500	1067	10.00	0.196	4.98	26,474	800	0.520	220	1518	57
Golden Gate	4200	1280	8.94	0.192	4.88	27,572	800	0.520	220	1518	83
Mackinac	3800	1158	10.75	0.196	4.98	12,580	380	0.245	220	1518	95
Great Belt	5325	1624	9.00	0.211	5.37	18,648	655	0.424	228	1570	69*
Akashi-Kaikyo	6528	1991	10.00	0.206	5.23	36,830	1226	0.791	260	1800	141

*Including the center clamp.

$$p_b = PN/L; \; p = p_{cable} + p_b$$

George Washington (4 cables):
Dana et al. (1933) reported design loads and cable parameters.
Golden Gate (2 cables):
Strauss (1937) reported design loads and the cable parameters.
Mackinac (2 cables):
The sources are secondary. According to Nishino et al. (1994) the Mackinac Bridge cables are built of wires similar to those of the Golden Gate Bridge, and were designed with the same allowable stress. At the Golden Gate $\sigma_{tensile}/\sigma_{max} = 220/78.17 = 2.8$. At Mackinac that ratio would obtain $T_{max} = 29,857$ kips. However $T_{max} = 32,258$ kips (144 MPa) has also been reported. This higher value was used, obtaining

$$\sigma_{tensile}/\sigma_{max} = 2.6.$$

Great Belt (2 cables):
Gimsing (1998) reported $\sigma_{tensile} = 1570$ MPa (228 ksi) and maximum cable force of 331 MN (74,106 kips) at the saddle towards the side span, corresponding to $\sigma_{max} = 780$ MPa (113 ksi). The resulting ratio $\sigma_{tensile}/\sigma_{max} = 2$ is "in compliance with the requirement for a minimum safety factor (corresponding to 2.2 on the characteristic load)".
Akashi-Kaikyo (2 cables):
HSBA (1998) reported $\sigma_{tensile} = 1800$ MPa (260 ksi) and a stress factor of 2.2, resulting in $\sigma_{max} = 818$ MPa (119 ksi).

The values in Table 3 are obtained by Equations 3 and 4 herein for the specified maximum vertical load p uniformly distributed along the span L, as shown in Figure 6.

The tension forces T^b_{max} and T^b_{min} in Table 4 result from a vertical load $p_b = p - p_{cable}$, distributed uniformly along the span L. $P = p_b L/N$ is the average force transmitted to the cable by the suspenders as shown in Equation 5. The values of σ_{max} and σ_{min} are enumerated in Table 3. The corresponding bending stresses induced at the cable bands by the suspenders are obtained by Equation 22 and enumerated in Table 5. As Equation 22 indicates, $\sigma_{b\,min} > \sigma_{b\,max}$.

The relationship between σ_b/σ, σ, and P/T values, obtained in Equation 24 is plotted in Figure 11. The respective positions of the representative bridges are numbered as in Table 5.

Gimsing (1997) points out that the maximum stress ($\sigma_{max} = \sigma + \sigma_b$) is the same, whether all parallel wires bend as a compact section or in individual "layers". In the "fixed" case however, only the "outer layers" are subjected to σ_{max}. In a circular section, there are fewer wires in the "outer layers". Under free slippage, all wires experience the same σ_{max}. The secondary bending stress σ_b is independent of the dimensions of the "string" and hence, the results apply to cables, strands, and wires. They depend on the tension σ in the "string", and the

Table 3.   Parameters of representative bridges based on the maximum design stress in the cable σ_{max}.

Bridge	P_{cable}		p		T_{max}		T_{min}		σ_{max}		σ_{min}	
	k/ft	KN/m	k/ft	KN/m	kips	MN	kips	MN	ksi	MPa	ksi	MPa
Williamsburg	0.754	11	4.95	72	9,832	44	9,000	41	44	304	40	276
Manhattan	0.951	14	8.10	118	14,800	66	13,547	62	53	365	48	334
George Washington	2.717	40	12.60	184	59,200	264	55,200	246	74	511	69	476
Golden Gate	2.717	40	12.60	184	65,045	291	59,373	266	78	538	72	497
Mackinac	1.290	19	5.48	80	32,258	144	27,983	125	85	587	74	511
Great Belt	2.224	33	11.30	165	74,106	331	67,619	302	113	780	103	711
Akashi-Kaikyo	4.164	61	16.60	243	146,013	652	135,456	605	119	818	110	800

Table 4.   Values of p_b, T^b_{max}, T^b_{min}, P and T/P for the representative bridges.

Bridge	p_b		$T_{b\,max}$		$T_{b\,min}$		P		$P/T_{b\,max}$	$P/T_{b\,min}$
	k/ft	KN/m	kips	MN	kips	MN	kips	MN		
Williamsburg	4.20	61	8,342	38	7,636	34	87	0.39	0.0104	0.0113
Manhattan	7.15	104	13,065	58	11,959	53	130	0.58	0.0100	0.0120
George Washington	9.90	145	46,514	208	43,371	194	608	2.71	0.0131	0.0140
Golden Gate	9.90	145	51,107	228	46,650	208	500	2.23	0.0098	0.0107
Mackinac	4.20	61	24,723	111	21,447	96	168	0.75	0.0068	0.0078
Great Belt	9.10	132	59,678	265	54,454	243	702	3.12	0.0117	0.0129
Akashi-Kaikyo	12.40	182	109,070	487	101,184	452	574	2.55	0.0053	0.0057

Table 5.   Estimated bending stresses in the cables of the representative bridges.

Bridge	$\sigma_{b\,max}$		$\sigma_{b\,min}$		$\sigma_{b\,max}/\sigma_{max}$	$\sigma_{b\,min}/\sigma_{min}$
	ksi	MPa	ksi	MPa		
1. Williamsburg	11.0	75	11.42	79	0.25	0.29
2. Manhattan	11.6	80	12.09	84	0.23	0.25
3. George Washington	17.0	117	17.60	122	0.22	0.26
4. Golden Gate	13.4	92	14.00	97	0.17	0.20
5. Mackinac	9.4	65	10.00	69	0.11	0.14
6. Great Belt	19.2	132	20.00	138	0.17	0.20
7. Akashi-Kaikyo	8.5	58	8.78	61	0.07	0.08

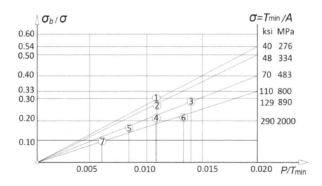

Figure 11.   Relationships between σ_b/σ, σ, and P/T.

Figure 12.   Inspection and re-compaction of a 100-year-old cable.

spacing of the suspenders. As a result, "secondary" bending at cable bands appears more significant at older bridges, such as the one illustrated in Figure 12, than at the most recent record holders in the field. Figure 12 shows wedges driven in the cable to allow for inspection of wires and re-compaction of the cable, following inspection.

5   CONCLUSIONS

The proposed technique provides a simple and realistic estimate of the bending stresses occurring at cable bands in the parallel wires of suspension bridge cables due to the tension in the suspenders. It relies on the basic parameters describing the cables of suspension bridges, such as the cable sag, main span length and number of suspenders. For the cases presented herein they are obtained from the bridge designers and owners, as specified in the references. The model can be applied to any set of suspension bridge data, under the critical simplifying assumption that loads are uniformly distributed along the horizontal, and are due either to the weight of the cable or to tension from the suspenders. The estimates can be taken into account in investigating more complex and non-linear phenomena beyond the scope of this study including: local effects, such as the condition between the saddle and the first cable band, discussed by Gimsing (1997); non-uniform distribution of tension among cable strands; non-uniform loads along the bridge span; stress states in cracked wires; local stress effects of the wires upon exiting from the cable bands; eccentricity at the transfer of the suspender load; ultimate cable strength.

   Applied to some familiar bridges, the method obtains σ_b/σ ranging from relatively low to over 25%. The difference parallels the evolution of suspension bridge design and construction over the 20th century. Between the completion of the Brooklyn Bridge in 1883 and that of the Akashi-Kaikyo in 1998, the diameter of cable wires has grown from 0.192 in (4.88 mm) to 0.21 in (5.37 mm) and the strength has increased from 160 ksi (1.1 GPa) to 260 ksi (1.8 GPa). Prefabricated Wire Strands have replaced the air spinning method. Corrosion is no longer retarded by inhibitors, but minimized by dehumidification. The design ratio of $\sigma_{tensile}/\sigma_{max}$ has dropped from over 4 at the Williamsburg (1903) and Manhattan (1909) Bridges to 2.2 at the Great Belt and Akashi-Kaikyo. Thus the management of recently constructed and century old suspension cables differ distinctly. Higher wire strength leads to higher working stresses and improves the overall stiffness, concurrently reducing the ratio σ_b/σ. As one consequence, at the 1624 m span of the Great Belt, suspenders can be spaced 24 m apart without causing significant secondary bending. On the other hand, ductility and tolerance to corrosion reduce as $\sigma_{tensile}$ increases. The sensitivity to fretting and fatigue is greater. At a ratio of $\sigma_{tensile}/\sigma_{max} = 2.2$, secondary effects cannot be ignored and deterioration is inadmissible.

With age, the assessment of the older suspension bridge cables gains importance for many reasons. Secondary stresses may be one of them. Assuming that wires alternate between perfect fixity under the cable bands and perfect slippage between them is extreme, but usefully reminds that bending is maximized at the band's edge. Broken wires at some older bridges have been found at cable bands, on the cable underside (also known as "6 o'clock"). Since the same areas are also most susceptible to corrosion and other local effects, pitting and stress of different origins contribute inseparably. Pre-existing micro-cracks further complicate the assessment.

The effects of compaction are reconsidered during the rehabilitation and in-depth inspection of suspension cables, such as the one in Figure 12. Suspenders are replaced at roughly 40–50 year cycles. Monitoring by various non-destructive (some of them invasive) methods is explored. Recent studies show that dehumidification can be effective at older bridges as well. These operations are costly, but inevitable over the hundreds of years of service.

REFERENCES

Dana, A., Andersen, A. & Rapp, G. 1933. George Washington Bridge: design of superstructure, *American Society of Civil Engineering,* Transactions, Paper No. 1820, New York.

Bowden, E.W. & Seely, H.R. 1933. George Washington Bridge Construction of the Steel Superstructure, Paper No. 1823, Transactions, American Society of Civil Engineers, Vol. 97, pp. 243–328.

Gimsing, N.J. 1997. *Cable Supported Bridges*, 2nd Ed., John Wiley & Sons, Ltd., New York.

Gimsing, N.J., Editor (1998) *The East Bridge,* A/S Storebæltsforbindelsen, Copenhagen.

Gjelsvik, A. 1991. The Development Length of an Individual Wire in a Suspension Bridge Cable, *Journal of Structural Engineering, American Society of Civil Engineering,* Vol. 117, No. 4, April, 1991, pp. 1189–1201.

Honshu-Shikoku Bridge Authority (HSBA) 1998. The Akashi-Kaikyo Bridge. Design and Construction of the World's Longest Bridge, Japan.

Irvine, M. 1981. *Cable Structures*, MIT Press, Cambridge, Ma.

Lee, M. & Kim, H.-K. 2016. Angular change and secondary stress in main cables of suspension bridges, *International Journal of Steel Structures,* June 2016, Vol.16, Issue 2, pp. 573–585.

Mayrbaurl, R. & Camo, S. 2004. Guidelines for the inspection and strength evaluation of suspension bridge parallel-wire cables. Report 534, National Cooperative Highway Research Program, Transportation Research Board, Washington, D.C.

Nishino, F., Endo, T. & Kitagawa, M. 1994. Akashi-Kaikyo Bridge under construction. *7th Structures Congress*, Atlanta, GA.

Raoof, M. & Huang, Y.P. (1992) Wire Recovery Length in Suspension Bridge Cables, Journal of Structural Engineering, Vol. 118, No.12, American Society of Civil Engineers, New York.

Steinman Boynton Gronquist & Birdsall, 1988. Williamsburg Bridge cable investigation program final report, in association with Columbia University, New York State Department of Transportation, New York City Department of Transportation.

Strauss, J.B. 1937. The Golden Gate Bridge. Report of the Chief Engineer to the Board of Directors of the Golden Gate Bridge and Highway District, California.

Williamsburg Bridge Technical Advisory Committee 1988. Technical Report to the Commissioners of Transportation of the City and State of New York, Howard Needles Tammen & Bergendorf, June 30.

Structural health monitoring of bridges

Chapter 18

Structural health monitoring of a historical suspension bridge

G.W. William
AECOM, Morgantown, West Virginia, USA

S.N. Shoukry
West Virginia University, Morgantown, West Virginia, USA

M.Y. Riad
Sargent and Lundy, LLC, Chicago, Illinois, USA

ABSTRACT: This paper describes the development of an innovative, cost-effective system for real-time condition assessment of the Market Street Bridge after the completion of its rehabilitation. The 547-m long suspended steel bridge was originally constructed in 1905, and was undergone major rehabilitation work that was completed in November 2011. The sensory system deployed in this study continuously monitors the global structural displacements of the bridge towers under operating and environmental conditions. A detailed 3D finite element model was developed for the suspended part of the Market Street Bridge and the model response was validated versus the field measured data. A damage identification technique was developed through comparison of vibration characteristics of structural elements between two different states. One represents the initial and often undamaged state and the second represents the deteriorated state after damage. A nonlinear trend of the first natural frequency of the bridge structure was observed once the towers reach certain inclinations, which could be attributed to the change in bridge stiffness due to damage or plastic deformation. This system was found to be a cost-effective remote monitoring system that provides the foundation for an enhanced structural bridge management.

1 INTRODUCTION

Structural Health Monitoring (SHM) is increasingly considered a valuable tool to increase the safety of structures, but also to optimize their life-cycle cost (Bakht et al. 2011, Xu and Xia 2012). As a matter of fact, monitoring a structure such as a bridge can lead to early damage detection and thus plan maintenance and rehabilitation work before the structure becomes deficient. Bridge structures have been monitored for multiple reasons. New or existent structures can be monitored to ensure safety, verify design assumptions, plan maintenance and operational activities, check integrity, and also prioritize rehabilitation or extension. Accurate knowledge of the behavior of structures is becoming more important as new structures become lighter and also an increasing number of existing structures is required to remain in service far beyond their initially planned service life (Inaudi et al. 1999, Inaudi 2009, Fraser et al. 2010).

Sensory systems which monitor the geometry and deformations of large civil structures are not new. Accelerometers, strain gauges, linear variable displacement transducers (LVDT), and total stations are familiar tools to many professionals involved with structural monitoring (Gastineau et al. 2009, Enckell 2011). Recently, Global Positioning Systems (GPS) have been demonstrated as a feasible alternative for monitoring the dynamic behaviour of tall buildings, towers, dams, and bridges (Erol 2004, Roberts et al. 2005, Ogaja et al. 2007,

Radhakrishnan 2014). The GPS system was deployed on several major suspension bridges including Tsing Ma Bridge in Hong Kong (Wong et al. 2001), Akashi Kaikyo Bridge in Japan (Nakamura 2000), Jiangyin Bridge in China (Leica 2003) and Golden Gate Bridge in San Francisco, USA (Turner 2003). The major challenges reported for the GPS technology were its low sampling rate for dynamic measurements of about 20 Hz as well as its high cost.

Aligned with the efforts above, this paper presents development of a cost effective monitoring system to monitor the structural deformation of the towers of the historical Market Street Bridge in West Virginia. The system is simple to use and costs much cheaper than the counterpart GPS alternative. The system could be expanded to include more sensors to extract and monitor the bridge's dynamic characteristics.

2   MARKET STREET BRIDGE

Owned and managed by the West Virginia Department of Transportation (WVDOT), the Market Street Bridge is a suspension bridge over the Ohio River connecting Market Street in Steubenville, Ohio with West Virginia Route 2 in Follansbee, West Virginia as shown in Figure 1. The bridge was constructed in 1905 with an overall length of 547 m (1,794 ft) and a roadway width of approximately 6.7 m (22 ft). The bridge has three cable suspension spans over the river stiffened by a welded Warren through truss. The main span length is 210 m (700 ft). The west approach consists of two deck girder spans and five spans of the original riveted steel through 132-m long (432 ft) truss, which is a quadrangular Warren with verticals. The cables are suspended from two steel towers that rise approximately 63 m (210 ft) from cut

Figure 1.   Elevation view of Market Street Bridge—looking west.

stone piers. The substructure consists of cut stone piers, concrete stub abutments, and both concrete and steel bents. Not only was the bridge called ornamental on its opening, but also it was a utilitarian structure that provided an important transportation link between Ohio and West Virginia. Construction of this bridge represented the expansion of communication and markets to new places with ease that could not have been conceived in the era of ferrying. The Bridge also uniquely represents an ever-evolving continuum of transportation history in its original use for streetcars, incidental use by pedestrians and finally, acquiescence to the automobile. The bridge remains of high value and importance to local citizens and businesses on both sides of the Ohio the river, from both historical and practical perspectives.

In 2009, the overall condition of the bridge was rated as poor. The towers controlled the load rating. The riveted tower columns are laced and braced with curved gusset plates. Previous strengthening repairs added full height plates on the transverse faces of the towers by stitch welding. Pack rust was causing these plates to bow away from the tower columns and breaking the stitch welds. Various bracing gussets had deteriorated with through holing and broken lacing bars. The base of the tower columns had isolated areas of section loss. The suspension system was generally in fair condition; however, a few locations had issues. The stiffening truss connections to the towers had expanded into the towers and in one location lost bearing due to pin wear. The cable clamps at the suspenders had deteriorated caulking and corrosion. Pack rust had formed behind many of the eye-bar connections to the floorbeams and at isolated locations caused concern for the adequacy of the eye-bars, clevises and pins. Uplift anchors and wind-locks had advanced loss of section. Deterioration to the floor system, stringers and floorbeams had caused through holing in several locations and reduced capacities near the posting level. As a result of inspection findings, the bridge load posting was reduced to 5 tons and major rehabilitation was recommended to keep the bridge in service (Lewellyn et al. 2012). Due to its historic nature, the WVDOT planned repairs to the bridge that included structural repairs to the truss and towers. The rehabilitation work that began in 2010 also included the installation of period lights along the outside cables of the bridge and a new paint and color scheme. The bridge was reopened to traffic on December 7, 2011.

Due to the poor rating of the Market Street Bridge structure before the repair work, the WVDOT requested a system to be designed and installed to reliably monitor the bridge response as it is open to traffic. This system is intended to supplement routine visual inspections due to the fact that the bridge was painted with a thick elastic coating during rehabilitation that would mask any new cracks or defects that may develop until they penetrate fully through the paint. Therefore, there was a necessity for innovative monitoring system that provides a tool for the real-life structural condition assessment of the repaired bridge and ensures that its deformations fell within the design limits. This was achieved by a monitoring system that was developed and used to monitor the inclination of the Market Street Bridge towers. Damage to a significant structural element such as a hanger or a truss member may produce an immediate out of the ordinary behavior of the plotted trends from the inclinometers.

3 INSTRUMENTATION SYSTEM

The instrumentation plan developed for the continuous monitoring of the Market Street Bridge towers utilized Inclinometers developed based on Micro Electro-Mechanical Systems (MEMS) technology that were installed side-by-side with vibrating wire inclinometers as shown in Figures 2 and 3. The implementation of the two inclinometer systems was necessary in order to validate the data from each system versus the other system, thus avoiding the withdrawal of incorrect conclusions regarding the bridge tower movements. Additionally, a laser imaging station was utilized to provide intermittent measurements of the changes in the profile of each tower. For this purpose, a set of reflectors was installed on

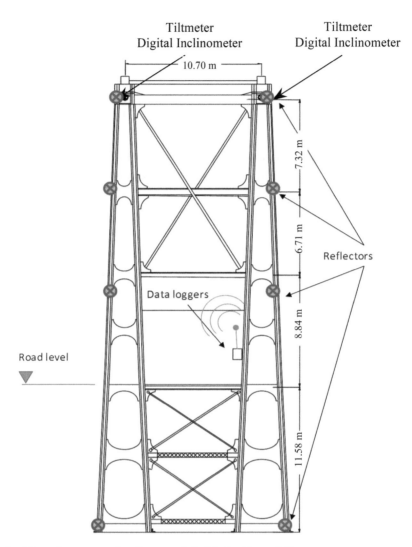

Figure 2.   Figure instrumentation diagram of the Market Street Bridge towers.

each tower as shown in Figure 2. Laser scanning was performed periodically on both towers to monitor the changes in the alignment and profile of each tower. The data from the vibrating wire inclinometer data loggers are manually collected together with the results of the laser scanning.

The MEMS sensor is a versatile 6-channel instrument that provides continuous data records of three-axis acceleration, two-axis inclinations, and temperature. This sensor allows for collecting and analyzing static and dynamic structural information easier than in the traditional configurations using separate sensing and data acquisition devices. The sensor also features a high-speed data acquisition board and digital communication into one small packaging. The accelerometer offers a bandwidth of DC to 200 Hz with a sampling frequency of 2,000 samples per second and a resolution of 50 µg. The inclinometer acquires data at 10 samples per seconds with a resolution of 50 µ° and a range of ±15° (Inaudi et al. 2011). It is worth mentioning here that this sensor is particularly adapted to the monitoring of extreme conditions, because it combines acceleration measurements

Reflector

Vibrating Wire Inclinometer

MEMS Inclinometer/
Accelerometer

Figure 3.   Measurement of tower inclination at the highest point.

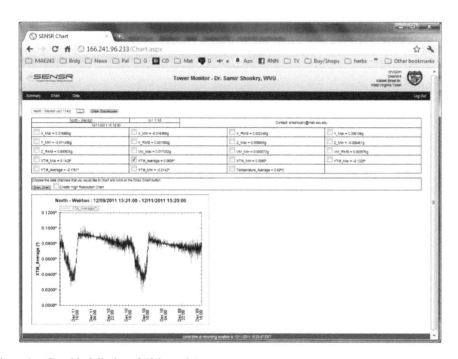

Figure 4.   Graphical display of 48-hour data.

that can identify actions during the event itself and static tilt measurements, which can be the result of irreversible deformations due to damage. In other words, acceleration readings provide information on the bridge response due to traffic loadings, while inclination readings provide slow response due to environmental changes and accumulated irreversible deformation.

Each sensor was mounted on a steel bracket that was welded to the bridge tower as shown in Figure 3. The data acquisition systems are powered continuously through AC power provided to the bridge. Each data acquisition is housed in a weather-proof enclosure and mounted on the bridge tower with accessibility at the traffic level. Cables connecting the sensors to their corresponding data acquisitions are directed along the steel members and hidden from exposure inside the bridge towers.

The data acquisition system for the MEMS sensors was fitted with cellular modem to provide an internet-based monitoring system to remotely monitor the bridge structural response from anywhere. A customized server is used to control the data acquisition system, collect data, store time histories, and display the data in real time for the last 48 hours of monitoring. The system automatically takes the measurements, creates the reports, and posts them to the internet. Figure 4 illustrates the web page created for West Virginia tower with the instantaneous sensors readings. This system was found to be a cost-effective remote monitoring system that provides the foundation for an enhanced management program that directly supports:

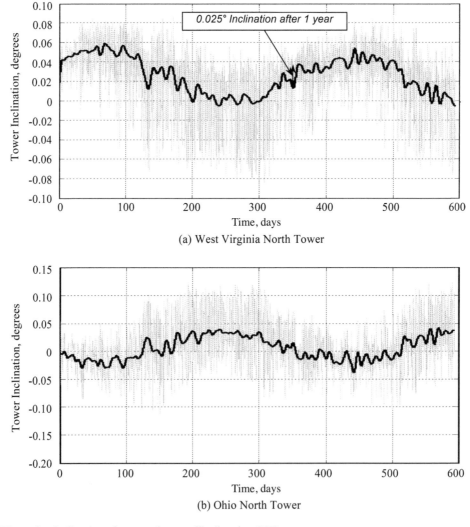

(a) West Virginia North Tower

(b) Ohio North Tower

Figure 5.   Inclination of towers along traffic direction (XT).

1. Enhanced inspection coverage—The system fills in the gaps between periodic inspections by providing continuous trend data and critical measurements that are not discernible by periodic visual inspections.
2. Real-time management—The system continuously monitors and reports structural responses, which enables producing constant awareness of the structure's response to all operating loads and environmental conditions.
3. Early detection—The sensors can detect different structural responses and conditions. This advanced detection capability enhances coverage and increases the opportunities for timely responses.

The data measured from the sensors had to be monitored for an initial twelve-month period to allow for the bridge to undergo expansion and contraction during a full seasonal temperature variation cycle. The limitation inherit in displaying the collected data for only

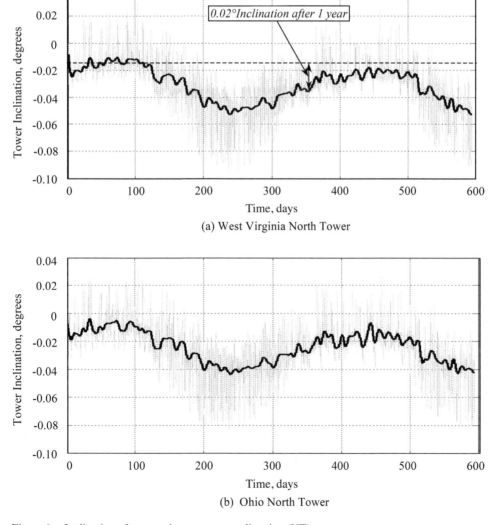

(a) West Virginia North Tower

(b) Ohio North Tower

Figure 6.   Inclination of towers along transverse direction (YT).

the last 48-hour period hinders the ability of the bridge engineer to inspect the full time history of the tower inclination. Therefore, it deemed essential to develop user-friendly software to communicate with the data acquisition system, update the data and display the full time histories of the measured bridge response on an internet website. Such software was developed and written in MATLAB and an internet website (http://marketstbridge.mae.wvu.edu) was established to provide secure access to the full-time history data. This aided the bridge engineers in their analysis of the data and helped them to draw useful conclusions about the bridge condition.

Figure 5 illustrates the time history plots for the inclination of the north side of the towers along the traffic direction (XT). Comparison of the two plots indicates that the West Virginia and Ohio towers incline toward each other as the bridge contracts in winter and away from each other as the bridge expands in summer. Although both towers undergo the same amount of inclination due to seasonal temperature change, Ohio tower was found to have higher range of inclination as daily temperature changes between day and night. The measured data also indicate that the north side of West Virginia tower, shown in Figure 5(a), accrued a permanent inclination of 0.025° after twelve months of monitoring. However, due to the limited monitoring duration, it is not possible to predict whether such a deformation would increase or decrease with time.

In the transverse direction, both north and south sides of each tower are inclining in the same direction as could be seen in Figure 6. It could be also noticed that West Virginia tower developed a permanent inclination after one year.

4   DAMAGE IDENTIFICATION

A three dimensional finite element model was developed for suspended spans of the Market Street Bridge as shown in Figure 7. The model includes detailed main structural elements of the steel bridge including the truss members, deck floor, cantilever walkway, bridge towers, suspension main chords and suspension cables. The geometry, section properties and

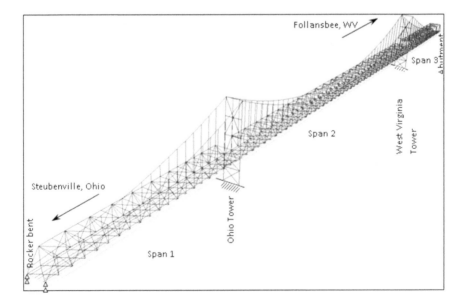

Figure 7.   Three dimensional model of the suspended bridge spans.

elevations are taken from the construction sheets as well as the various rehabilitation documents of the bridge structure. In total, the bridge model contains 7075 nodes, and 8862 elements. Among those are 5980 Hermitian beam elements representing the steel bridge members in the stiffening truss members, upper and lower lateral bracing members, stringers and floor beams, as well as the main members of the towers, 2568 shell elements representing the open deck floor and sidewalk, and 314 3-D truss elements representing the main suspender cables and hangers.

The main towers are fully fixed to the piers in the real structure, therefore the base of both towers in the model are restricted for all displacements and all rotations. Hinged supports were assigned for both ends of the model where the rocker bents exist at the Ohio side, and at the east abutment on the West Virginia side. Also hinged supports are specified at both ends of the main suspension cables. Rigid links were specified to connect the deck and the supporting steel structure to the towers. This arrangement takes into account the large dimension of the tower width, while conveying the weight of the structure to the towers as it is the case in the field.

The kinematics control to run the program solver is specified for large displacements and rotations, and small strains. The solution process uses the sparse equation solver while the nonlinear analysis uses the full Newtonian method for the iteration scheme. The iteration tolerances have been specified to use the force convergence criteria where the contact force tolerance is 0.05 and the minimum reference contact force as well as the force (moment) tolerances are 0.01. The reference force and moment are set to 1 and the maximum number of iteration was set at 15 trials to complete the solution.

The finite element model is subjected to two types of loading configurations. The first loading configuration consists of the own weight of the entire structure through gravity loading, in normal temperature conditions of 21°C (70°F). The suspension cables are also pre-stressed with initial strains that are applied to simulate the bridge condition in the field providing minimum bending moments and deflections on both towers and bridge deck. The second loading configuration after applying the own weight of the structure and the pre-stressing tension on the suspension cables is variation of ambient temperatures. The entire bridge model has been subjected to temperature loading ranging from -1°C to 38°C (30°F to 100°F) with reference temperature of 21°C. The values of inclinations at the highest point of each tower are recorded for each temperature. The difference in inclination from that at the reference temperature value of 21°C is plotted versus the temperature. Figure 8 illustrates this

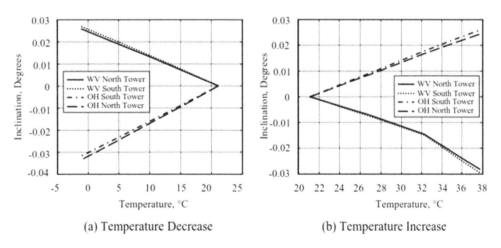

(a) Temperature Decrease   (b) Temperature Increase

Figure 8.   Relative inclination of towers vs temperature (Reference position at 21°C).

relationship where it can demonstrated that the temperature increase tend to bring the top end of the towers away from one another.

Validation of the Market Street Bridge FEM is conducted in two different ways. The vertical deflection values of the bridge model subjected to gravity load is compared to field measurements of the bridge deck profile in absence of traffic loading. Also field measurements of the inclination of both West Virginia and Ohio towers in response to seasonal ambient temperature changes are compared to FEM results subjected to similar temperature change values.

Table 1 lists a comparison between the levels of the bridge deck profile predicted by the FEM in response to gravity load, and corresponding values measured from the field. The values listed in Table 1 are relative to the level of the bridge deck at the rocker bent. The average discrepancy between FEM levels and those measured in the field are calculated and found to be within 4.1%.

In a similar manner, the inclination of the towers in the FEM is recorded at each temperature and compared to those collected from the field at corresponding temperature values. Values of the inclination of the towers at different temperature values are identified from each sensor installed on the bridge structure. The mean value of the time history is used in order to filter out the effect of diurnal oscillations and any dynamic effect that traffic loads might have on the towers behavior. Figure 9 illustrates the change of

Table 1.   Comparison of FEM versus field measurements of bridge deck profile due to own weight.

Location	FEM (m)	Field measurement (m)	Difference (%)
Mid-Span 2	4.176	4.145	0.74
Quarter-Span 2	4.328	4.359	0.70
WV Tower	4.084	4.075	0.22
Mid-Span 3	4.663	4.481	4.08

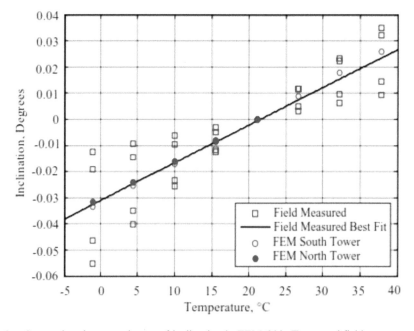

Figure 9.   Comparison between change of inclination in FEM Ohio Tower and field measurements.

inclination as recorded from time-history field measurements of the Ohio Tower versus ambient temperature change, taking 21°C as the reference temperature. A linear trend line is drawn across the data points to obtain the average relationship between those two variables. Similar relationships could also be established between seasonal temperature values and the change in inclination of the Ohio tower, as recorded from the instrumentation installed at both North and South Ohio towers. The Comparison between the mean values of inclinations collected from the field and those recorded from the FEM shows that the latter can predict a close response to that of the field. The discrepancy between measured field values and those from the FEM can be attributed to multiple factors that contribute to the behavior of bridge and cannot be reproduced in the FEM. Examples of such factors is the effect of wind, effect of traffic loading, condition of the expansion joints that might add constrains to expansion and contraction, and the real values of the steel material properties in the field. Because the field data have been collected from time histories through an entire year, it is hard to identify the effect of traffic loading for example on the tower inclination especially in absence of records of the traffic patterns on the bridge at this stage. Also there is no means to identify the effect of wind gust on the tower inclination since no weather station is available at the bridge site. For these reasons, the mean value of the field tower inclination is used for this comparison, which it is believed that this could neutralize most of those unknown effects. The percentage errors between mean measured inclination values and those maximum values recorded from the FEM were calculated and summarized in Table 2.

An interesting observation in Figure 9 is that the fluctuations in the measured tower inclination corresponding to temperatures below 21.1°C (70°F) than those for high temperatures. This can be explained by the fact that as the bridge superstructure contracts at low temperatures, the tension force in the main suspender decreases increasing its sag. As a result, the towers experiences higher inclinations under operational conditions.

The data presented in Table 2 indicate that the average discrepancy between the average change in inclination values from the field and those recorded from the FEM of the Ohio towers amounts 6.95%, while that of the West Virginia towers amounts 27.68%. The average percentage error between the field measurements and those values recorded for the response of both West Virginia tower and Ohio tower combined is calculated to be 17.31%. Given the complexity of the bridge structure and the varying levels of deterioration and section losses existed in various members, the changes in member properties due to the rehabilitation work in each tower among other factors that contribute to the towers inclination in the field and the level of details in the FEM, this amount of error in the overall response of the FEM to temperature change is considered acceptable.

Table 2.   Change of towers inclinations versus ambient temperature change.

Temperature °C (°F)	Ohio Tower			West Virginia Tower		
	Mean field (Deg.)	FEM (Deg.)	Difference (%)	Mean field (Deg.)	FEM (Deg.)	Difference (%)
−1.11 (30)	−0.03265	−0.0335	2.60	0.04345	0.0269	38.09
4.44 (40)	−0.02465	−0.0254	3.04	0.03259	0.0205	37.10
10.0 (50)	−0.01664	−0.0171	2.76	0.02172	0.0138	36.46
15.6 (60)	−0.00864	−0.00866	0.23	0.01085	0.00699	35.58
21.1 (70)	0	0	0.00	0	0	0.00
26.7 (80)	0.00737	0.00884	19.95	−0.01089	−0.0073	32.97
32.2 (90)	0.01537	0.0178	15.81	−0.02176	−0.0148	31.99
37.8 (100)	0.02338	0.026	11.21	−0.03263	−0.0296	9.29

In order to use the results of the Finite element model in identifying signs of concerns in the behavior of the bridge structure, a particular parameter representative of the bridge behavior needed to be identified as an indicative of the behavior of the entire bridge structure rather than the response of particular bridge element at the local level. This philosophy was established when it was decided to monitor the inclination of the bridge towers, which are the main supporting elements of the bridge, in a manner that would indicate the overall behavior of the bridge superstructure at the global level rather than at the local level. Therefore, it was decided to relate the inclination of the bridge towers to the natural frequency of the bridge superstructure. Identification of modal vibration parameters such as natural frequencies and mode shapes of structural systems can be achieved experimentally through various instrumentation based studies and also through finite element modelling studies. As damage occurs in structural systems, their performance is adversely affected, so as their mechanical properties. Based on this concept, damage identification can be achieved by a comparison of vibration characteristics of structural elements between two different states. One represents the initial and often undamaged state and the second represents the deteriorated state after damage. The basis of most damage detection methods is that damage will alter the measured dynamic response of the system that is dependent on mass, stiffness and energy dissipation properties of the system (Fisher 1984, Sohn et al. 2003).

The natural frequency of the bridge structure as calculated from the Finite Element model was used to identify the degree of linearity in the response of the bridge as the tower change their inclination in response to temperature change. Figure 10 illustrates the relationship between the change in tower inclination from the initial state at regular temperature of 21 °C and the first natural frequency of the bridge superstructure as calculated from the finite element model due to temperature change.

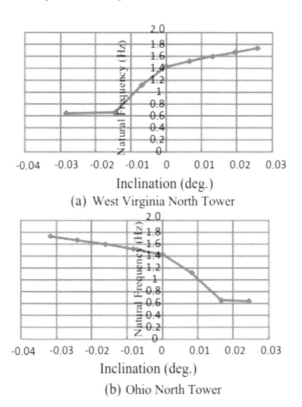

(a) West Virginia North Tower

(b) Ohio North Tower

Figure 10.   Change in inclination of towers versus bridge first natural frequency.

The plots in Figure 10 indicate that the first natural frequency changes linearly as temperature decreases. The first natural frequency increases from 1.43 Hz to 1.74 Hz linearly while the change in inclination of West Virginia Tower increases from zero at 21°C to 0.027° at -1°C. On the other hand, as temperature increases, the change in the first natural frequency is also linear till a certain point, then a non-linear behavior is observed in the trend. For the West Virginia tower, the non-linear behavior appears as the change in tower inclination reaches -0.014°. For the Ohio tower, this non-linear behavior is reached when the change in tower inclination reaches 0.017°.

The nonlinear trend of the first natural frequency only indicates nonlinear values of the stiffness of the bridge superstructure once the towers reach those corresponding inclinations. The first natural frequency of any structural system depends only on natural properties of the structure, consisting of the stiffness of the system and its mass. Since the mass of the bridge superstructure could be considered as a constant value, the change in natural frequency is only attributed to the change in stiffness. Thus, it can be concluded that the stiffness of the bridge superstructure becomes nonlinear once the mentioned values of tower inclination are reached, thus a plastic deformation of the members is likely to occur.

5   SUMMARY AND CONCLUSIONS

A sensory system was developed and deployed in this study to continuously monitor the global structural inclination of the historical Market Street bridge towers under operating and environmental conditions. Four 6-channel sensors were installed on top of the tower to provide continuous data records of three-axis acceleration, two-axis inclinations, and temperature. The data acquisition system was fitted with cellular modem to provide an internet-based monitoring system to remotely monitor the bridge structural response anywhere in the world. A user-friendly software was developed in order to communicate with the data acquisition system, update the data, and display the full time histories of the measured bridge response on an internet website, which aids bridge engineers to analyze the data and draw useful conclusions about the bridge condition. The system instrumentation system can fill in the gap between periodic inspections by providing continuous trend data and critical measurements that are not discernible by periodic visual inspections. However, the system was not intended to replace periodic visual inspection. The trend data should be used as a quantitative screening method for the need for an in-depth inspection.

Based on the field measured data collected during the monitoring period from November 8, 2011 and June 24, 2013, the following conclusions could be made:

1. In the traffic direction, both West Virginia and Ohio towers incline toward each other as the bridge contracts in winter and away from each other as the bridge expands in summer. Although both towers undergo the same amount of inclination due to seasonal temperature change, Ohio tower was found to have higher range of inclination as daily temperature changes between day and night.
2. The field measured data indicated that the north side of West Virginia tower accrued permanent inclinations of 0.025° and 0.02° in the longitudinal and transverse directions respectively, after twelve months of monitoring. However, due to the limited monitoring duration, it is not possible to predict whether such a deformation would increase or decrease with time.
3. The 3DFE model results as well as the field measured data were used to develop a damage identification technique through comparison of vibration characteristics of structural elements between two different states. One represents the initial and often undamaged state and the second represents the deteriorated state after damage.
4. A nonlinear trend of the first natural frequency of the bridge structure was observed once the towers reach inclination of 0.015° due to bridge expansion, which could be attributed to the change in bridge stiffness due to out-of-plane lateral deformation of the truss members.

5. Periodic evaluation of the structural performance both over long time scales as well as after a damaging event is always necessary.

ACKNOWLEDMENTS

The research work presented in this paper was funded by West Virginia Department of Transportation. The authors acknowledge in particular the support and encouragement of Mr. Marvin Murphy and Mr. Jimmy Wriston. Acknowledgements are also extended to the support and cooperation of the Engineers of the Engineering Division and District 6.

REFERENCES

Bakht, B., Mufti, A.A. & Wegner, L.D. 2011. *Monitoring Technologies for Bridge Management*. Multi-Science Publishing Co Ltd.

Enckell, M. 2011. Lessons Learned in Structural Health Monitoring of Bridges Using Advanced Sensor Technology. Ph.D. Dissertation, Department of Civil and Architectural Engineering, Royal Institute of Technology (KTH), Denmark.

Erol, S., Erol, B., & Ayan, T. 2004. A General Review of the Deformation Monitoring Techniques and a Case Study: Analyzing Deformations Using GPS/Levelling. *Proc., XXth ISPRS Congress*, Vol. XXXV, part B7, Istanbul, Turkey, 622–627.

Fisher, J.W. 1984. *Fatigue and Fracture in Steel Bridges*. John Wiley and Sons, Inc., New York, USA.

Fraser, M., Elgamal, A., He, X., and Conte, J.P. 2010. Sensor Network for Structural Health Monitoring of a Highway Bridge. *J. Comput. Civ. Eng.*, 11(1): 11–24.

Gastineau, A., Johnson, T. and Schultz, A. 2009. Bridge Health Monitoring and Inspections - A Survey of Methods. Research Report No. MN/RC 2009-29, Minnesota Department of Transportation, St. Paul, Minnesota, USA.

Inaudi, D., Casanova, N. & Vurpillot, S. 1999. Bridge Deformation Monitoring with Fiber Optic Sensors. *Proc., IABSE Symposium: Structures for the Future—The Search for Quality*, Rio de Janeiro, Brazil, 475–482.

Inaudi, D. 2009. Overview of 40 Bridge Structural Health Monitoring Projects. *Proc., 26th Annual Int. Bridge Conf., Engineers' Society of Western Pennsylvania*, Pittsburgh, 343–350.

Inaudi, D., Favez, P., Belli, R., and Posenato, D. 2011. Dynamic Monitoring Systems for Structures under Extreme Loads. *Applied Mechanics and Materials* 82: 804–809.

Leica Geosystems TruStory. 2003. Jiangyin Bridge, China: Monitoring with GPS RTKT Technology. http://www.leica-geosystems.com.

Lewellyn, M., Whited, D., & Juszczak. 2013. Revitalization of an Ohio River Suspension Bridge. *Proc. 29th Annual Int. Bridge Conf., Engineers' Society of Western Pennsylvania*, Pittsburgh, 564–574.

Nakamura, S. 2000. GPS Measurement of Wind-Induced Suspension Bridge Girder Displacements. *J. Struct. Eng.* 126(12): 1413–1419.

Ogaja, C., Li, X., & Rizos, C. 2007. Advances in Structural Monitoring with Global Positioning System Technology: 1997–2006. *J. of Applied Geodesy jag* 1(3): 171–179.

Roberts, G.W., Brown, C., & Meng, X. 2005. Deflection Monitoring of the Forth Road Bridge by GPS. *Proc., ION GNSS 18th Int. Technical Meeting of the Satellite Division*, Long Beach, California, USA, 1016–1021.

Radhakrishnan, N. 2014. Application of GPS in Structural Deformation Monitoring: A Case Study on Koyna Dam. *J. Geomatics* 8(1): 48–54.

Sohn H., Farrar C., Hemez F., Shunk D., Stinemates D., & Nadler B. 2003. *A Review of Structural Health Monitoring Literature: 1996–2001*. Report No. LA-13976-MS, Los Alamos National Laboratory, Los Alamos, New Mexico, USA.

Turner, L. 2003. *Continuous GPS: Pilot Applications-Phase II*. Final Report F-2001-OR-05, FHWA/CA/IR-2003/05, California Department of Transportation, Sacramento, California, USA.

Wong, K., Man, K., & Chan, W. 2001. Monitoring Hong Kong's Bridges: Real-Time Kinematic Spans the Gap. *GPS World* 12(7): 10–18.

Xu, Y.L., and Xia, Y. 2012. *Structural Health Monitoring of Long-Span Suspension Bridges*. Taylor and Francis Group, Boca Raton, Florida, USA.

Chapter 19

Protection of existing structures using health monitoring

A. Ramakrishna & R. Mankbadi
Hardesty & Hanover, LLC, New York, USA

ABSTRACT: A common challenge faced by foundation engineers while working in urban settings is the maintenance and protection of adjacent structures. The standard practice to mitigate damage caused by adjacent foundation construction work is to limit the existing structures' vibrations within acceptable safe constraints. In spite of these efforts, numerous existing structures have suffered damage from adjacent foundation construction activities even in full compliance to the standard practice. In recent years, Structure Heath Monitoring has been recognized as an effective approach for the protection of existing structures from ground excitation caused by adjoining foundation construction work. This paper presents case histories on protecting existing structures from adjacent foundation construction activities using Structure Heath Monitoring.

1 INTRODUCTION

At urban project sites, foundation construction activities can create soil movements that may cause damage to existing adjacent structures. Assessing the potential damage from planned construction activities is a difficult task due to uncertainties of the existing structures' conditions.

The standard practice to mitigate damage of existing structures from adjacent foundation construction work is to limit the existing structures' vibrations within acceptable safe limits. The majority of assessment resources are spent on characterizing the subsurface soil type and amount of energy transferred into the ground with little or no consideration given to condition of structure. This approach has had a fair share of success, but has also led to costly litigation. This paper presents an approach to safeguard existing adjacent structures from the ground excitation caused by foundation construction activities by assessing condition of structure, herein referred as *health monitoring*.

2 STRUCTURAL RESPONSE

The structural response to ground excitation depends on soil and structure types. A structure can sustain damage if construction induced waves propagate through the structure or soil cause excessive elastic or plastic deformation. A structure will respond to these vibrations at its natural time period (ω_n), which is correlated to two system properties: mass (m) and stiffness (k). The time in seconds that it takes to complete one cycle of vibration can be referred to as its natural time period (ω_n).

The stiffness of a structure is dependent on its intrinsic properties and the stiffness of soil supporting the structure. It can be inferred that the stiffness of a structure (k) is positively correlated to the structure's ability to resist the motion from vibration: the stiffer the structure the smaller the deformation and higher the natural frequency. Therefore, potential damage to a structure from ground excitation can be mitigated by ensuring structural response falls within the elastic deformation per structure's natural frequency through Structure Heath Monitoring. This approach assumes an elastic deformation not plastic deformation response of the supporting soil induced by the ground excitation.

3   HEALTH MONITORING

The Structure Heath Monitoring offers valuable insight on analyzing bridge movement in real time. When subjected to ground excitation, each structure has its own unique response that is dictated by its materials, design, and condition (the mass and stiffness of its members, and their connections to each other and to the earth). This unique response is referred to as the "dynamic signature". The Heath Monitoring reads the dynamic signature of a structure using three highly sensitive accelerometers: one up/down sensor and two horizontal sensors (north/south and east/west).

4   CASE STUDIES

4.1   *The MBTA Saugus Drawbridge No. S-05-040 (Ramakrishna et al. 2013)*

Introduction – The MBTA (Massachusetts Bay Transportation Authority) Saugus Draw-bridge No.S-05-040 (A5G) is located near the mouth of the Saugus River, north of Boston Harbor, in Saugus, Massachusetts. It is part of a multi-span bridge that crosses the Saugus River between the towns of Lynn and Saugus, as shown in Figure 1.

The bridge has eastbound and westbound tracks which accommodate simultaneous traf-fic on both tracks. The Bridge was built in 1911 and the structure includes a plate girder bascule span over the Saugus River on the MBTA Commuter Rail. The original piers consist of granite block resting on timber piles. Due to the concern of degradation, the piers went through an underpinning effort in 1987. Continued concern over the condition of rest pier led MBTA to engage Diversified Technology Consultants (DTC), who in turn hired Hardesty and Hanover, LLC (H&H) to inspect the condition of the rest pier. During the condition evaluation H&H noticed that the original granite construction, which was no longer intended to support loads from the structure but was intended to offer lateral stability to drilled pin pile, had deteriorated significantly. Many of the stones had fallen from the face of the pier revealing that the drilled pin piles were severely corroded and indicating that the 1987 underpinning effort may not be performing as intended. Refer to Figure 2 for Rest Pier observed "as-inspected" condition during evaluation stage.

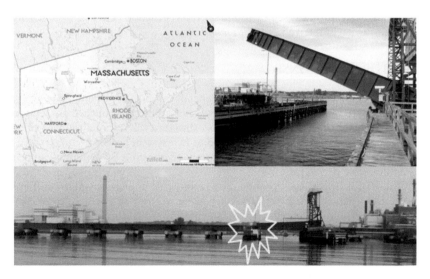

Figure 1.   Project location map.

Figure 2.   Rest pier "as-inspected" condition.

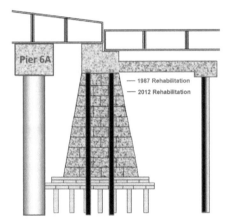

Figure 3.   Rest pier interim repair.

The subsequent load rating analyses of the rest pier indicated the need for rehabilitation of the rest pier. The MBTA proposed to perform remedial action in two stages. The first stage (2012 Rehabilitation), which is covered in this paper, has been identified as the remedial/ interim repair. The MBTA considers this to be an emergency repair. The second stage of repairs is permanent in nature and will take place at a later date.

It was decided that a temporary Pile Bent (Pier 6A) would be constructed in front of the rest pier to share the load. The remedial repairs did not address the condition of Pier 6 (rest pier) directly, but rather improved the capacity of the pier by reducing the load and providing an alternate load path for the lateral loads. The recommended temporary pile bent consisted of six (6) 30-inch diameter, open-ended, steel pipe piles driven to refusal with a precast concrete cap. Refer to Figure 3 for details.

Safeguard of Existing Structure – The construction of the temporary pile bent required installation of piles in close proximity to the existing in-service, structurally deficient rest pier. The design team chose Structural Health Monitoring to assess the health of structure and establish a pile installation protocol based on the structure's response to soil movements induced during pile installation. The dynamic structure monitoring approach included:

1. Continuous monitoring of structure's dynamic response (i.e., acceleration and tilt),
2. Assessing tolerable structural response using mathematical models,
3. Tracking change in dynamic response during construction operation, and
4. Alerting design team on any unstable structural response.

For this project, the instruments were mounted at two locations on the rest pier, as shown in Figure 4. The position of the sensors was chosen to maximize the acceleration and tilt information to be acquired during the monitoring. Selected monitoring positions were considered to be the most sensitive in order to capture the overall movement of the structure, as well as, differential movement from two sides of the support. The accelerometers were remotely connected to main data collection/analysis/communication equipment of the Structurocardiogram (SKG), which was located on the deck adjacent to the rest pier with access for any necessary maintenance. The on-site measurements of acceleration were processed using proprietary digital signal processing (DSP) technique, which alerted the design team and the bridge owner of any unusual structural response.

The observed structural response prior to the commencement of pile driving was simulated using mathematical models (SAP2000/FB-Pier). The "as-inspected" condition of structure was compared with the "1987-retrofitted designed" to estimate the tolerable movement and tilt the structure could withstand without sustaining permanent damage or undesired permanent movement. The analyzed loads cases consisted of dead load, wind (per AREMA), and live load. Live load induced from braking and traction forces were also included. The superstructure loads were applied to the pier at the bearing locations. To ensure integrity of the load path from superstructure to substructure, a rigid connection link was made between the pile cap and adjacent elements. Refer to Figure 5 for mathematical model details.

Based on the mathematical model, the following Structural Monitoring, Notification, and Action Plan was developed to mitigate the concerns associated with undermining the foundation of existing structure in service.

1. "Warning" if structure average tilt exceeds 0.08 degrees from the base. This equates to approximately 0.5^2 of horizontal movement at the bearings.
2. "Alert" if structure average tilt exceeds 0.28 degrees from the base. This equates to approximately 1.76^2 of horizontal movement at the bearings.
3. Upon receipt of a "Warning", the Contractor responsible to reduce pile impact forces to prevent further "Warnings" and to prevent average tilt values from progressing towards the "Alert" level.
4. Upon receipt of an "Alert", the Contractor shall be instructed to halt pile driving operations and use alternate means to install the piles.

Production piles were driven using an impact hammer with maximum rated energy of 107.7 kips-ft. Per project requirement Pile Dynamic Analyzer (PDA) test was performed on all the six (6), 30-inch diameter steel pipe pile. Since acceleration data obtained from the sensors could not be converted to vibration in real-time during the installation of piles, the hammer-operated stroke was adjusted based on the field measured "tilt' criteria recorded

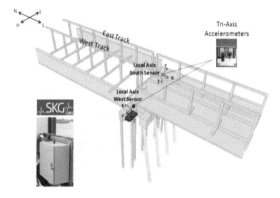

Figure 4.   Real time dynamic structural monitoring.

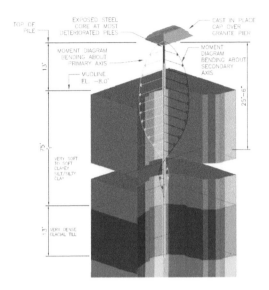

Figure 5.   Soil structure analyses.

by the instruments. All the foundation piles were installed successfully without incident. On completion of the temporary pile bent construction, Pier 6 pile rating was upgraded.

4.2   *Garden State Parkway (GSP) over the Great Egg Harbor Bay and Drag Channel (Hardesty & Hanover 2012, and Jeary and Winant Company 2014)*

Introduction – The project described herein is the New Jersey Turnpike Authority (NJTA) replacement of the existing southbound bridges carrying the Garden State Parkway (GSP) over the Great Egg Harbor Bay (Structure No. 28.0S) and Drag Channel (Structure No. 28.5S). The overall condition of the existing structures is poor. Both structures exhibit substantial deterioration and require replacement to eliminate structural deficiencies, address substandard geometry, and maintain coastal evacuation capabilities. Refer to Figure 6 for Project Location Map.

The new 28.0S structure consists of 21 spans of precast, prestressed concrete bulb tee girders varying in lengths between 148 and 250 ft. The substructure consists of reinforced concrete piers, which are comprised of three, six-foot diameter concrete columns and a capbeam supported on a concrete footing and 30-inch solid precast prestressed square concrete (PPSC) piles. The abutments are reinforced concrete cantilever walls, which are supported on 24-inch square, prestressed concrete piles. The new 28.5S structure will consist of 10 AASHTO Type III 77-ft precast, prestressed concrete I-beam spans made continuous for live load with joints only at the abutments. The substructure of the Drag Channel Bridge (DCB) consists of reinforced concrete pile bent piers supported by ten 30-inch solid PPSC piles arranged in a staggered layout. The abutments are similar to the 28.0S structure.

The alignment of the proposed replacement structures necessitates the installation of approximately fifty (50) new piles in close proximity to the piles supporting the existing bridges. At a few locations, the existing piles were battered outwards or towards the new piles as shown in Figure 7.

For this project, 4 tri-axial accelerometers were placed on the bridge at Pier 3 (close to the southern end), Pier 32, Pier 36 and Pier 45 (close to the Drag Island), as shown in Figure 8. The accelerometers were mounted on the western side of the pile cap also shown in Figure 8. The sensors are all oriented so that the X and Y axes follow the longitudinal and transverse directions of the bridge, respectively. Two monitoring systems were mounted on the barrier of the catwalk.

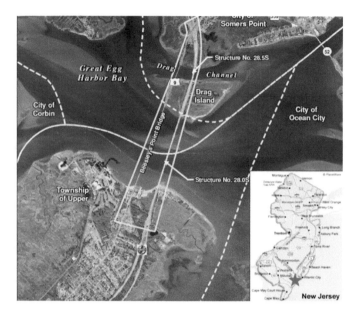

Figure 6.   Project location map.

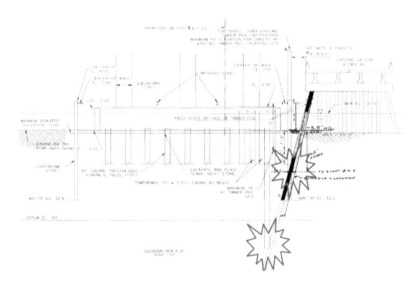

Figure 7.   Planned pile driving activities.

The monitoring for the Great Egg Harbor Bridge has been continuous during foundation work that is in close proximity to the existing structure. Based on the mathematical model three levels were developed, as shown in Table 1.

Spectral Analysis was performed to identify the structural modes of the bridge. Hourly Spectra, as well as longer term Spectra were used in the system identification process. A basic model was built to assist identifying the modal frequencies and mode shapes, which form the baseline signature of the bridge. Hourly Spectra are automatically computed and updated on the project central website and compared with the baseline to detect any changes in the frequencies of resonance of the structure that could be indicative of any potential loss of stiffness.

Figure 8.   Monitoring and data collection system and sensor locations.

Table 1.   Threshold set based on fixity assumption.

Warnings	Horizontal movement (inch)	Pier 03	Pier 32	Pier 36	Pier 45	Pier 03	Pier 32	Pier 36	Pier 45
		Changes in degrees for warnings (°)				Changes in inches/inch for warnings in/in			
Early warning	0.325	0.0273	0.0309	0.0309	0.0383	0.0007	0.0005	0.0005	0.0007
Warning to contractor	0.75	0.0629	0.0712	0.0712	0.0884	0.0017	0.0011	0.0012	0.0015
Stop work	1.0	0.0839	0.0949	0.0949	0.1179	0.0023	0.0014	0.0017	0.0021

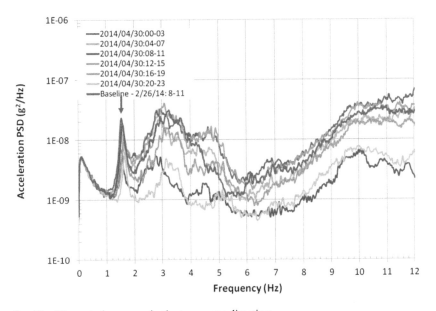

Figure 9.   Pier 03 spectral response in the transverse direction.

The Spectra recorded during construction activities from the sensors on the bridge were superposed on the baseline curve. The Spectra computed from all the four piers were examined and compared with the baseline signature. The principle modes of vibration as identified in the baseline found to have remained at, or close to, the baseline level, which suggests that the overall modal stiffness of the structure remained unchanged or very close the baseline during pile driving activities.

As an example, Figure 9 shows the East-West (transverse) sway mode of the Pier is identified at 1.61 Hz in the baseline, as highlighted by the red arrow. A slightly higher vibration level was observed during the day due to higher traffic intensity or construction activities in the vicinity. It returned to the ambient level at night. The frequency of the first fundamental mode remained close to the baseline during the day, which signifies that the structure remained stable.

During the installation of piles, the hammer-operated stroke was adjusted based on the field measurements and caution was exercised to ensure that the threshold set level of "Early Warning" was not exceeded. All the foundation piles were installed successfully without incident.

5 CONCLUSIONS AND SUMMARY

The Structure Heath Monitoring of measuring structure's "dynamic signature" in real time offers valuable insight on structure's behavior to the ground excitations. The "Dynamic signature" refers to structure's unique response that is dictated by its materials, design, and condition.

The method of limiting structural damage from ground excitation by regulating its maximum tilt/horizontal movement using Real Time Dynamic Structural Monitoring is effective in mitigating damages.

The information reported in this paper will advance as a conduit to gain a further understanding of the notion that different structures respond differently to the same given vibration because of their own unique intrinsic properties. Moreover, this concept stands in stark contrast to the standard parctice and will influence the future approach taken to protect adjacent structures from damage incurred by vibration.

The discussion and conclusion presented in this paper applicable to cited case histories. Further documentation of Real Time Dynamic Structural Monitoring for urban construction in other projects is greatly needed in order to better understand and advance the state of practice for mitigating damage to adjacent existing structures.

ACKNOWLEDGEMENTS

The authors are thankful to their employer Hardesty & Hanover, LLC (H&H) for the support and commitment during the preparation of this manuscript. Jeary & Winant Company (STRAAM Corporation), New York provided structure monitoring service and performed Spectral Analysis.

REFERENCES

Hardesty & Hanover, LLP: *Geotechnical Foundation Design—Replacement of GSP Bridge NOS. 28.0S and 28.5S Mile Post 27.0 to 28.8, NJ,* October 2012.
Jeary and Winant Company: *Structural Integrity Assessment Reports 1 thru 5—Continuous Monitoring of Great Egg Harbor Bay Bridge,* 2014.
Ramakrishna, A., A. Coates, and R. Mankbadi., *Foundation Design and Construction Challenges of 100-Year Old Saugus River Railroad Bridge Emergency Rehabilitation,* International Bridge Conference, 2013, Pittsburgh.

Chapter 20

Data-to-decision framework for monitoring railroad bridges

S. Alampalli & S. Alampalli
Prospect Solutions, LLC, Loudonville, NY, USA

M. Ettouney
Mohammed Ettouney, LLC, West New York, NJ, USA

J.P. Lynch
University of Michigan, Ann Arbor, MI, USA

ABSTRACT: Structural Health Monitoring as a nondestructive evaluation tool can be vital for effective bridge maintenance. The advent of wireless sensors and associated technologies has made data collection relatively easy, making continuous monitoring of structures a feasibility. However, it leads to large volumes of data in different formats. This paper presents the data-to-decision framework where data from commercial remote sensing and spatial information technologies are used in a decision-making tool chest (DMT) software with built-in predictive analytical models for making cost-effective predictive maintenance and management decisions. The framework is applied on a selected sub-network of the Union Pacific railroad. The near real-time streams of semi-structured data collected from remote sensors, past visual inspections, and observations from structural analysis are stored in a central data repository. DMT uses this data to assess bridge condition, derive the load capacity, compute reliability indices for bridge components, and provide bridge management recommendations.

1 INTRODUCTION

Our Rail Network is an important segment of the United States' transportation portfolio and has revolutionized transportation and catalyzed economic development. The U.S. freight rail network consists of 140,000 rail miles operated by seven Class I railroads, 21 regional railroads, and 510 local railroads (FRA, 2016). The Class I railroads own and maintain over 61,000 bridges. In addition, 40 Class II and 509 Class III railroads own and maintain over 15,000 bridges (GAO, 2007). The railroad traffic on Class I railroads has increased dramatically along with the freight volume. These changes have not only increased the strain on key bridges but also made them more important. According to a bridge survey completed in 1993 by the Federal Railroad Administration (FRA), more than half of the nation's railroad bridges were built before 1920 (FRA, 1999). However, with the increased weight, freight cars are approaching the design load limits of older bridges. At the same time, fatigue occurring in some components of older bridges from repeated heavy freight train operations is causing bridge failures and leading to serious service disruptions within the national rail network. Hence, the railroads need to increase their investment in inspection, maintenance, and replacement to keep existing railroad lines serviceable (Palley, 2013).

FRA regulations require that all railroads have comprehensive bridge safety management programs, which guide all bridge safety efforts and includes specific requirements concerning railroads' methods of inspection, evaluation and structural work (AAR, 2017). Current bridge management programs in the railroad industry largely rely on visual inspections with bridges inspected on an annual or multi-annual basis. These inspections include careful examination of each component of the bridge for corrosion or cracks in trusses, decking, and other components (UP, 2016). Post-inspection analyses are then used in assessing the load carrying capacity of the inspected bridge to derive condition ratings for each bridge span. Unfortunately, visual inspection methods are known to introduce subjectivity to the evaluation process thereby hindering reliable structural management (Moore et al., 2001). In spite of these shortcomings, visual inspections have proven to be an effective bridge management tool. However, the railroad industry simultaneously recognizes the role of commercial remote sensing and spatial information (CRS&SI) technologies that can provide an effective means of collection and processing structural performance data for making more informed and quantitative decisions associated with assessing the condition of bridges (Moreu et al., 2012).

Structural Health Monitoring (SHM) is a damage identification strategy and can be a vital tool in bridge management to improve the safety and maintainability of critical structures. Automated bridge management systems and CRS&SI technologies can play a critical role in monitoring railroad bridges, which have many components of SHM paradigm. SHM not only involves the collection of data from number of sensors but also conversion of sensor data to physical parameters, data analysis, and comparison of data with expected behavior of the structure derived from analysis or design. This comparison provides information related to structure deterioration or damage along with its location on the bridge. Finally, information can assist in predictive and just-in-time maintenance, and make appropriate decisions for managing the infrastructure.

Various forms of SHM have been used on bridge structures for many years, but lately computer-based technologies have been coupled with sensing technologies to provide information for evaluating structural integrity, risk, and reliability in order to ensure optimal maintenance and safe bridge operations. Nevertheless, these sensing technologies create large streams of data, which require new and improved information technology (IT) tools for analysis. Big data analytics is one such process that examines large data sets containing a variety of data formats for uncovering useful information (McKinsey Global Institute, 2016). Bridge management can leverage big data analytics in identifying bridge problems and monitoring problematic bridges. This initiative is currently driven in the "Health Assessment and Risk Mitigation of Railroad Networks exposed to Natural Hazards using Commercial Remote Sensing and Spatial Information Technologies" project, in which the authors are involved. This paper discusses a data-to-decision framework for monitoring railroad bridges that uses historic data, such as bridge inventory and visual inspection data, along with automatically generated unstructured sensor data. In addition, big data analytics provide insights to bridge condition that can improve efficiency.

2   DATA-TO-DECISION FRAMEWORK

The main objective of this project is to integrate data from CRS&SI technologies to create a novel data-driven decision-making framework that empowers the railroad industry to monitor bridge response to dynamic train loads, environmental loads, and hazard loads along with assessing and managing risks associated with the aging bridge inventories. The project is funded through the USDOT Research and Innovative Technology Administration (RITA) CRS&SI Program. This project is dedicated to the collection of sensor data, conduct data analysis, and create a data-driven decision-making framework (see Figure 1). The data-to-decision (D2D) framework consists of three components, namely: 1) Automated wireless sensing and communication technologies to remotely monitor bridges and collect data; 2) A central data repository to store previously created data such as structural information, finite element models, visual inspection data; automatically generated sensor data; and management metrics; 3) The Decision-Making Toolchest software with built-in analytical models to evaluate bridge condition and provide recommendations to bridge owners.

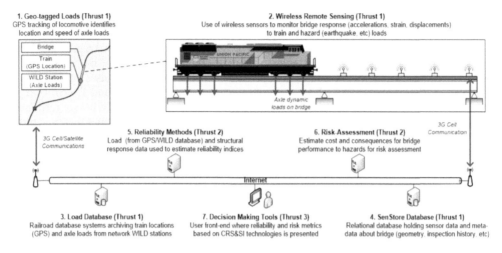

Figure 1.   Health assessment and risk mitigation of railroad networks using CRS&SI.

Figure 2.   Harahan Bridge and Parkin Bridge.

2.1   *Sensing technologies*

In this project, CRS&SI system for bridges consist of a sensor system, data acquisition system, and data transmission system. The system is deployed on the Harahan and Parkin bridges, owned and operated by the Union Pacific Railroad (see Figure 2). The Harahan Bridge is a 2550 ft. long five-span cantilever through truss bridge that carries two rail lines and a pedestrian walkway across the Mississippi River between West Memphis, Arkansas and Memphis, Tennessee. The Parkin Bridge is a 56 ft. short-span bridge located in Parkin City, Arkansas.

The Harahan Bridge, located in the New Madrid fault zone, is exposed to natural hazards such as weather, scour or flooding events from the Mississippi River along with man-made disasters such as barge collisions. These hazards contribute to the deterioration of individual members of the bridge with a possibility of multi-component bridge subsystems not interacting properly. In order to better maintain the Harahan Bridge and to develop a monitoring system, the Parkin Bridge is selected to act as a trigger and an additional weigh station for the sensing system before trains reach the Harahan Bridge. The instrumentation on the Harahan Bridge addresses the following sensing functionalities: 1) global modal properties; 2) lateral loads (i.e., earthquake, wind and barge collision); 3) vertical loads; 4) floor system interaction with bottom chord of truss; 5) relative tautness of parallel eyebar chord elements of the truss;

6) local free body diagram to identify subsystem interaction; and 7) vehicle-bridge interaction between the bridge and passing trains. The objective of the Parkin Bridge's sensing system is to attain extraction of modal properties, monitoring and identification of train loads, over-height vehicular collision detection, and structural health condition diagnosis.

The design and implementation of the sensing system is done in two phases. First, a Finite Element Model (FEM) of the bridge is created for both the Harahan and Parkin bridges utilizing SAP/CSiBridge and ABAQUS software to analyze and quantify bridge responses (modal, strain, displacement, relative stiffness, etc.) to various dead, live, and dynamic moving loads. The results from FEM analysis, structural analyses, and design documents are used in developing an optimal sensing strategy and instrumentation plan for the bridge. Phase two includes the installation of sensors on bridges. This phase also includes a calibration of the system, testing to check the operation of the system, and replacement of malfunctioning sensors. The objective of the sensing system is to measure the structural health under daily service conditions. Currently, the sensing system on the Harahan Bridge consists of 17 sensors, which includes 6 uniaxial accelerometers to measure eyebar relative tautness, 3 tri-axial accelerometers to measure global model properties, and 8 strain gauges to measure axial strains for fatigue assessment and floor system assessment. The Parkin Bridge sensing system consists of 6 uniaxial accelerometers and 2 tri-axial accelerometers for global modal analysis and model calibration, and 16 strain gauges to measure girder moments for load estimation and transverse response to vehicle collisions. Furthermore, a specific train axle detection system using two piezoelectric elements is used to detect the instance when each axle of a train enters and exits the bridge. Data communication is based on base station triggers, where each train event triggers system results for both accelerometers and strain gauges. Data is passed through cyberinfrastructure to be stored in a central data repository.

2.2   *Central data repository*

The advent of inexpensive wireless sensors and their associated electronics, higher computing power, and wireless communication technologies has made data collection relatively easy and economical, making continuous monitoring of structures a feasibility. However, it leads to large volumes and different formats of data (both structured and unstructured), which needs to be gathered, aggregated, stored, and analyzed to find useful information. The big data analytics has a potential to play a major role in this end. Big data analytics is the process of collecting, organizing, and analyzing large sets of data to discover useful information (Bollier, 2016). Big data tools and technology has the potential to improve decision-making, minimize risks, and reveal valuable insights that are crucial in SHM. In addition, multidisciplinary team and multi-organizational involvement can provide a technical foundation to unlock the potential from massive data resources.

The framework for big data analytics in this project (see Figure 3) is similar to a traditional analytics project in bridge management. However, the difference lies in executing various processes. In a traditional analytics project, the analysis can be done with a stand-alone system. Whereas big data analytics requires large data repositories, distributed processes, various platforms, and tools to gain insights for making better data-driven informed decisions. While the algorithms and models are similar to traditional analytics, big data analytics tools are complex, programming intensive, and require multiple skill sets.

Dataset in this project includes bridge properties, current and past visual inspections, sensor data, structural analysis models such as Finite Element Models, Wheel Impact Load detector (WILD) data, and management metrics. The visual inspection data provides the condition of different components and bridges. WILD data provides per-car information on axle loads, speed, number of cars, and number of locomotives that form a basis for estimating the imposed train load on bridges instrumented. Sensor measurements are used to diagnose the health condition of the bridge itself, which ensures the reliability and safety of the bridge service. In addition, sensor data is used to identify the dynamic axle loads as well

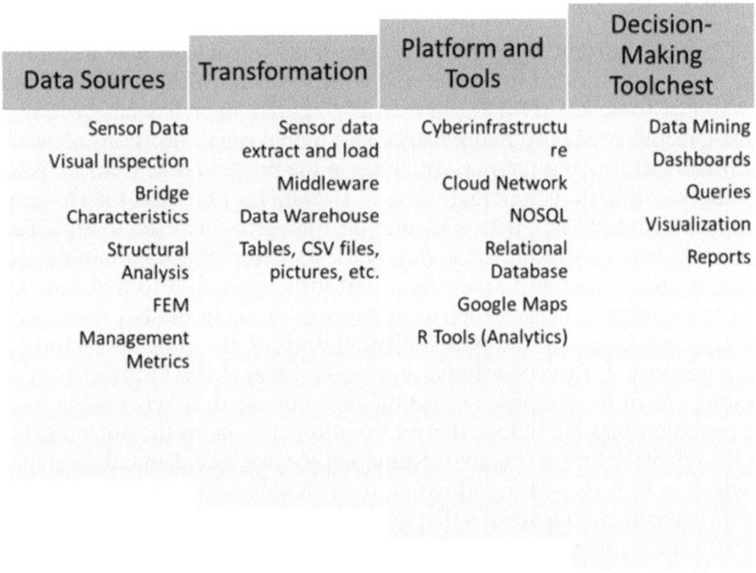

Data Sources	Transformation	Platform and Tools	Decision-Making Toolchest
Sensor Data	Sensor data extract and load	Cyberinfrastructure	Data Mining
Visual Inspection	Middleware	Cloud Network	Dashboards
Bridge Characteristics	Data Warehouse	NOSQL	Queries
Structural Analysis	Tables, CSV files, pictures, etc.	Relational Database	Visualization
FEM		Google Maps	Reports
Management Metrics		R Tools (Analytics)	

Figure 3.   Applied architecture of big data analytics in SHM.

as dynamics properties (i.e., stiffness and damping of suspension systems) of trains, which in turn provides a valuable basis for studying the bridge-vehicle interaction. The two key data sources generating vast amount of data are from WILD stations and sensors. However, edge computing is applied on large volumes of real-time data collected from sensors at edge of the data-to-decision framework. For example, acceleration data from tri-axial accelerometers are used in a Fourier analysis to obtain the natural frequencies (for the transverse and vertical directions) and to automate the modelling of the corresponding mode shapes of the instrumented span. Also, peak picking algorithms are used on acceleration data to extract the modal frequencies of the individual eyebars, which in turn are used to quantify irregularities in the axial load distribution among eyebars. Strain data collected on the eyebars and vertical hangers are passed through a low-pass filter to acquire the static response of the bridge and used in calculating the dynamic load factor (ratio of the maximum dynamic strain to the maximum filtered static strain). Cross-correlation is used on strain data to determine the speed and direction of a train travelling on the bridge. Additionally, rainflow—counting algorithm is applied to measured strain response sequences in order to conduct fatigue analysis for each component of the bridge. Reliable and appropriate processed data are stored in a powerful, scalable, flexible data repository for analysis and visualization.

In this project, given the heterogeneity of data, many alternative designs were considered to store information with the key requirements for the system being scalability, consistency, and usability. Finally, Microsoft SQL Server—a relational database—has been used as a reliable data repository. The relational database links structural analytical models to physical bridge components; sensors to physical bridge components; and results to related components. The created database schema is structured and normalized to ensure that data are stored efficiently and optimized for highest performance. The design provides the capability to disclose knowledge and reliable hidden patterns; crosscheck various datasets; and validate and uncover relationships within data. For example, visual inspection constitutes a major basis for decisions regarding the performance of bridges. So, the visual inspection data is linked with sensor data in order to provide information regarding limit states of bridges below failure through different ratings.

2.3   *Decision-Making Toolchest*

One of the most important aspects of big data is its impact on how decisions are made and results are communicated in a clear and concise form. In this project, a well-designed Decision-Making Toolchest (DMT) powered by big data analytics has been developed for making cost-effective predictive maintenance and management decisions in ensuring bridge safety and other performance metrics. Analytics in the project range from statistics—such as moving averages, correlations, and regressions—to complex functions such as graph analysis. In addition, predictive models, networks, and decision trees are used to anticipate behavior and events. Predictive models utilize a variety of statistical and data mining techniques to study near real-time sensor and historical visual inspection data to make predictions. The outcome of these analyses does not tell what happens in the future but forecasts what might happen in future because of the probabilistic nature of the predictive models. However, DMT take a step further from descriptive and predictive analytics to prescriptive analytics by recommending one or more courses of action and showing the likely outcome of each decision, so the decision-maker can take this information and act on the outcome. For example, data from the reference bridge is used in estimating the live load demand on all bridges in the network, which in turn is used in estimating the risk values for the network. This information assists in recommending actions that are simple as a visual inspection (within a certain time frame) for ascertaining the loss of capacity to all the way recommending an emergency closure of a route/bridge for certain rail traffic. Results are provided to decision-makers through a rich palette of visual dashboards.

Decision-Making Toolchest software is developed in .NET environment using latest technologies. DMT uses big data along with built-in predictive analytical models to provide bridge management recommendations based on owner-defined thresholds and feasible actions to infrastructure owners through visual dashboards. DMT also includes user friendly input screens, allowing users to add or manipulate data that is stored back into the database. Analysis in DMT can be done at three different levels—Network, Bridge, and Component levels. Analysis results are presented in a tabular or graphical form making it easy to understand.

Network Level: The effectiveness of a resilient infrastructure or enterprise depends upon its ability to anticipate, absorb, adapt to, and/or rapidly recover from a potentially disruptive event (NIAC, 2009). DMT incorporates a resilience model at the network level. A Resilience model depends on robustness, resourcefulness, recovery, and redundancy, which is computed using sensing information, visual inspection, and other objective/subjective data in the form of resilience issues stored in the central repository. Selected network topology and its resilience assessments are presented, as shown in Figure 4. Assessments provide infrastructure owners the ability to reduce the magnitude and/or duration of disruptive events.

Bridge Level: Analysis at the bridge level can be done by selecting a bridge from the data repository. Structural information, sensors installed on the bridge, historical inspection, design (COOPER E80) axle loads and train axle loads associated (see Figure 5) with the selected bridge are presented for easy access and analysis. COOPER E80 simulates the live load from a train, usually on a bridge. It is a series of point loads (simulating locomotive wheels) followed by uniform distributed load of 8 kips per linear foot of track (WisDOT, 2017). Standard Cooper E10 loading is a 2–8–0 locomotive equivalent, 10,000 lbs per driving axles. Whereas E80 is the current standard design criteria, 80,000 lbs per driving axle (i.e. E10 X 8).

Component Level: The third level of analysis is done at component level, where analysis can be done by selecting any component associated with the bridge. Inspection findings and sensor events related to sensor installed on the component can also be accessed for the selected component. Load rating is one of the analytical models, at the component level, integrated into the DMT software.

The estimation of dynamic load on the bridge from a train is based on measures of its static axle weights using calibrated monitored bridges. The analysis utilizes CRS&SI data

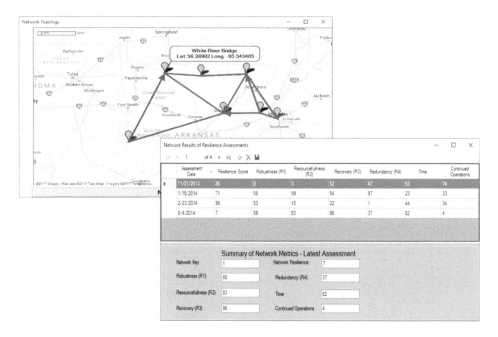

Figure 4. Network topology and resilience assessment.

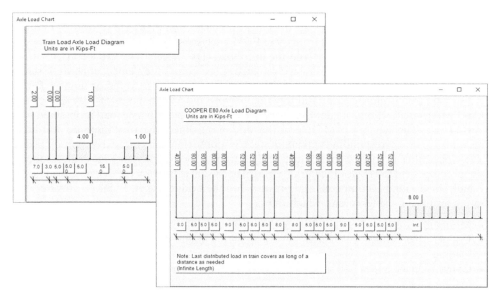

Figure 5. Train axle load and COOPER E80 axle load.

(axle weights, locomotive locations, and bridge responses) to make estimates of the imposed dynamic load for rail bridges in a bridge network and estimates the residual capacity of the monitored bridge systems in order to develop a reliability analysis (i.e., calculation of reliability indices) of bridge components and spans. Load rating is based on capacity and demand. Load rating for a single event or based on all events is computed and presented to

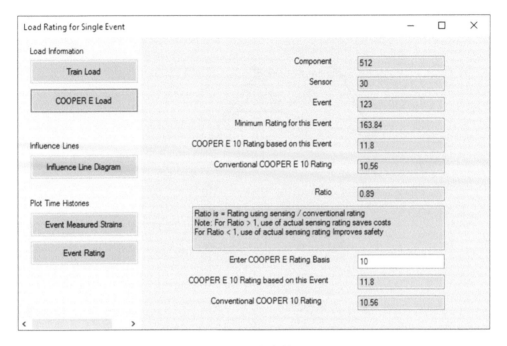

Figure 6.   Load rating for a single sensor event on the bridge component.

bridge owners in DMT. First, bridge drawings and inspection plans are studied and utilized in FEM models. Second, results from these FEM analysis, stored in the database, are used to estimate the bridge load capacity and the dead load (DL) effects. The live load (LL) effects corresponding to COOPER loading is calculated either from finite element analysis or sensor data measured during the train passage. Similarly, the dynamic load factor (impact factor) is also estimated from American Railway Engineering and Maintenance-of-Way Association (AREMA) specifications or captured using sensor data. The train axle load is obtained from WILD data. The dynamic load factor is the ratio of maximum dynamic response (unfiltered response) to the corresponding maximum static response (low-pass filtered response). Hence, a low-pass filter is applied to measure bridge strain response for removing high-frequency signal component (50 Hz for the Parkin Bridge) needed in extracting the dynamic load factor. Finally, load rating is computed and presented to the bridge owner. Figure 6 provides load rating results for a single event, which shows

- Minimum rating—calculated using LL demand and DL based on event axle data,
- Cooper E80 rating—calculated by extrapolating event based data to COOPER E80 loading,
- Conventional Cooper E10 rating—based on Train LL demand and DL for COOPER E10 loading, and
- Ratio—computed using sensing and conventional rating.

3   CONCLUSIONS

Use of SHM has steadily increased in bridge management as a means to establish effects of load through long-term monitoring of bridges and other structures. However, SHM has generated a variety of data, driven by visual inspection, sensors, analytical models, and management metrics. Data-to-decision framework involving big data analytics in SHM can supplement routine visual inspection of bridges by improving value and reducing uncertainty. In this project, the developed data-to-decision framework uses a set of advanced technologies designed to work with heterogeneous data. The framework holds the potential to discover patterns and indicators for decisions needed for effective bridge maintenance. In addition, the data-driven decision-making software with a broad range of IT technologies along with analytical models can provide a better understanding of bridge conditions, and assist in making predictive maintenance and management decisions.

ACKNOWLEDGEMENTS

The authors acknowledge the support of the U.S. Department of Transportation (USDOT)/ The Office of the Assistant Secretary for Research and Technology (OST-R) for funding this research and the support of the Union Pacific. The views, opinions, findings and conclusions reflected in this paper are the responsibility of the authors only and do not represent the official policy or position of the USDOT/OST-R, or any State or other entity.

REFERENCES

AAR. 2017. Railroad Bridge Safety. Association of American Railroads, https://www.aar.org/Bridges.

Bollier, D. 2016. The Promise and Perils of Big Data. The Aspen Institute, https://assets.aspeninstitute. org/content/uploads/files/content/docs/pubs/The_Promise_and_Peril_of_Big_Data.pdf.

FRA. 1999. Audit Report: FRA's Interim Statement of Policy on the Safety of Railroad Bridges. TR-1999-077. Washington, D.C.

FRA. 2016. Freight Rail Today, http://www.fra.dot.gov/Page/P0362.

GAO. 2007. Railroad Bridges and Tunnels: Federal Role in Providing Safety Oversight and Freight Infrastructure Investment Could Be Better Targeted. U.S. Government. Accountability Office. Washington, D.C. GAO-07-770.

McKinsey Global Institute. 2016. Big data: The next frontier for innovation, competition, and productivity, http://www.mckinsey.com/business-functions/business-technology/our-insights/big-data-the-next-frontier-for-innovation.

Moore, J., Phares, B., Graybeal, B., Rolander, D., & Washer, G. 2001. Reliability of Visual Inspection for Highway Bridges. Federal Highway Administration. Washington D.C. FHWA-RD-01-020.

Moreu, F., LaFave, J., & Spencer, B. 2012. Structural Health Monitoring of Railroad Bridges—Research Needs and Preliminary Results. Structures Congress 2012. Chicago, IL.

NIAC. 2009. Critical Infrastructure Resilience Final Report and Recommendations. National Infrastructure Advisory Council. Washington, DC.

Palley, J. 2013. Freight Railroads Background, Federal Railroad Administration. Washington, D.C.

UP. 2016. Union Pacific Protects Bridge Integrity with Rigorous Inspections. Union Pacific, http://www.up.com/cs/groups/public/@uprr/@newsinfo/documents/up_pdf_nativedocs/pdf-bridge-safety-fact-sheet.pdf.

WisDOT. 2017. Railroad Structures. Wisconsin Department of Transportation Bridge Manual, http://wisconsindot.gov/dtsdManuals/strct/manuals/bridge/ch38.pdf.

Modeling of bridges

Chapter 21

A numerical model of lateral response of a drilled shaft adjacent to a caisson foundation

L. Wei & D. Ha
Jacobs Engineering Group Inc., Boston, MA, USA

S. Patel
Jacobs Engineering Group Inc., Clark, NJ, USA

ABSTRACT: Currently the p-y method is widely used in industry to model the pile behavior under lateral loads. Appropriate p-multipliers and y-multipliers are usually applied to the "backbone" p-y curves in order to account for group effect or any other possible influence on the pile lateral behavior. At Piers 76 and 77 of Pulaski Skyway, deep foundations with groups of drilled shafts are proposed to replace the existing caisson foundations. The existing caisson foundations will be left in place and hence may influence the drilled shaft behavior. In this study, a numerical model was developed to investigate the possible effect on drilled shaft lateral behavior due to the existing caisson. A 2D analysis was performed using the commercial finite difference program FLAC. A plain strain condition was assumed in a homogenous soil layer and the shaft was subjected to a lateral force directed towards the caisson. The lateral response behavior was evaluated in both soft clay and medium stiff clay. The effect of interface roughness (rough or smooth) between shaft-soil and caisson-soil were also investigated. First, the lateral behavior of drilled shaft was calibrated with the "backbone" p-y curves. Next, the p-y curves were compared between "with caisson" and "without caisson" conditions. It is found that the soils exhibit a slightly stiffer behavior at small shaft deflections with the caisson presence, particularly for a rough soil-structure interface. However, the soil ultimate resistance may be lower with the caisson than without the caisson, particularly for a smooth soil-structure interface. It is also observed that the displacement level to fully mobilize the soil resistance with the caisson is significantly less than that required to fully mobilize the resistance without the caisson. For a full-range p-y curve, therefore it may be necessary to apply both p and y multipliers. Nevertheless, under service conditions that the shaft deflection is small (e.g. less than 2.5 cm), such effect is minor and the "backbone" p-y curves may still be applicable with the caisson presence.

1 INTRODUCTION

1.1 *Project background*

The Pulaski Skyway is a vital link in the Northern New Jersey/New York Metropolitan transportation network, carrying over 67,000 vehicles a day. The Skyway has been in operation for approximately 80 years and is near its project useful life. The goal of the current ongoing project is to rehabilitate the Skyway; address the existing structural deficiencies along the nearly 5793 m of this structure; make safety improvements where possible; and achieve a life expectancy of at least 75 years before another major rehabilitation is necessary. See Figure 1 for photo of the Pulaski Skyway.

As part of the rehabilitation design effort, a comprehensive coring and testing program revealed that the pier and foundation concrete is in a severely deteriorated condition, thus warranting replacement of the piers/foundations in lieu of repairing them. It was accepted that

Figure 1.   Aerial photo of the Pulaski Skyway (looking east).

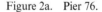

Figure 2a.   Pier 76.   Figure 2b.   Pier 77.

new construction would be the best way to provide the expected 75 year service life required for the rehabilitated structure. Many of the pier columns are being replaced with a new reinforced concrete shell, which would be isolated from the existing pier columns. The new foundations will involve a new pile cap and drilled shafts over and outside the limits of the existing caisson foundation, some of which are 24 m long, 9 m wide and 30 m deep, and are to remain.

1.2   *Purpose of study*

At Piers 76 and 77 of Pulaski Skyway, deep foundations with groups of drilled shafts are proposed to replace the existing caisson foundations. However, the existing caisson will remain and loads it previously carried from the pier columns will be removed and transferred to the new deep foundations. The distance between the edge of the caisson and the center of drilled shaft is 1.45 m in the bridge longitudinal direction. The proposed drilled shaft is 1.52 m in diameter, so the caisson-to-shaft distance (caisson edge to shaft center) is slightly less than 1D (D is the drilled shaft diameter). The caisson is about 18.3 m long and 7.0 m wide, extending all the way to the top of bedrock. The anticipated lateral response of the proposed drilled shaft may be affected by the adjacent existing caisson. Figures 2a and 2b shows an existing photo of the existing Pier 76 and 77. Figure 3 shows a sketch of the proposed foundation for Pier 77.

The purpose of this study is to simulate the drilled shaft lateral behavior, while taking into consideration the effect of the adjacent caisson. Currently the p-y method is widely used in industry to model the pile behavior under lateral loads. Therefore, this numerical model is created to capture the soil-pile interactions in terms of p-y curves with the caisson nearby. It is anticipated that the original p-y curves (the condition without caisson effect) may need to be modified based on the numerical modeling results to account for the caisson effect.

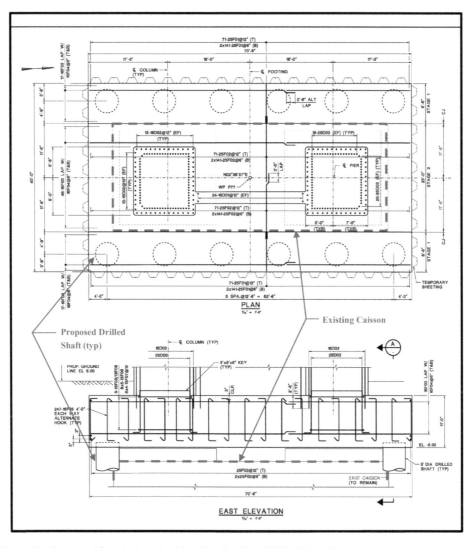

Figure 3.   Proposed foundation plan/elevation for Pier 77 (Pier 76 similar).

2 SUBSURFACE CONDITIONS

2.1 *Geology*

Based on surface geology maps of New Jersey, the Pulaski Skyway lies entirely within the Piedmont Plateau subdivision of the Appalachian geographic province. The surface geology primarily consists of man-made fill overlies marine tidal marsh and estuarine deposits containing brown to black deposits vary from sand to clay with minor amounts of gravel and contain abundant organic material. Where not overlain by man-made fill, the marine tidal marsh typically consists of 0.6 to 1.5 m of organic soil followed by stratified silty clays and clays. These silty and clayey soils are usually highly compressible with low densities (PB 2013).

2.2 *Soil stratigraphy*

Numerous borings and cone penetration tests (CPTs) were performed in the vicinity of the proposed piers (Arora, 2017). The soils at the site generally have a fill layer about 3.0 m below grade overlying a soft to very soft organic clay layer with an approximate thickness of 3.0 m. Below the organic layer are the soft to medium stiff silty clays extending more than 30 m below grade. This thick silty clay layer is notably varved with fine-grained sand and silt (PB 2013). Loose to medium dense sand layers are also encountered at varying depths at some of the boring locations and the thickness is usually less than 2.7 m. Below the thick varved silty clay layer are the glacial tills overlying the bedrock. The groundwater is typically about 1.5 to 2.1 m below grade.

The proposed foundation at piers 76 and 77 generally consists of a group of drilled shafts extending into the bedrock. The top of the drilled shaft is within the organic clay layer. It is anticipated that the shaft lateral behavior is largely governed by the organic clays and the varved silty clays. Laboratory test results from the organic soils indicated that the organic content values ranged from 2 to 45 percent and the natural water contents varied from 19 to 309 percent (PB 2013). The standard penetration test (SPT) indicates the blow counts were mostly "weight of hammer". The laboratory and CPT results generally indicate that the undrained shear strength of the organic clay is in the range of 14.4 and 19.2 kPa.

Laboratory test results of the varved cohesive soils indicated natural water contents typically between 20 and 35 percent (Arora, 2017). The Liquid Limit (LL) is typically less than 50 and the Plasticity Index (PI) generally ranges between 6 and 23. The standard penetration test (SPT) indicates the blow counts were mostly "weight of hammer" to less than 10 blows per foot. The laboratory and CPT results generally indicate that the undrained shear strength of the varved silty clay ranges between 24 and 48 kPa.

3 NUMERICAL MODEL

3.1 *Introduction*

When a shaft or pile is subject to a lateral force, passive wedges form in front of the pile, which provide the most lateral resistance to the pile (Ashour et al. 1998). The overall pile lateral response is a 3-dimensional behavior and would be ideally addressed by a 3D numerical modeling. In the strain wedge method, it is assumed that a constant shape of a passive wedge may be appropriate in a homogenous soil sublayer to simulate the lateral resistance (Ashour et al. 1998). At a great depth below ground in cohesive soils, the failure is more represented by plastic flow of the soil around the pile as it deflects laterally (Randolph & Houlsby, 1984). In this study, the shaft top is located about 4.3 m below ground and therefore a plain strain condition is assumed when modeling the shaft lateral behavior. A 2D finite difference analysis was performed using the commercial program FLAC (Version 7.0). Since the on-site soils are predominantly cohesive (very soft to medium stiff) in the upper depths where maximum pile deflection occurs, the caisson effect was investigated in both the soft organic layer and the medium stiff clay layer.

3.2 Model overview

Figure 4 shows the 2D numerical model without the adjacent caisson. The concrete pile is 1.5 m in diameter with an elastic modulus of 25,000 MPa with a Poisson's ratio of 0.15. The soil properties used in the numerical modeling are summarized in Table 1. The soils were modeled as linear-elastic perfectly plastic material using Mohr-Coulomb yielding criterion. Both the pile and soil were divided into quadrilateral elements generated automatically by FLAC. Interface elements were introduced to simulate the soil-pile interactions along the pile surfaces. A 15 m by 15 m soil region is fixed (no movement) to represent the far-field boundary conditions.

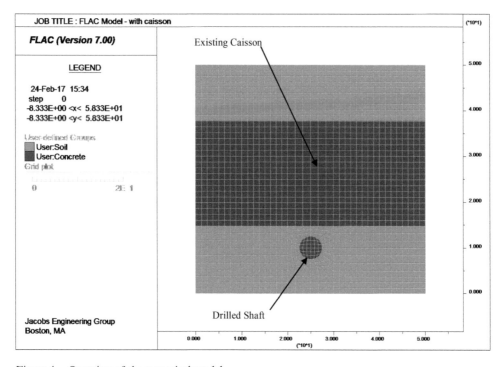

Figure 4.   Overview of the numerical model.

Table 1.   Soil properties in the numerical model.

Soil Layer No.	Soil Type	S_u (kPa)	E (kPa)	v	ε_{50}
1	Soft Organic Clay	10.6	1680	0.49	0.02
2	Medium Stiff Clay	26.9	8832	0.49	0.01

S_u: Undrained shear strength
E: Elastic modulus
v: Poisson's ratio
ε_{50}: Axial strain at 50% peak strength

3.3   *Model calibration*

The "Matlock" soft clay (Matlock 1970) model was used as "backbone" p-y curves for both soil layers 1 and 2. Soil layer 1 represents a "soft organic clay" condition with undrained shear strength less than 24 kPa. Soil layer 2 represents a "medium stiff clay" condition with undrained shear strength between 24 and 48 kPa. The soil properties presented in Table 1 were used to calibrate the lateral behavior to closely match the "backbone" p-y curves, as shown in Figure 5 and Figure 6. In this process, an incremental lateral force was applied to the pile and the corresponding pile movement at each load level was recorded. Figures 7 and 8 show the soil displacement pattern under lateral load of the pile in soil layers 1 and 2, respectively. Both figures show the soils in front of the shaft tend to move normally away from the pile circular surface which is consistent with a passive wedge assumption (Ashour et al. 1998), while the soils behind the shaft move more similar to a plastic flow around pattern (Randolph & Houlsby, 1984). By recording the applied force versus the measured pile deflection, the pile lateral behavior in terms of p-y curve is captured by the numerical model. In this process, it is assumed that the shaft surface is rough and that no soil slips on the shaft surface. Next, the existing caisson was placed at 1.45 m from the pile (edge of caisson to center of pile) and the above procedure was repeated with the same soil properties summarized in Table 1.

3.4   *Soil-structure interface*

There are two soil-structure interfaces in the numerical model: the soil-shaft interface and the soil-caisson interface. Interface elements were introduced to model the soil-structure interaction on the interface. In FLAC (Itasca 2011), the interface elements are zero thickness elements to transfer shear and normal stresses between the soils and the structures. The interface shear strength is governed by the Mohr-Coulomb failure criterion. Three interface conditions were considered: smooth, rough, and medium rough. The smooth condition

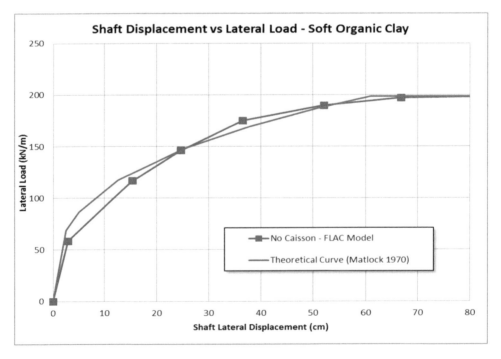

Figure 5.   Calibration of p-y curve in soft organic clay.

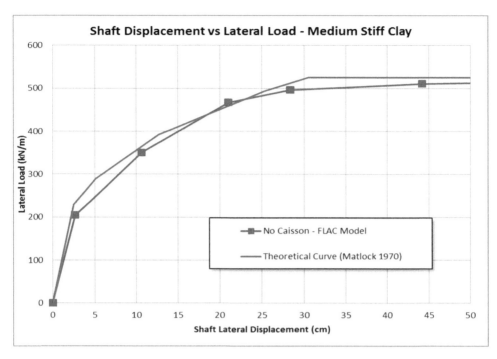

Figure 6.   Calibration of p-y curve in medium stiff clay.

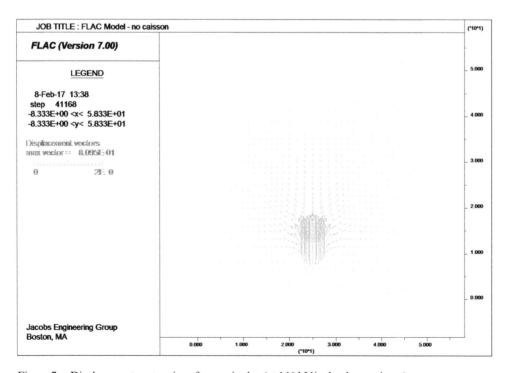

Figure 7.   Displacement vectors in soft organic clay (at 146 kN/m load, no caisson).

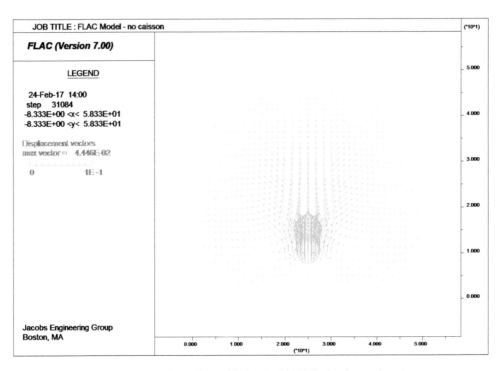

Figure 8.   Displacement vectors in medium stiff clay (at 146 kN/m load, no caisson).

simulates the case that the soil is free to move along the structure surface at any mobilized shear stress level. The rough condition simulates the case that the soil will "stick" to the surface and no relative movement will occur between the soil and the structure. The medium rough condition represents a more realistic interface behavior and an adhesion (c_α) value of $c_\alpha = 0.55S_u$ was assumed. This estimate is based on the recommended axial friction resistance for drilled shafts (Brown et al. 2010).

4   NUMERICAL RESULTS

4.1   *Caisson effect*

Figures 9 and 10 show the soil displacement pattern under lateral load of the pile with the caisson presence in soil layers 1 and 2, respectively. Compared to Figures 4 and 5 without the caisson, the soil displacement pattern is significantly different and the soils are getting more "squeezed" laterally out of the space between the pile and the caisson due to the presence of the caisson. The resulting p-y curves due to the existing caisson effect are shown in Figures 11 and 12 for soil layers 1 and 2, respectively. The analysis shows that the effect of existing caisson on the drilled shaft lateral behavior strongly depends on the mobilized lateral movement of the shaft.

The existing caisson has two contradicting effects on the lateral behavior of the shaft. The existing caisson is rigid and will not move as the shaft is deflecting towards the caisson. Therefore, the presence of the caisson would tend to stiffen the soils and provide higher resistance against shaft deflection. However, for continued increases in shaft deflections, the soils between the shaft and caisson become thinner and thinner, and the soils become more and more susceptible to be squeezed laterally out of the space between the two structures. When the shaft lateral deflection is small, the "stiffening" effect is dominant and the soil

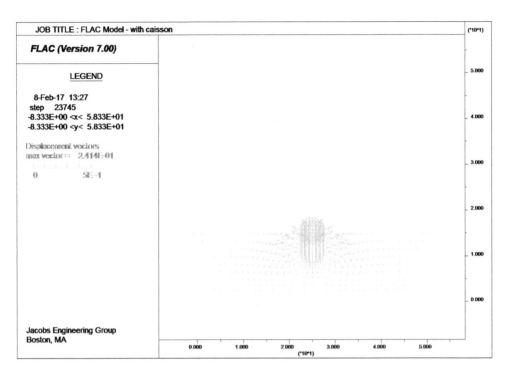

Figure 9.   Displacement vectors in soft organic clay (at 146 kN/m load, with caisson).

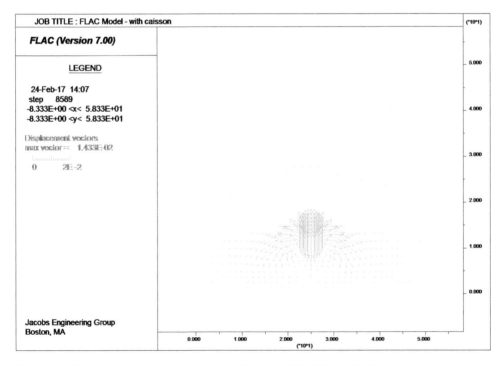

Figure 10.   Displacement vectors in medium stiff clay (at 146 kN/m load, with caisson).

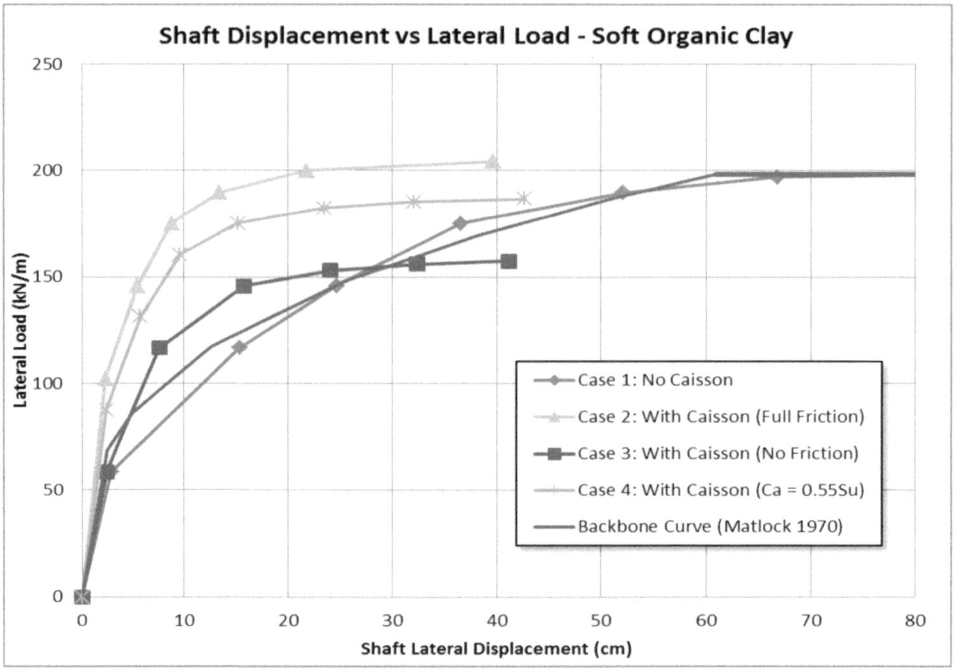

Figure 11.   Comparison of p-y curves in soft organic clay.

lateral resistance is actually higher than the condition without the caisson. When the shaft lateral deflection increases further, the "squeezing" effect becomes more and more significant and the soil lateral resistance will tend to decrease. Both "stiffening" and "squeezing" effects are obvious in Figures 11 and 12.

4.2   *Soil-structure interface effect*

The results also show that the soil-structure interface roughness has a significant impact on the shaft lateral behavior when the caisson is present, particularly at large deflection levels. When the interface is smooth, the soils are easier to get "squeezed" out of the space and hence the lateral resistance to the shaft becomes less. When the interface is rough, the soils are less susceptible to get "squeezed" out of the space and hence the lateral resistance to the shaft increases. A more realistic interface condition is between "smooth" and "rough", and the resulting soil lateral resistance is in between "smooth" and "rough" as well.

4.3   *Lateral behavior effect*

Based on Figures 11 and 12, the ultimate soil resistance is approximately 0%, 10% and 20% less than the fully mobilized resistance without the caisson, for rough, medium-rough, and smooth interface conditions, respectively. On the other hand, the required soil displacement to fully mobilize the ultimate resistance with the caisson is much less than the condition without the caisson. In the "soft organic clay" layer, the ultimate soil resistance is mobilized at about 25 cm lateral deflection with the adjacent caisson while it requires about 60 cm to fully mobilize the soil resistance without the caisson. In the "medium stiff clay" layer, the ultimate soil resistance is mobilized at about 12 cm lateral deflection with the adjacent caisson while it requires about 30 cm to fully mobilize the soil resistance without the caisson.

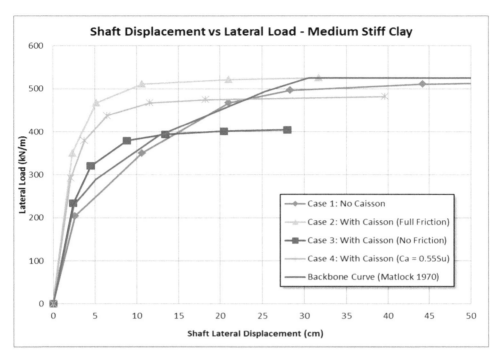

Figure 12.   Comparison of p-y curves in medium stiff clay.

The above observation tends to modify the "backbone" p-y curves with both p and y multipliers due to the caisson effect. However, when the lateral deflection is small (e.g. less than 2.5 cm), the p-y curves are only slightly stiffer than the "backbone" p-y curves. For a smooth interface condition, the p-y curves with caisson effect are almost the same as the "backbone" curves when the lateral deflection is less than 2.5 cm.

4.4   *Soil stiffness effect*

The above observations are consistent between "soft organic clay" and "medium stiff clay". This indicates that the soil stiffness does not have a significant impact on the caisson effect with both soil properties presented in Table 1. This may further suggest that the caisson effect is about the same when the soil consistency is between very soft to medium stiff (e.g. undrained shear strength less than 48 kPa).

5   CONCLUSIONS

This paper presents the lateral behavior of the drilled shaft due to a nearby caisson foundation. Currently the p-y method is widely used in industry to model the pile behavior under lateral loads. As a result, such caisson effect was investigated in terms of p-y modifiers. The following conclusions may be drawn based on the numerical model with an approximate caisson-to-shaft distance (caisson edge to shaft center) of 1D (D is the shaft diameter):

1. For very soft to medium stiff clays, the caisson effect is not sensitive to the actual soil strength or stiffness values. This appears to be the case for soil undrained shear strength less than 48 kPa and an E/S_u ratio between 150 and 300.

2. The caisson effect strongly depends on mobilized lateral movement of the shaft. The presence of caisson has two contradicting effects on the lateral behavior of the shaft. When the caisson movement is small, the caisson would tend to stiffen the soils and provide higher resistance against shaft deflection. When shaft deflection becomes large, the soils between the shaft and caisson become thinner, and the soils are more susceptible to be squeezed laterally out of the space between the two structures. When the shaft lateral deflection is small, the "stiffening" effect is dominant and the soil lateral resistance is higher than the condition without the caisson. When the shaft lateral deflection increases further, the "squeezing" effect becomes more and more significant and the soil resistance will tend to decrease.

3. The interface roughness between the caisson and soils has a significant impact to the ultimate soil resistance. When the interface is smooth, the soils are easier to get "squeezed" out of the space and hence the lateral resistance to the shaft becomes less. When the interface is rough, the soils are less susceptible to get "squeezed" out of the space and hence the lateral resistance to the shaft increases.

4. Based on the numerical results presented in this study, the caisson will affect both the ultimate soil resistance and the displacement level to fully mobilize the soil resistance. A p-multiplier of 0, 0.9 and 0.8 is suggested for a perfectly rough, medium rough and perfectly smooth caisson surface, respectively. In all above cases, a y-multiplier of 0.42 is suggested to account for the less mobilized displacement level at the ultimate resistance due to the caisson presence.

5. For a proper designed foundation, the lateral deflection of piles or shafts is usually small, e.g. less than 2.5 cm. At such displacement levels, the soils may exhibit a slightly stiffer behavior, particularly for a rough caisson interface, when compared to a no caisson condition. However, such effect is minor and the impact of existing caisson to the drilled shaft lateral behavior may be neglected at small deflection levels (e.g. less than 2.5 cm).

REFERENCES

Arora and Associates, P.C. 2017. *Geotechnical Foundation Report for Pulaski Skyway, Piers 76&77, Township of Kearny, Hudson County, NJ.*

Ashour, M., Norris, G. & Pilling, P. 1998. Lateral loading of a pile in layered soil using the strain wedge model. *J. Geotech. Geoenviron. Eng.*, 124(4): 303–315.

Brown, D.A., Turner, J.P. & Castelli, R.J. 2010. *Drilled Shafts: Construction Procedures and LRFD Design Methods.* FHWA-NHI-10-016, May 2010.

Itasca Consulting Group Inc. 2011. *FLAC: fast lagrangian analysis of continua. Version 7.0 [computer program].* Minneapolis, Minn.

Matlock, H. 1970. Correlations for design of laterally loaded piles in soft clay, *2nd Offshore Technology Conference*, Vol. I: 577–594. Houston, TX.

Parsons Brinckerhoff (PB), 2013. *Rehabilitation of the Pulaski Skyway: Geotechnical Data Report for Seismic Evaluation, submitted to New Jersey Department of Transportation.*

Randolph, M. F. & Houlsby, G. 1984. The limiting pressure on a circular pile loaded laterally in cohesive soil. *Geotechnique*, 34 (4): 613–623.

Chapter 22

Widening an existing multi-cell box structure with shallow depth steel girders

L. Rolwes
HNTB Corporation, St. Louis, MO, USA

ABSTRACT: The Mississippi DOT's plans to widen northbound I-55 at the interchange with I-220 near Jackson, MS were impeded by various geometric constraints, the most challenging being the limited vertical clearance under an existing CIP multi-cell box structure. Shallow depth steel plate girders were found to be the only viable solution to accommodate the maximum span length within the limited structure depth. An in-depth analysis using 3D FEM methods was conducted to assess the interaction between the existing and new structures. The relatively flexible steel structure was found to deflect and twist away from the existing resulting in large bending stresses within the slab. Further analysis demonstrated that the use of haunched plate girders would limit stresses in the deck to acceptable limits while maintaining minimum clearance to the roadway below. This solution allowed the DOT to overcome the impasse and proceed with their plans to widen this important transportation corridor.

1   INTRODUCTION

The stretch of northbound I-55 between State Street and Natchez Trace Parkway near Jackson, MS necks down to two lanes, causing traffic congestion during peak hours. The Mississippi DOT's plans to add additional lane capacity to alleviate the congestion were impeded by various geometric constraints related to widening of the existing cast-in-place, multi-cell, box-type structure that carries NB I-55 over the northbound ramp from I-220. The existing five-span structure (49'-72'-100'-71') is situated on a curved horizontal alignment, passing over the northbound ramp from I-220 (I-220 NB Ramp) and under the westbound ramp from I-55 (I-55 WB Ramp) as shown in Figure 1. The bridge section is comprised of a 5'-3" inside shoulder, two 12-foot lanes, and an 11'-3" outside shoulder. A section through the existing bridge is shown in Figure 2. The total width of the bridge is 43'-4", including bridge railings. It was designed for HS20-44 vehicle loading.

Several alternatives for widening the structure inside and/or outside of the existing alignment were developed and are summarized in Table 1. A comparative study was conducted considering both roadway and bridge implications and costs. Though inside widening could easily be accomplished with prestresssed girders, it required extensive roadway and drainage modifications to accommodate a lane shift. Work within the medians was also anticipated to be costly due in part to the need for retaining structures to accommodate the grade difference between the southbound and northbound lanes, as well as the associated traffic impacts in both directions. Widening to the outside was therefore found to be the most cost effective solution with the least impact to traffic. The proposed structure could be constructed outboard of the existing with limited lane closures and no lane shift. Some minor roadway reconfiguration would be required before and after the bridge. Total construction cost was estimated to be about $3.5M, of which $1.2M was figured for widening of the bridge.

It was found that widening to the outside, however, would require the use of shallow depth steel plate girders to maintain the minimum vertical clearance to the I-220 NB Ramp.

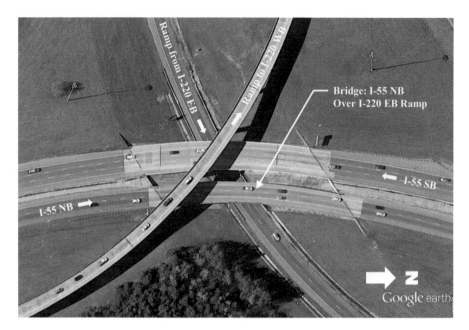

Figure 1.   Aerial view of I-55 and I-220 interchange.

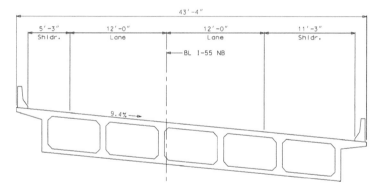

Figure 2.   Typical section through existing multi-cell CIP box structure.

The large difference in flexibility between the existing and proposed structures was viewed as a potential issue for long term durability, particularly of the slab.

 This paper discusses the various constraints leading to selection of steel girders as well as the modeling approach for evaluating the behavior of the widened structural system. This model served as the basis for a parametric evaluation of various structural constraints to hone in on an optimal solution resulting in acceptable stress levels in the slab.

2   LAYOUT AND CONSTRAINTS TO WIDENING

2.1   *Proposed widened section*

The proposed roadway section for widening to the outside of the existing alignment consists of three (3) 12-foot lanes and a 12-foot outside shoulder. The existing inside shoulder width

Table 1.   Summary of proposed I-55 widening options.

Option	Inside widening	Outside widening
56'–1" 5'-3" Shldr. / 12'-0" Lane / 12'-0" Lane / 12'-0" Lane / 12'-0" Shldr. ←BL I-55NB	NA	Structure: 3-haunched, steel PG's 8" CIP slab Girder Depth: 33" (min.) Spans: Match Existing Width: 12'-9"
56'–10" 6'-0" Shldr. / 12'-0" Lane / 12'-0" Lane / 12'-0" Lane / 12'-0" Shldr. 3'-9" ←BL I-55NB	Structure: 2-Type IV P/S girders 8" CIP slab Girder Depth: 54" Spans: Match existing Width: 9'-0"	Structure: 1–36" P/S LA girder 8" CIP slab Girder Depth: 36" Spans: Match existing Width: 4'-3"
56'–10" 6'-0" Shldr. / 12'-0" Lane / 12'-0" Lane / 12'-0" Lane / 12'-0" Shldr. ←BL I-55NB	New Structure	Structure: 11 steel PG's 8" CIP Slab Girder Depth: 33" (min) Spans: 20', 90'-130'-90', 50' Width: 56'-10"

of 5'-3" would be maintained such that the overall width of the widened structure would be 56'-1", which is 12'-9" wider than the existing bridge.

The critical geometric constraints for widening to the outside were the limited vertical clearance over the I-220 NB Ramp and the proximity of a column on the I-55 WB Ramp flyover at the south approach to the bridge. Each of these will be briefly discussed.

2.2   Vertical clearance

Both the width and depth available for the widening were limited by the vertical clearance requirements over the I-220 NB Ramp. The point of minimum clearance at the edge of shoulder on the ramp occurs within span 3 of the existing bridge, as shown in Figure 3. Plans indicated a clearance of 16.3 feet, while field measurements suggested a lower clearance of about 16.0 feet. The minimum clearance required at this location is 16.0 feet. To achieve this, the proposed structure depth would need to be shallower than the existing CIP box, decreasing with width of widening to accommodate a −9.4% cross slope on the deck and an upward grade of 5.6% at the underpass roadway. Based on these constraints, the maximum potential beam depth was 33 inches for the desired roadway width.

2.3   Horizontal clearance

Widening of the roadway section at the south approach to the bridge was limited by horizontal clearance to the adjacent column of the I-55 WB Ramp flyover. The distance from existing edge of shoulder to the column is about 13 feet as shown in Figure 3. This would accommodate a shoulder width of about 10 to 11 feet, depending on the protective measures required for the column, which would then taper to a 12-foot shoulder to be carried across the bridge.

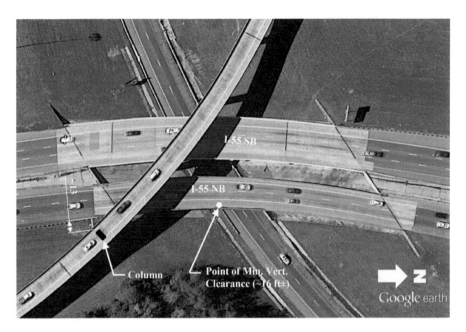

Figure 3.   Plan view showing point of minimum vertical clearance at existing I-55 NB structure and horizontal clearance to existing I-220 WB flyover ramp column.

3   STRUCTURAL ALTERNATIVES

The limitation on beam depth precluded the use of prestressed I-girders, which would have been ideal for the 100-foot span. Three alternative structure types were therefore developed for evaluation: (1) steel plate girders; (2) steel box beams; (3) prestressed box beams. Each of these are briefly discussed below.

3.1   *Steel plate girders*

This alternative consists of three closely spaced 33-in deep steel plate girders with a composite concrete deck. The structure would be connected to the existing CIP box by a closure-pour between the two slabs. No diaphragms would be provided between the existing exterior web and adjacent plate girder.

The span to depth ratio of the girder is 28 at span 3, which is within the suggested limits given in Section 10.5.2 of the **AASHTO LRFD** Bridge Design Specifications (AASHTO 2014). Preliminary design indicated a workable solution meeting both strength and serviceability requirements could be achieved with the limited beam depth.

The flexibility difference between the existing and proposed structures posed some concern for this alternative. The stiffness of the plate girder structure would be about 2 to 3 times less than the existing CIP box, resulting in an uneven response of the two structures under vehicle loading. This behavior would likely cause large transverse stresses to develop in the deck at the joint between the existing slab and closure pour with serviceability implications for both structures.

MDOT's recent experiences on a similar project substantiated this concern. A similar type bridge had been widened with precast, prestressed concrete (PPC) I-beams. Within several years after completion of the work, spalling began to develop along the construction joint in the slab, resulting in exposed rebar and the need for frequent patching. An assessment of the structure by HNTB demonstrated that the flexibility difference between the existing and widened portions of the structure was a contributing cause (HNTB 2015).

3.2 *Steel box girders*

A steel box beam solution improves upon the plate girder alternative as the torsionally stiffer section would help reduce the uneven response of the two structures. Practical considerations, however, made this alternative unfeasible. The minimum depth generally recommended for steel box beams for construction and maintenance access is 60 in, which exceeds the limiting depth of 33 in. No further consideration was therefore given to this alternative.

3.3 *Prestressed concrete girders*

Building on the box beam concept, consideration was given to a prestressed box beam alternative, which would provide even greater stiffness benefits over both the plate girder and steel box beam alternatives. Unlike the steel box beams, however, access for inspection is not required such that shallow depths are possible.

Preliminary design was conducted for a section comprised of two 48"x33" prestressed box beams with a composite concrete slab. A satisfactory strand pattern with f'$_c$ equal to 8,000 psi and 0.6" diameter strands could not be achieved, however, for the maximum design span of 100 feet. No further consideration was therefore given to this alternative.

3.4 *Selection of structural system*

The only viable structural alternative for full widening to the outside was therefore found to be a steel plate girder system. Further analysis was conducted on this option to study and quantify the impacts to the existing structure.

4 EVALUATION OF STEEL PLATE GIRDERS

4.1 *Modeling approach*

The new portion of the structure was assumed to consist of a 13'-11" slab on steel plate girders tied to the existing slab through a 2'-0" closure pour. A typical section through the widened structure, which has an overall width of 56'-1", is shown in Figure 4. The first interior plate girder was set at the edge of the proposed slab to minimize the distance to the existing exterior CIP box web. The exterior girder was set 3'-3" from the fascia to accommodate slab drains.

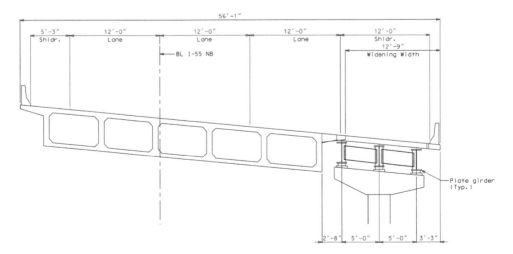

Figure 4.   Typical section through widened structure with steel plate girders.

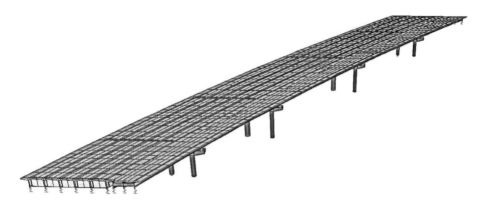

Figure 5.   CSI Bridge model of existing CIP box and plate girder structures.

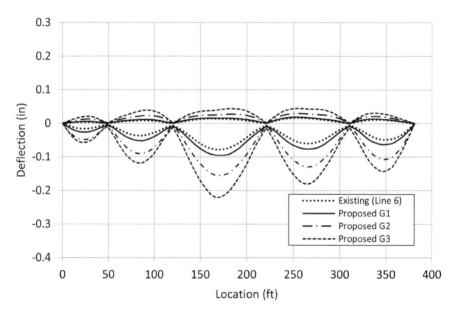

Figure 6.   Comparison of single lane live load deflections at proposed girders and existing web line.

A 3D finite element model of the widened structure was created using CSi Bridge. An iso-metric view of the model is provided in Figure 5. The existing CIP box was modeled with shell elements. The webs of the steel girders were modeled with shell elements while the flanges were modeled with frame elements. Shell elements were also used for the slab. Intermediate bent footings and abutments were assumed to be rigid supports. The existing bridge was modeled using plan dimensions with no reductions for wear or damage. Gross section properties were used assuming no cracking. Load analysis was limited to dead loads and vehicle loading.

4.2   *Modeling results*

The behavior of the structure under live load was first evaluated by placing a single lane of HS20-44 vehicle loading on the plate girder portion of the structure and comparing displace-ments between girders. Figure 6 shows the variation of live load deflection with location for the three girders and the existing exterior web. The girders are numbered G1 to G3 starting from the inside (adjacent to the existing CIP box). Deflections along girder G1 are similar to those along the exterior web. Those at the remaining girders increase with distance from the existing structure. The largest deflections occur in span 3 where the existing web deflection is about 0.08 in and the deflection at girder G3 is about 0.22 in. For comparison, the maximum girder deflection for the widened structure acting independently of the existing is about 0.8 in.

The live load deflection pattern indicates a certain twisting of the deck within each span due to the uneven response of the two structures. With this twisting, large moments develop in the closure pour slab that cause tension at the top surface normal to centerline of bridge. An area plot of factored transverse bending moments throughout the proposed deck is shown in Figure 7. The existing structure is not shown. The contour scale corresponds to a moment range of 0 kip-ft/ft to -20 kip-ft/ft. Areas of larger moments are clearly distinguishable near the center of each span along the joint between the existing slab and closure pour. The largest magnitudes are on the order of 20 kip-ft/ft, which is approaching the capacity of the section (21 kip-ft/ft). Moment and shear capacities of the existing CIP structure were also evaluated per AASHTO Standard Specifications (AASHTO 1996) and found to be adequate for the increased demands in the widened condition.

4.3   *Parametric study*

With such a high demand to capacity ratio, it was felt prudent to consider measures to reduce the bending moments in the closure pour in order to mitigate potential serviceability issues along the construction joint in the slab. Several parameters were evaluated in the CSi model with respect to their influence on deflections and thus stresses in the slab. These included: number of girder lines, connectivity between existing and proposed structures, girder depth, and girder spacing.

Seven different trial concepts were developed, which are summarized in Table 2. Each concept incorporates one or more of the study parameters. The impact of each trial on slab bending stresses at the construction joint is also shown in the table. Girder spacing (Trials 1 and 2) had minimal impact while the addition of struts connecting the bottom flange of G1 to the web of the CIP box (Trial 4) generally resulted in a 25% decrease in slab stresses. This came at the cost of increasing the cross frame member forces such that larger members and connections would be required. The addition of a girder line with or without struts (Trials 5 and 6) had a similar impact. Lastly, haunched girders (Trial 7) decreased slab bending stresses without any increase in cross frame forces. This solution is described in more detail in the next section.

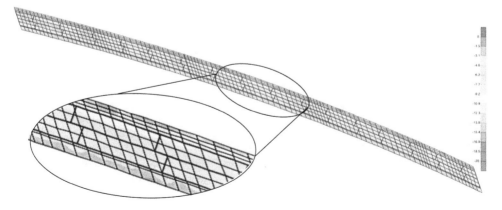

Figure 7.   Contour plot of transverse bending moments in new slab with partial blowup at span 3.

Table 2.   Summary of parametric summary trials.

Trial description	Change in slab stresses	Comments
1 Move G1 closer to CIP box web	None	
2 Vary steel girder spacing to shift more stiffness towards exterior girder	None	
3 Provide maximum web depth at G1 and decrease depth at G2 and G3 based on clearance	None	
4 Add struts connecting exterior web of CIP box to G1	25% decrease	Large forces developed in the cross frames requiring larger members and connections
5 Add a 4th girder	25% decrease	
6 Add a 4th girder and struts	25% decrease	Large forces developed in the cross frame requiring larger members and connections
7 Increase web depth; haunch girders in span 3 to meet clearance	25% decrease	

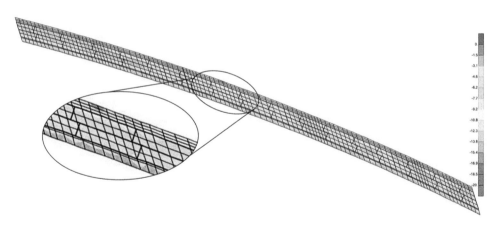

Figure 8.   Greyscale contour plot of transverse bending moments in new slab with haunched plate girder with partial blowup at span 3.

4.4   *Haunch plate girder solution*

The haunched plate girder solution consisted of increasing the depth of the girder webs to 60 inches throughout and haunching them to 30 inches in span 3 to meet the clearance requirements. This configuration provided the greatest benefits as demonstrated in Figure 8, which shows a revised area plot of transverse bending moments throughout proposed deck. Moments were substantially reduced in spans 1, 2, 4, and 5 in comparison with the original concept (refer to Figure 7). Moments in span 3 were moderately reduced to a maximum of about 15 kip-ft/ft. This effect was attributed to the increased stiffness of the adjacent spans, which resulted in less live load deflection throughout all spans, including span 3. As demonstrated previously, there is a direct connection between deflection of the plate girders and slab moments at the joint.

5   SUMMARY AND CONCLUSIONS

FEM modeling of the widened structure demonstrated the impacts of differential stiffness between the existing and proposed structures on stresses in the slab. The model identified the construction joint in the slab as a critical location. Through a parametric study, haunched plate girders were found to be a viable structural solution for widening. The increased stiffness provided by the variable depth girders was shown to reduce live load deflections in all spans, including the span with the least girder depth, thereby mitigating the stress within the deck to acceptable levels.

Final details will need to address specifics of the closure pour including material specifications and development of existing and new reinforcing steel. Details similar to those used for longitudinal joints between full depth precast slab panels are envisioned.

The proposed solution made it possible for MDOT to move forward with this needed widening project. Final design is slated to begin this summer (2017) with construction letting in 2018.

REFERENCES

American Association of Highway and Transportation Officials (AASHTO) (7th ed.) 2014. *AASHTO LRFD Bridge Design Specifications*. Washington D.C.: AASHTO.
American Association of Highway and Transportation Officials (AASHTO) 1996. *Standard Specifications for Highway Bridges, Division IA Seismic Design*. Washington D.C.: AASHTO.
HNTB 2015. *US 78 Bridge No. 64.7, 65.0, 65.0B Deck Repairs—Condition Assessment and Analysis Report*; prepared for MDOT.

Chapter 23

3D structural modeling of stringer-bent connections on Gowanus Expressway Viaduct, New York

X. Wei, A. DeVito & W.S. Najjar
WSP USA, New York, USA

ABSTRACT: During inspections of Gowanus Expressway Viaduct, cracks in stringer-end copes, stringer webs, and stringer to floor-beam connection angles have been found on a regular basis. In order to understand the behavior of the structural system and determine remaining fatigue life, a WSP/URS Joint Venture performed analytical and experimental studies of the stringer to steel bent connections on the viaduct for NYSDOT.3D finite element models, including a global viaduct model and a detailed connection model, were developed. To validate the modeling approach, diagnostic load testing on one typical span of the viaduct was conducted. Comparisons were made between analytical and experimental results for the rotational stiffness values for stringer–bent connections, stresses in the connection components, and stringer member forces. It is shown that the models can predict the structural responses well, and the analysis results can be used to estimate the remaining fatigue life of the connection components. The analytical study is the primary focus of this paper.

1 INTRODUCTION

Built in the late 1930s and early 1940s, the Gowanus Expressway Viaduct carries traffic with an elevated roadway through the Borough of Brooklyn, New York City. During inspection of the 322 span 3.8 mile steel framed elevated viaduct of Gowanus Expressway, cracks have been found on a regular basis in the stringer-end copes, stringer webs, and stringer to bent connection angles. To evaluate the structural deficiencies and predict the remaining fatigue life of structural components, a WSP/URS Joint Venture performed analytical and experimental studies of the stringer to steel bent connections on the viaduct for NYSDOT. This paper summarizes the analytical study and its findings.

Three-dimensional (3D) finite element models were created in CSiBridge software (CSiBridge 2014) for four representative bents (Figure 1). The models include (1) a global model with full-size structural components (e.g., columns, beams and slabs) of the four spans to estimate the member forces of the viaduct; and (2) a detailed connection model (local model) with connection angles, bolts/rivets, partial concrete deck, stringers, and columns, to estimate the rotational stiffness of the stringer to column connections and calculate stresses in the structural components.

Although originally designed as pure shear connections, empirical evidence suggests that the stringer to bent connections behave with partial fixity. The magnitude of the partial fixity, i.e., the rotational stiffness of the connections, is a critical input parameter in the global model to determine member forces and for subsequent structural component stress calculations. Considering interactions between viaduct deck slabs and steel components and between angles and column web, the connection rotational stiffness is determined by the detailed connection model for a typical stringer to bent connection. To ensure that the detailed CSiBridge model accurately represents the structural behavior, diagnostic load testing on one typical span of the viaduct was conducted with dump trucks, whereby strains and displacements in

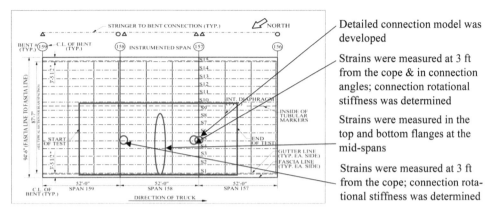

Detailed connection model was developed

Strains were measured at 3 ft from the cope & in connection angles; connection rotational stiffness was determined

Strains were measured in the top and bottom flanges at the mid-spans

Strains were measured at 3 ft from the cope; connection rotational stiffness was determined

Figure 1.   Bridge framing and testing instrumentation plan.

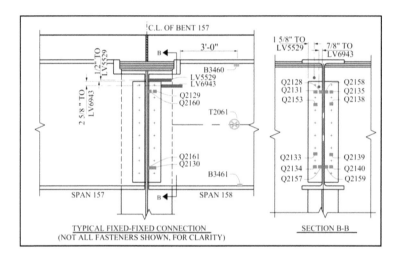

Figure 2.   Testing instrumentation at Stringer S5 end at Bent 157 in Span 158.

selected steel elements were measured. Calculated stringer–bent connection rotational stiffness values were then calibrated based on experimental tests.

Subsequently, the connection rotational stiffness values were applied to the global model to obtain member forces at the stringer ends. To calculate the stresses in the stringer to column connection angles, member forces at the stringer ends were applied to the detailed connection model. Analytical and experimental results for member forces in the stringers and stresses in the connection components were compared to confirm the accuracy of the analytical models. With the accuracy of the CSiBridge models confirmed, remaining fatigue life of the connection components was calculated.

2   DIAGNOSTIC LOAD TESTING OF A TYPICAL SPAN

This study focuses on four typical spans. Each span measures 52 feet in length and consists of a non-composite reinforced concrete deck, steel stringers, steel bent cap beams, and columns. The stringers are connected to steel bents through simple web-shear connections. This part

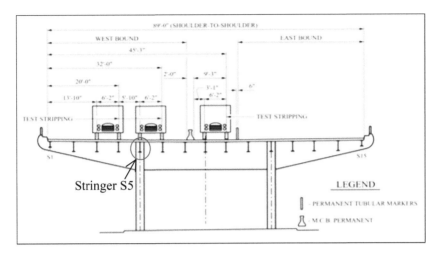

Figure 3. Truck configuration of test runs 3 & 4 (looking north).

of the viaduct was originally built in 1940s with seven stringers. In the 1960s, a cantilevered bent cap supporting four stringers was installed on both sides of the existing bent to expand the roadway. Currently, the viaduct carries seven lanes.

For this study, diagnostic load testing at one typical span (Span 158) was conducted on October 23, 2013 to validate the analytical approach (WSP 2013). While dump trucks were driven over the span, strains and displacements in selected steel elements were recorded. Figure 1 shows the bridge framing plan between bents 156 and 159, as well as the testing instrumentation plan. As shown, strain histories were recorded at all stingers in span 158, i.e., at the mid-spans of stringers, 3 feet from the cope in both ends of Stringer S5, the stringer-end cope of S5 at bent 158, and connection angles attached to the end of S5 at bent 157. Figure 2 shows the details of testing instrumentation at stringer S5 and at bent 157 in span 158. A total of twelve test runs were conducted. Figure 3 shows the truck configuration of two of the twelve test runs. From the strain recorded history, connection rotational stiffness values were determined for both Stringer S5 ends in span 158. Testing results also showed partial composite behavior between the stringer and concrete deck on the top of the stringer.

3 DETAILED CONNECTION MODEL (LOCAL MODEL)

3.1 Modeling

A detailed connection model was developed for the end of Stringer S5 to Bent 157 connection to determine the connection's rotational stiffness. A photo of the connection is provided in Figure 4. The analysis results were compared to the test results to validate the rotational stiffness.

As shown in Figures 5 and 6, angles, stringers, and columns are modeled as shell elements, bridge decks are modeled as solid elements, and interactions between the bridge deck and stringer top flange/column are modeled with link elements.

Joint restraints were applied at the column top surface and along the column edge, as shown in Figure 6, to eliminate the deformation of the column, which is captured in the global model analysis. In addition, bolts and rivets are modeled as beam elements, interactions between bolts and angles/column web are represented by gap elements, and the partial composite behavior is considered by using a smaller concrete modulus of elasticity calculated based on the results from the diagnostic tests discussed in Section 2. A partial length of

Figure 4.   Photo of Stringer S5 end at Bent 157 in Span 157 (Connection details at the backside similar but without the supporting bracket).

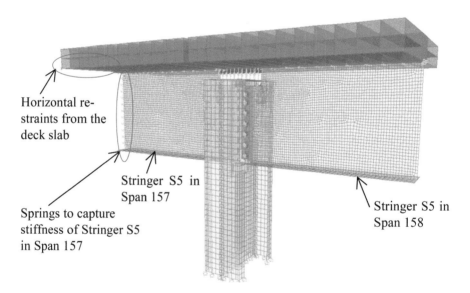

Figure 5.   Detailed connection model for the calculations of connection rotational stiffness and stresses in connection components.

Stringer S5 measuring 6 ft in length was modelled on each side of the bent (Bent 157). The support stiffness at the cut section of Stringer S5 in Span 157 was determined by assuming a simply supported beam of Stringer S5. The aforementioned cut section of Stringer S5 in Span 158 is a free end and loaded with loads (shear, moment and axial forces) determined from global model analysis. Contact between part of a connection angle and the bent web may theoretically be lost when part of the angle is subjected to a pulling force from the stringer. Due to this behavior, compression only springs are used to connect the angles to the bent web to accurately represent the interaction between the angles and the bent web. The springs are shown by the green lines in Figure 6.

3.2 *Connection rotational stiffness: Analysis vs. testing results*

Applying a vertical concentrated load at the cut section of the stringer in the detailed connection model, the moment, corresponding rotational angle, and subsequent rotational stiffness can be determined for the stringer to bent connection. Stiffness results from the analytical method and the empirical tests are compared in Figure 7, with the corresponding values pro-

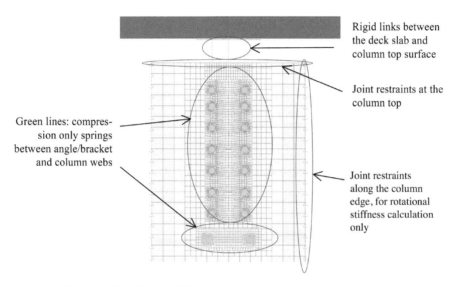

Figure 6.   Modeling details of Stringer S5 end to bent connection.

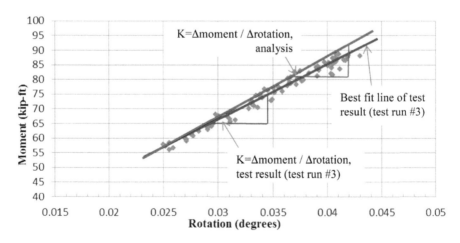

Figure 7.   Connection rotational stiffness = $\Delta_{moment} / \Delta_{rotation}$, Stringer S5 end at Bent 157 in Span 158.

Table 1.   Numerical rotational stiffness values, S5 end at Bent 157 in Span 158 (kip-ft/degree).

	Analysis	Test
Connection rotational stiffness	2053	1867

vided in Table 1. It can be concluded that the analytical and test results are very close, thus validating the assumed modeling techniques and the boundary conditions, and proving that the finite element analysis approach is effective for the calculation of connection rotational stiffness.

The same approach can be applied to other connection locations to determine the corresponding rotational stiffness values. Modifications to the detailed connection model may be needed to consider various geometries and boundary conditions.

4   GLOBAL MODEL

4.1   *Modeling*

A global model representing the global behavior of four representative spans (Spans 157 to 160) was developed to calculate the member forces of the viaduct. The spans in the model are structurally independent extending from expansion joint to expansion joint on the bridge. In the model, steel components (i.e., columns, cap beams, stringers and diaphragms) are modeled as beam elements, bridge decks are modeled as shell elements, and the interactions between the decks and steel components are modeled with link elements. Figure 8 shows the extruded views of the model.

Analysis of the global model was focused on the span tested (Span 158), so that member forces from the global model can be compared with forces measured in the diagnostic test. At Span 158, finer deck elements are used and grillage beams transverse to the stringers were added to the model (Figures 9 and 10). Based on maximum stringer-flange strain values measured from the diagnostic tests, the distance to the neutral axis from the bottom of each instrumented stringer was calculated, and then the stiffness of the deck-beam section.

Figure 8.   Global model of the bridge, spans 157 through 160.

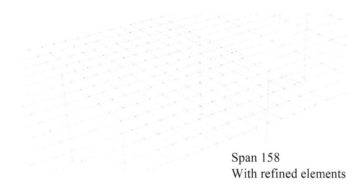

Span 158
With refined elements

Figure 9.   Refinement of global model at Span 158.

Lane over stringer S5 for fatigue analysis

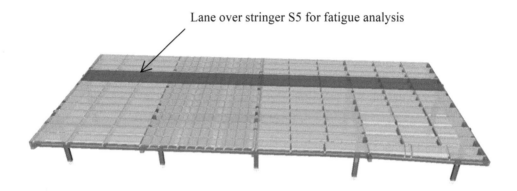

Figure 10.   Extruded view of the refined global model.

Table 2.   Connection rotational stiffness values used in the global model (kip-ft/degree).

Stringers	S2	S3	S4	S5	S6	S7
Span 157, bent 157	6000	6000	6000	6000	6000	6000
Span 158, bent 157	3500	4400	4400	2200	2200	3500
Span 158, bent 158	3500	4400	4400	2600	3500	3500

The stiffness of the partial composite deck-beam section is about 2 to 3 times the stiffness of the steel section alone. In the model, the contribution of partial composite behavior between the concrete deck and steel frame was accounted for by increasing steel member stiffness values.

The rotational stiffness values of Stringer to bent connections (see Table 2) were applied to the global model. Among these values, the stiffness of Stringer S5 end at bent 157 was determined by local model analysis as discussed in Section 3. Other values are from test results or estimated based on geometries and analysis/testing results. This is because no local models were developed for these locations.

Truck loading and spacing configurations, similar to Test Runs 3 and 4 of the diagnostic tests discussed in Section 2, were applied to the global model. For simplification, it is

Table 3.   Comparison of maximum moment*
in Stringer S5 in Span 158 (kip-ft).

Bents	158	157
Analysis	−82.9	−83.7
Test	−88.8	−82.0
Difference	−7%	2%

*Negative signs indicate negative moments.

Table 4.   Comparison of Mid-span moments of
Stringers S2 through S7 in Span 158 (kip-ft).

Stringers	S2	S3	S4	S5	S6	S7
Analysis	130	217	223	264	302	211
Test	113	189	205	242	308	223
Difference	15%	15%	9%	9%	−2%	−5%

assumed that the wheel spacing is the same as the stringer spacing. This spacing configuration results in conservative member forces.

4.2   *Member forces, analysis vs. test results*

Tables 3 and 4 show the comparison between analysis and test results for the maximum moments at the ends of Stringer S5 and at the mid-span of various stringers, respectively. It is indicated that, at the location studied (Stringer S5 ends), the difference is small and the model predicts member forces well. The larger differences at mid-span is likely due to the stringer end rotational stiffness assumed.

5   LOCAL MODEL ANALYSIS FOR MEMBER STRESSES

Member end forces, obtained from the global model analysis for the end of Stringer S5 at Bent 157 in Span 158, were applied to the detailed connection model discussed in Section 3 to calculate member stresses. For the stress calculation, boundary conditions were updated in the detailed connection model, e.g., removing the restraints at the column top surface and along the column height and adding horizontal springs at the column top surface to capture the horizontal restraints from the deck and stringers.

5.1   *Stresses in stringer copes*

Stress analysis of the detailed connection model gives a maximum localized tensile stress of 4.7 ksi at the end cope of Stringer S5 at Bent 157 in Span 158. It is noted that there is no experimental stress/strain measurement for this location because of corrosion. Strain measurement in the cope of the other end (at Bent 158) of the same stringer indicates a maximum tensile stress of 8.65 ksi. The higher stress is likely due to the higher connection rotational stiffness.

5.2   *Stresses in connection angles*

Analysis and test results of stresses in connection angles of Stringer S5 at Bent 157 in Span 158 are shown in Figure 11. Corresponding values are also provided in Table 5, together

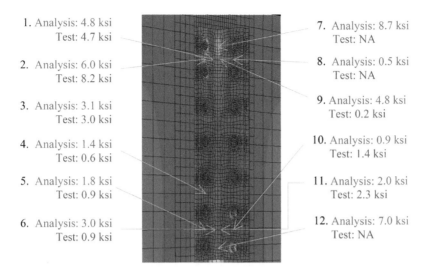

1. Analysis: 4.8 ksi
   Test: 4.7 ksi

2. Analysis: 6.0 ksi
   Test: 8.2 ksi

3. Analysis: 3.1 ksi
   Test: 3.0 ksi

4. Analysis: 1.4 ksi
   Test: 0.6 ksi

5. Analysis: 1.8 ksi
   Test: 0.9 ksi

6. Analysis: 3.0 ksi
   Test: 0.9 ksi

7. Analysis: 8.7 ksi
   Test: NA

8. Analysis: 0.5 ksi
   Test: NA

9. Analysis: 4.8 ksi
   Test: 0.2 ksi

10. Analysis: 0.9 ksi
    Test: 1.4 ksi

11. Analysis: 2.0 ksi
    Test: 2.3 ksi

12. Analysis: 7.0 ksi
    Test: NA

Figure 11.   Stresses in the connection angle.
Note: 1. The maximum stress measured in the test is 8.2 ksi in the connection angle. 2. The above stress comparisons are also presented in the table below, together with corresponding labels of strain gages.

Table 5.   Stresses at various strain gage locations on angles. Test (runs 3 & 4 Max./Min. stresses) vs. Analysis results (Max. stresses) (ksi).

Strain gages	Angles attached to column web						Angles attached to stringer web			
	Top		Bottom				Top		Bottom	
	2158	2131	2153	2133	2134	2157	2129	2160	2130	2161
Analysis	4.8	−6.0	−3.1	−1.4	−1.8	−3.0	0.5	4.8	−0.9	−2.0
Test	4.7	−8.2	−3.0	−0.58	−0.91	−0.9	–	−0.2	−1.4	−2.3

with corresponding strain gage labels shown in Figure 2. Comparison indicates that the stress values from the finite element analysis and the test are reasonably close. It is noted that the maximum stress values in the connection angle are 8.2 ksi measured from the test and 8.7 ksi calculated with the analytical method, respectively, and the difference is 6%.

6   FATIGUE ANALYSIS

Criteria for fatigue analysis used in this study are from AASHTO LRFD Bridge Design Specifications (AASHTO LRFD 2010) and AASHTO Manual for Bridge Evaluation (AASHTO BEM 2011) Section 7 "Fatigue Evaluation of Steel Bridges".

An AASHTO fatigue truck, Figure 12, was applied to the global model on a single lane over stringer S5. A moving truck load analysis was performed with the lane shifted transversely to obtain maximum negative moment in Stringer S5 to bent connection. Forces (moment, shear, and axial force) at Stringer S5 end from the global model analysis were then applied to the detailed connection model to obtain fatigue stresses in the connections. Fatigue stresses (without the impact factor) in the connection angle are shown in Figure 13. The maximum fatigue stress in the base metal of connection angle for Stringer S5 is 4.66 ksi, and the stress in the net section of the connection angle is less than 2 ksi. In addition, the maximum fatigue stress at the Stringer S5 cope at Bent 157 in Span 158 is 2.6 ksi. Considering a fatigue impact factor of 1.35,

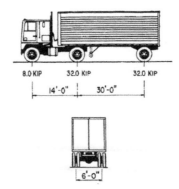

Figure 12.   Fatigue truck.

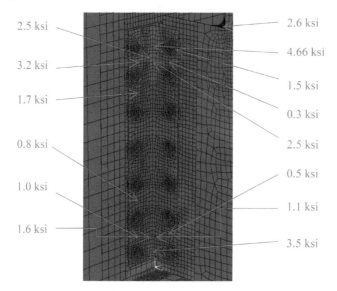

Figure 13.   Fatigue stresses in the connection angle.

Table 6.   Nominal fatigue resistance for infinite life (AASHTO LRFD) (ksi).

Category*	Fy = 36	Fy = 33
A	24	22
B	16	14.7
C	10	9.2

*Category A: base metal, except non-coated weather steel. Category B: (1) base metal, non-coated weather steel; (2) gross section of high-strength bolted joints, slip-critical; (3) net section of high-strength bolted joints, bearing type. Category C: (1) member with re-entrant corners at copes, cuts, block cuts. (2) for evaluation only, net section of rivets connected joints.

determined from the test discussed in Section 2, and a factor of 1.5 for maximum stress, the corresponding maximum fatigue stresses are 9.4 ksi, 4 ksi and 5.3 ksi in the base metal of the connection angle, the net section of the connection angle, and the cope of Stringer S5, respectively.

The fatigue thresholds based on AASHTO LRFD are shown in Table 6. Connection angle and stringer cope fatigue life values are compared with the fatigue thresholds in Table 7.

Table 7. Check of infinite fatigue life, Stringer S5 end, Bent 157 (ksi).

Location	Max. Stress	Category	Stress Limit	Check
Stringer flange cope	5.3	C	9.2	OK
Angle, base metal	9.4	B	14.7	OK
Angle, net section	<4.0	C	6.4	OK

Calculations indicate that the connection possesses an infinite fatigue life. Therefore, additional fatigue analysis for remaining fatigue life was not performed for the Stringer S5 to bent connection at Bent 157 in Span 158.

7 CONCLUSIONS

In this study, finite element models were developed for four typical spans of the Gowanus Expressway viaduct. The models include a global model with full-size primary structural components for the global behavior of the viaduct and a detailed connection model with connection components for calculations of connection rotational stiffness and stresses in connection components.

Results from analytical methods and tests were compared for the connection rotational stiffness values, connection component stresses, and stringer member forces. It is shown that the models can predict the structural responses accurately, validating the assumed modeling techniques and model boundary conditions. It is concluded that the finite element analysis is effective and can be applied to other connection and span locations with various geometries. There are several key factors that contributed to the effectiveness of the finite element analyses. In the local model, the key factors include the compression only springs for the interactions between connection angle and column web, the reduced concrete modulus of elasticity for deck-beam partial composite behavior, and the support stiffness at the cut section of partial stringer S5. In the global model, the key factors include the rotational stiffness values determined in the local model analysis and the deck-beam stiffness considering partial composite behavior.

Furthermore, fatigue analysis was performed for the Stringer S5 to Bent 157 connection in Span 158. Results determined that the connection possesses an infinite fatigue life for AASHTO fatigue loading conditions. The study is based on a typical simple shear connection subjected to the typical loading assumed.

ACKNOWLEDGMENT

The authors would like to express sincere appreciation to NYSDOT for supporting this study and permitting presentation and publication of this paper. Many of our WSP USA colleagues had made considerable contributions to this work. Rakesh Murthy and Carolyn Mariano developed the computer models. Satrajit Das led the field testing. Paul Nietzschmann, Philippe Bousader, and Ronald Paproski of WSP USA, and coworkers from AECOM gave valuable guidance. We are very grateful to all of them.

REFERENCES

AASHTO LRFD 2010. AASHTO LRFD Bridge Design Specifications, 5th Edition (US), American Associates of State Highway and Transportation Officials.

AASHTO MBE 2011. The Manual for Bridge Evaluation, 2th Edition, American Associates of State Highway and Transportation Officials.

CSIBridge 2014. Computers and Structures, Inc.

WSP 2013. Field Instrumentation and Diagnostic Load Testing of Span 18 Superstructure Over 40th Street, Gowanus Expressway Viaduct (I-278).

Bridge history & aesthetics

Chapter 24

Modern art and New York City bridges

S. Rothwell

Stuart Rothwell and Associates Pty Ltd., Brisbane, Australia

ABSTRACT: The evolution of the modern bridge, from around 1850, both coincided with, and influenced the growth of modernist art movements. Many European avant-gardists used the new bridges as subject material for their revolutionary art. That trend continued in New York, as modern art permeated the city, particularly after the celebrated Armory Exhibition of 1913. The bridges of New York were a rich source of inspiration for a plethora of twentieth century artists. This paper discusses some bridges of New York City, both the iconic and unsung, the people who built them, the artists who were inspired by them, and the paintings they produced.

1   TIME IS SO SHORT AND THE SUBJECTS SO MANY—AN INTRODUCTION

> *The only works of art America has given*
> *are her plumbing and her bridges*
> Marcel Duchamp (1917)

When the Sydney Harbour Bridge was under construction in the 1920's, young Australian artists rushed to the water's edge to capture the thrill of the epic event in a series of paintings that hallmark the birth of modern art in Australia (Rothwell 2013). The emergence of the bridge and the art were contemporaneous, and had the iconic bridges of New York City been built at the same time a plethora of American artists could have captured the dynamism of the moment. As it was, modernist art arrived in New York to bridges that were virtually all in-situ, some for a generation or two. Consequently, the impact of modern art, both in fact—and its effect on society, played out on already mature structures; a notable exception, perhaps, being the 1915 etching *The Hell Gate Bridge*, by Joseph Pennell (1857–1926), whose comment about a life in art is encapsulated in the title of this introduction. Born in Philadelphia, and trained at its Academy of Fine Arts, Joseph Pennell spent most of his adult life in London, but he did return to America occasionally, including 1915, when he observed '*nothing in this world was so marvelous as to watch, day by day, last summer, the great arch of the Hell Gate Bridge growing from each side of the river till the two arms met in the middle and the subject ceased to exist*' (Pennell 1916).

In his celebrated history of modern art, the expatriate Australian critic Robert Hughes (1938–2012) referred to the Brooklyn Bridge as '*the New World's answer to the Eiffel Tower*' (Hughes 1980). The grandeur of the bridge evident in Pennell's delightful sketchbook study of the bridge, shown in Figure 1, indicates why such a view might prevail. There is a certain irony at play here though. After all, when the Parisian tower was completed, in 1889, the bridge in New York had been open for six years, and the authentic '*answer*' had, in fact, been the eponymous Ferris Wheel constructed in 1893 for the World's Fair in Chicago. Nevertheless, Hughes had a point. He was, of course, referring to the emotional response that each structure evoked in the people who built them and the populations that embraced them. It's no surprise, then, that of all the river crossings in New York City, the Brooklyn Bridge is the most painted, artistically.

Figure 1.   Joseph Pennell *Brooklyn Bridge*, 1921, Brooklyn Museum.

From a family of Quakers, Pennell witnessed—as a press illustrator—the carnage at Verdun during WW1, which had such a cathartic effect on him that in 1917 he returned to America and eventually settled in Brooklyn Heights, from where it is likely he produced *Brooklyn Bridge*. On his deathbed he asked to be moved nearer the window for one last gaze at the scene.

No Australian sees the spandrel arch Hell Gate Bridge without being immediately reminded of the bridge over Sydney Harbour. Consulting engineer for the bridge at Hell Gate was Gustave Lindenthal (1850–1935), fresh from his controversial commissionership at the city's bridge department. Other notable engineers who worked on the project were Othmar Ammann (1879–1965) and David Steinman (1886–1960). The architect Henry Hornbostel (1867–1961) was also involved (Dupré 1998).

2   WESTWARD HO AND MURMURINGS OF THE MODERN

William Louis Sonntag Sr (1822–1900) is now primarily remembered as a landscape painter, particularly his Arcadian panoramas of the American wilderness, and for much of his life he painted in a group now commonly referred to as the Hudson River School. It is often observed that he reflected mid-century sensibilities about manifest destiny. Although his son, William Louis Sonntag Jr (1869–1898), is '*better known today as an illustrator of New York urban life*' (Gerdts 1994), his 1895 painting of the Brooklyn Bridge, shown in Figure 2, has been portrayed as an icon of that providential ideal with the '*grandeur*' of the image suggesting that Sonntag Jr '*saw the Bridge as the apotheosis of America's westward march*' (Kachur 1983), and '*a metaphor for the vastness of the both the nation and its destiny* (Gerdts 1994).

Stylistically, the painting is a nocturne, a term coined by Pennell's friend and mentor, A M Whistler (1834–1903) for any painting of a night scene intended to evoke a '*dreamy, pensive*

Figure 2.   William L Sonntag Jr, *Brooklyn Bridge*, 1895, Museum of City of New York.

mood' and the most appropriate Whistler contribution to this style is *Nocturne in Blue and Gold; Old Battersea Bridge*, painted around 1875. Compositionally, Sonntag Jr practiced a diagonal axis technique, in which the eye is led from the right foreground to the left background, which is shown to full effect in his *Brooklyn Bridge*, where the eye moves from the steamer in the river to the oblique elevation of the bridge. As if to signal our starting point he has reflected the port-side light in the river and gently bathed the ship in the sunset. Sonntag Jr died before turning thirty, but he appears to have heard that muffled modernist drumbeat coming from Europe.

The Brooklyn Bridge had been opened in 1883. When building commenced in 1869, the bridge should have been the culmination—for John Roebling (1806–1869) - of a lifetime spent in the early development of the modern suspension bridge. Within weeks of approval being granted for construction of the bridge, however, John Roebling was dead from a site injury. Responsibility for completing the bridge passed to his son, Washington Roebling (1837–1926), who had already proved his abilities on the Cincinnati Bridge. When completed in 1866 that bridge had, at 1057 ft, the longest span in the world (Dupré 1998). Washington Roebling succumbed to a debilitating illness in 1872 after prolonged exposure to conditions in the pressurised caissons used to sink the tower foundations. Washington's wife, Emily Warren Roebling (1843–1903), then assumed the role of her husband's amanuensis. She conveyed his instructions to the assistant engineers at the site, until the bridge was complete (McCullough 1972).

3   THE AMERICAN IMPRESSIONISTS

The Australian historian Bernard Smith (1916–2011) has remarked that modern art *'developed out of French impressionism'* (Smith 1992), the origins of which can be traced to the work of Edouard Manet (1832–1883), who commenced his art training in 1850.

Figure 3.   Childe Hassam, *Brooklyn Bridge in Winter,* 1904, Telfair Museums, Savannah, Georgia.

Ironically, modernity—as a concept that today's structural engineer can appreciate—commenced around the same time, with the construction of large span railway bridges in wrought iron, designed on the basis of rational, mathematically based analysis and rigorous material testing. These developments were driven by the industrialisation and urbanisation shaking the world, which triggered revolutions in transportation and social mobility (Rothwell 2011) that the French Impressionists sought to convey in new ways of applying paint and using colour.

From the 1880's onwards American painters travelling and studying in Europe found inspiration in this new style and techniques, As they returned to New York towards the end of the nineteenth century, they found a city in transition and dedicated themselves to applying Impressionism in an attempt to capture the spirit of the age.

After his return from France in 1889, Childe Hassam (1859–1935) is the painter most identifiable as an American who devoted himself to the Impressionist idiom in New York, contemporaneously described as *'one who paints the life he sees about him, and so makes of his own epoch'* (Gerdts 1994).

Hassam painted Brooklyn Bridge several times (Figure 3) and contended that Paris and London *'had nothing to compare with our own Manhattan Island when seen in an October haze or an early morning mist from Brooklyn* Bridge' (Gerdts 1994).

In 1892 he had painted *Winter Day on Brooklyn Bridge*, a watercolour as viewed from a user's perspective, which predicts the focus of future modernists.

4   THE EIGHT

Early in the twentieth century a group of painters, centered around the theorist, Robert Henri (1865–1929), became frustrated by the closed shop and conservative values represented by the National Academy of Design. In 1908, an octet of these artists, subsequently dubbed

'*The Eight*', held an independent show of their work at the Macbeth Gallery to protest against the restrictive exhibition practices and narrow tastes of the Academy. Along with Henri, the other members of the group were William Glackens (1870–1938), John Sloan (1871–1951), George Luks (1867–1933), Everett Shin (1876–1953), Arthur B Davis (1863–1928), Maurice Prendergast (1858–1924), and Ernest Lawson (1873–1939).

Lawson was not an obvious radical. Undemonstrative, quietly spoken and self-effacing, he was in reality a refugee Impressionist, a style most of his colleagues amongst the Eight eventually derided. Nevertheless, his work had been rejected for an Academy exhibition, so he was at least philosophically aligned with the group.

Born in Halifax, Nova Scotia, Lawson arrived in New York in 1891, via Kansas, and studied under the American Impressionist, John Twachman (1853–1902). Between 1893 and 1896 Lawson was in France, where his Impressionist style was further developed, including a 'plein air' association with Alfred Sisley (1839–1899), the British Impressionist remembered today for his paintings of the Seine and its bridges. Importantly, for his initial years back in New York City, Ernest Lawson lived in Washington Heights, where he often painted the bridges over the Harlem River, as in *The Bridge* (c 1912), the High Bridge (viewed from the Bronx), which opened in 1842 as part of the Croton Aqueduct, Manhattan's first reliable supply of potable water.

The work of John B Jervis (1795–1885), who began his career on the Erie Canal, the aqueduct bridge was the city's first cross river bridge. Fayette B Tower (1817–1857), an assistant engineer on the aqueduct scheme, was a keen sketcher of the works carried out, and his drawing of the bridge was later reproduced by J. Napoleon Gimbrede (1820-?) and published in 1843.

The High Bridge, simply called the Harlem River Bridge by Jervis, was 1450 feet long, with 15 semicircular arch spans varying from 50–80 feet. Its clearance of 100 feet was a legislated allowance for future navigation. It was constructed from '*well dressed granite*'. In 1928, five arches were replaced with a single steel arch of 450 feet over the main waterway, to improve navigation.

Nevertheless, Lawson's *Spring Night* (see Figure 4), which depicts the nearby Washington Bridge over the Harlem River at 181st Street, painted in 1913 at the pinnacle of his career, is his most evocative work.

Figure 4.   Ernest Lawson, *Spring Night, Harlem River*, 1913, The Phillips Collection, Washington DC.

The Board of Commissioners of Central Park first mooted the idea for this road bridge connecting the Counties of New York and Westchester in 1868, and executive officer, Andrew H Green (1820–1903), envisaged *'a suspension bridge two thousand feet long'* (Hutton 1889). Land for the bridge was approved by the Supreme Court in 1876, and in 1881 W J McAlpine, Chief Engineer for the Department of Public Parks, prepared four alternative plans, including two suspension bridge options, although each with more modest spans of 800 feet. In 1883 the Park Commissioner invited designs from A.P Boller (1840–1912), Messrs Buck and McNulty, and the Wilson brothers. The Buck and McNulty design was favoured, but funds were elusive at the time.

In 1885, a special Commission was instituted with authority to construct the bridge. McAlpine was appointed Chief Engineer and a design competition was arranged. Seventeen entries were received, but only two premiums were awarded, the first to C. C. Schneider (1843–1916) and the second to Wilhelm Hildenbrand (1843–1908). Both entries were for twin arches, Schneider's of 450 feet and Hildenbrand's 540 feet. Both the options utilized trussed arches. The eventual design, as modified by the Union Bridge Company, replaced the arch bracing with web plates. Further amendments by Chief Engineer McAlpine and Theodore Cooper (1839–1919), by then consulting engineer, led to the plans included in the tender documents, with the span for each arch set at 510 feet. Hildenbrand and George McNulty had been prominent on the design and construction of the Brooklyn Bridge (McCullogh 1972), and the others were all eminent engineers of the day.

Eventually in 1886, a contract was awarded and the work was complete by March 1889. In 1902 Childe Hassan painted the bridge on a sunny winter's day, originally entitled *Washington Bridge*, but later erroneously renamed *High Bridge*. It would be a quarter of a century after it opened before Lawson gave it his memorable artistic voice.

Ernest Lawson was not done with bridges, but his melancholy painting *Brooklyn Bridge* (Terra Foundation of Modern Art), circa 1920, illustrates his reluctance to incorporate the modernist trends by then engulfing the art world. He became artistically divorced from the Eight, whose more modernist flame shone a little longer, until they too were consumed.

5   THE ASHCAN SCHOOL

The Eight consisted of artists *'more closely allied in friendship than belief'*, and of the original *'black gang'*, five—Henri, Luks, Glackens, Sloan & Shinn—essentially eschewed the Impressionist style that Lawson was so reluctant to relinquish, and painted in an alternative style that became known as the Ashcan School. This was more a matter of palette than technique. They were not really interested in colour, to the extent the Impressionists were, and their work was typified by a certain grittiness, portraying everyday images of the less salubrious aspects of modern urban life, painted in dark, almost muddy tones, loosely applied.

This association attracted artists like George Bellows (1882–1925) and Glenn O. Coleman (1887–1932), and several notable bridge paintings came out of the group, like Luks *Brooklyn Bridge* (1916), now in a private collection.

Although Coleman was born in Ohio and grew up in Indiana, his relatively short adult life was consumed with capturing the character and content of New York. Arriving there in 1905, he studied at the New York School of Art with Robert Henri and Everett Shinn. For the next twenty years he lived a life marked by financial hardship, punctuated by a sojourn in Cuba, building up an impressive body of work comprising paintings, drawings and lithographs, gleaned from a patient observation of the insalubrious world around him.

Although the popularity of urban realism, as championed by the Ashcan movement, waned after the Great War, Coleman persevered with his simple and lyrical, almost naïve style, depicting the everyday life of the city. When asked late in his somewhat short life to describe the subject of his work he merely replied, *"New York"*. His painting of the Brooklyn Bridge, shown in Figure 5, is a wonderful example of his mature work.

Figure 5.   Glenn O. Coleman *Bridge Tower*, 1929, Brooklyn Museum.

Painted well after the glory days of the Ashcan movement it retains that obsession with the working city. However, although Coleman's characteristic palette of burnt siennas, viridian greens, and lemon yellows remains from his earlier work, there is a flush of sunlight across the work that perhaps mirrors the more financially comfortable circumstances of his later years.

6   THE ARMORY

1913 was a big year for the art world in New York, although not because a young Edward Hopper (1882–1967) went down to the banks of the East River to paint the recently built bridge at Blackwell Island (see Figure 6). It was nevertheless fortuitous, as Hopper went on to be an artist of some considerable fame, now best known, of course, for paintings that appear to explore the solitude of the human condition. For a time he too was classified, by some critics, as part of the Ashcan group, perhaps because Robert Henri had been one of his teachers at the New York Institute of Art and Design. It was a label Hopper always rejected, claiming his urban scenes were painted, *"with not a single incidental ashcan in sight"*. Just the same, his Queensborough Bridge painting does contain a house in the middle ground, and the interplay of light and shadow on walls, porches and roofs became a particular fascination and subject matter throughout his career (Hodge 2009).

The Queensborough Bridge was designed and constructed between 1901 and 1909 in an atmosphere of almost Shakespearean drama, during a period that included the controversial Commissionership of the New York City Bridge Department by Gustave Lindenthal. While design of the bridge is normally attributed to Lindenthal and the architect Hornbostel,

Figure 6.   Edward Hopper *Queensborough Bridge*, 1913, Whitney Museum of American Art.

Lindenthal was only Commissioner from 1901 to 1904, and eventually the bridge had, like the Washington Bridge over the Harlem River, many fathers, so to speak. Originally conceived as a double cantilever bridge with suspended spans, by Richard S Buck, Lindenthal's principal contribution seems to have been modification of Buck's concept to a continuous cantilever. The rational for this may have been aesthetic, as continuity allowed the top chord to follow a smooth curve, similar in shape to the catenary of a suspension bridge. The result, however, was to make the structure highly indeterminate, and the analysis must have required considerably greater effort.

The real excitement in 1913 though was the art exhibition held at the Armory of the New York National Guard at the corner of Lexington Avenue and Twenty—Sixth Street, something of an ironic choice given the explosive reaction that resulted from it. Originally intended as another show by a group of independent American artists, calling themselves the Association of American Painters and Sculptors, its initial organisers included members of the Eight, but they struggled to find a suitable venue and raise finance. In a masterstroke, Jerome Myers, who was, in his own words, a man *'who painted ashcans and the little people around them'* approached Arthur B Davies, the most anomalous member of the Eight. Of a Pre-Raphaelite disposition, Davies was, in the words of Myers *'a painter of unicorns and maidens under moonlight'* (Hunter 1973).

Importantly, Davies was well connected to a bevy of wealthy women patrons. Artistically conservative he nevertheless had an eclectic interest in, and sympathy for the emerging overseas trends in painting and sculpture. He changed the focus of the exhibition to include a survey of modern European Art. Even though about three quarters of the exhibition was still work by American artists, not everyone was happy with this strategy.

Davies sourced work from Picasso, Matisse, Kandinski, Piabia, Cezanne, Van Gogh and Gaugin, as well as the Impressionists and their earlier influences, such as Corot and Courbet. The exhibition was a sensation and created an enormous public reaction, especially *Nude descending a staircase* by Marcel Duchamp (1887–1968), and changed the way Americans viewed art forever.

7   THE FUTURE REVEALED

Ironically, Davies avoided the Futurists. He was not an admirer and they had been difficult to deal with (Hunter 1973). Futurism, originating in the 1909 Manifesto of Fillipo Tommaso Marinetti (1876–1944), was the only modern art movement to come out of Italy. To be a futurist was

Figure 7.   Joseph Stella, *Brooklyn Bridge*, 1919, Yale University Art Gallery, New Haven.

to believe that technology—as enjoyed at the beginning of the nineteenth century—had created a new kind of man, one who eschewed the past and who, instead, was in love with 'the beauty of speed'. The futurists professed a hatred of history and memory; rather they were enamoured with danger, energy, fearlessness, courage, audacity and revolt. Their icon was the racing car. When you worship at the altar of speed, the principal problem is developing a painting technique to portray movement, which—for the futurists—generally tended towards abstraction.

Brooklyn Bridge had been in place for almost four decades before Joseph Stella (1877–1946), an Italian who immigrated to the USA in 1896, used this style to give the bridge a truly memorable artistic voice. Stella had visited Paris and been strongly influenced by the Futurist's vision of reality. This, combined with his innate religiosity and appreciation of Italy's cultural heritage, particularly Dante's *Divine Comedy*, led him to view the bridge as a spiritual experience, *'the shrine containing all the efforts of the new civilisation'*. In this he was not alone. The Russian Futurist poet Vladimir Mayakovsky (1893–1930) walked onto the bridge *'as a crazed believer enters a church'*, and a poem about the bridge by his American contemporary, Hart Crane (1899–1932), contains the line *'O harp and altar, of the fury fused'*. So, not surprisingly, Stella's images of the bridge, including the one shown in Figure 7, convey its structural elements as a subjective, transcendental, psychological sensation—a stained glass like effect that presaged the future move to more abstraction in art (Rothwell 2011).

Here at last was Robert Hughes' expression of the Brooklyn Bridge as America's emotional clarion call to the world.

8   291 AND A DYNAMIC AWAKENING

While the Armory show on 26th Street awakened the general public to the latest trends in modern art, this was not its first exposure to New York's more discerning art fraternity. When the Eight began to question the conservative and restrictive orthodoxy of the Academy, and held their 1908 exhibition, other subversives were also on the move. But, while Henri and his *'gang'* were merely pursuing a new approach to realism, the other group, forming around the photographer Alfred Stieglitz (1864–1946), was pursuing an altogether different agenda.

In 1906, Stieglitz, and his friend, Edward Streichen (1879–1973), opened a small gallery at 291 Fifth Avenue (it quickly became known simply as 291), initially as a showcase for innovations in photographic expression. Soon, however, it garnered a reputation as a venue for painting and sculpture.

In 1908, the same year the Eight held their Macbeth Gallery show, Steiglitz held two small, but—in retrospect—remarkable exhibitions. Streichen, who spent part of each year in France, was a conduit through which Steiglitz obtained both bodies of work; the first a series of watercolours by the sculptor Rodin, and the second a collection of Matisse drawings. Of equal importance, Stieglitz began providing opportunities for aspiring artists, out of favour with the new realism, to exhibit their work.

One of these was John Marin (1870–1953), who commenced his working life as an architect, but eventually turned to art. At the Pennsylvania Academy he trained under William Merritt Chase (1849–1916), the American Impressionist and esteemed teacher, and befriended the *'brilliant colourist'*, Arthur Carles (1882–1952), who became a close companion when they both lived in Paris. Marin was in Europe between 1905 and 1911, producing some respected etchings of celebrated monuments such as Rouen Cathedral. During this time he also became an accomplished watercolourist and in 1909 held his first one-man show at 291. Viewing his watercolours the art critic Paul Haviland believed them to be *'the best examples of the medium which have ever been shown in New York'* (Hunter 1973).

In 1911 Stieglitz held two more boundary breaking exhibitions at 291, the first of Cezanne's late watercolors, and then later in the year early Cubist drawings and watercolours of Picasso. Marin had already shown some bias towards abstraction, what the critic Charles Caffin referred to as *'a spiritualised version of form and colour'*, and the effect of these two exhibitions propelled him towards what Sam Hunter has called *'a dynamic new graphic*

Figure 8.   John Marin, *Brooklyn Bridge* ca. 1912, The Metropolitan Museum of Art, New York.

style' (Hunter 1973), clearly on display in his subsequent etchings and watercolours of the Woolworth Building and Brooklyn Bridge (see Figure 8).

This was the culmination of more than a decade of study and experience for John Marin; the heightened sensation of the Fauves, the structural distortion of Cubism, and the lightness of Cezanne's watercolours, all brought together in a Dr Frankenstein like enlivening of the Brooklyn Bridge as a dynamic entity.

9 GOODBYE TO ALL THAT

Georgia O'Keeffe (1887–1986), a close associate of John Marin, was born near Sun Prairie in Wisconsin. Interested in art, as a young girl she eventually studied at the Art Institute of Chicago. Always keen to extend her knowledge she studied the influential teachings of Arthur Wesley Dow (1857–1922) and imbibed his theory of composition, massed colour and harmony. Inspired by his philosophy she began to experiment with abstract charcoal drawings, which came to the notice of Alfred Stieglitz. She held various art-teaching positions until 1918, when Stieglitz encouraged her move to New York and become an artist. They married in 1924, although his philandering eventually reduced it to a union of carefully avoided conflict, which from 1928 led O'Keeffe to spend summers painting in New Mexico.

Stieglitz died in 1946, and Georgia O'Keeffe spent the next three years in New York settling his estate, after which she left to live permanently in New Mexico. Now best known for her paintings of enlarged flowers, desiccated skulls and southeastern landscapes, she was also a skilled illustrator of the urban scene, as evident in her series on New York City skyscrapers and skylines painted from the Shelton Hotel between 1925 and 1928.

And she may have left the best until last. O'Keeffe had shown no previous interest in bridges as subject matter. After all her view, *East River from the 30th story of the Shelton Hotel* appears to avoid them. Yet, as she was about to quit New York, in 1949, she painted a poignant abstract of the Brooklyn Bridge, which seems to wrap Stella and Marin up into a neat Dowian package; her parting gift to the city, Figure 9.

Figure 9.   Georgia O'Keeffe. *Brooklyn Bridge*, 1949, Brooklyn Museum.

It had been just over 100 years since the young engineer Fayette Tower had sketched The High Bridge at Croton. To him the city would have been unrecognisable; the bridges immense and unimaginable, and the art incomprehensible. The modernists had triumphed and the seemingly unrelated worlds of the artist and the bridge engineer had marched into a new world arm in arm.

10 CONCLUSIONS

There are sinewy threads running through the paintings of New York City bridges produced in the first half of the twentieth century and the limited commentary the artists produced about the work and their motivations. Firstly, there is a sense of pride and wonder, and a desire to somehow capture that spirit. Then there is also the struggle to break free of the constraints imposed by traditional institutions, tempered by the internecine disputes of those suspicious of European influences, and those eager to embrace them.

Eventually, the technological daring of New York's bridges was matched by equivalent artistic radicalism.

REFERENCES

Dupré, J. 1998. *Bridges*, Köln, Germany: Könemann.

Gerdts, W.H. 1994. *Impressionist New York*, New York: Artabras.

Hodge, S. 2009. *How to Survive Modern Art*, London, Great Britain: Tate Publishing.

Hughes, R. 1980. *The Shock of the New*, London, Great Britain: British Broadcasting Corporation.

Hunter, S. 1973. *American Art of the 20th Century*, London, Great Britain: Thames and Hudson Ltd.

Hutton, W. 1889. *The Washington Bridge*, New York: Leo Von Rosenberg.

Kachur, L. 1983. The Bridge as Icon. *The Great East River Bridge, 1883–1983*, New York: Brooklyn Museum.

McCullough, D. 1972. *The Great Bridge*, New York: Simon & Schuster.

Pennell, J. 1916. *Catalogue of an Exhibition of Etchings by Joseph Pennell*, New York: Frederick Keppel & Co

Rothwell, S. 2011. The Bridge Painters. *Engineering World*, **21**(1), February / March 2011, 14–25. Crows Nest, NSW, Australia: Engineers Media.

Rothwell, S. 2013. Artists' Portraits of the Sydney Harbour Bridge. *Durability of Bridge Structures*, 265–274. London, United Kingdom: Taylor and Francis Group.

Smith, B. 1992. *Australian Painting 1788–1990*. South Melbourne, Australia: Oxford University Press.

Chapter 25

The Hell Gate Arch Bridge in New York City

K. Gandhi

Gandhi Engineering, Inc., New York, USA

ABSTRACT: The Hell Gate Arch Bridge of the New York Connecting Railroad was dedicated 100 years ago on March 9, 1917. When constructed, it was the longest arch bridge in
the world with a span of 997.5 ft between centers of bearings and 1017 ft between the faces
of abutments. The Chief Engineer of this project was Gustav Lindenthal and working under
him were Othmar H. Ammann and David B. Steinman, two future giants of long span bridge
engineering in the United States. The rivalry developed between them on this project continued for the rest of their careers. This paper describes the development of this project and the
design and construction of this monumental bridge.

1 NEW YORK CONNECTING RAILROAD COMPANY

Although the Hell Gate Bridge was opened to traffic in 1917, its idea was conceived in 1892
by the incorporators of the Connecting Railroad Company (Connecting RR), namely, Oliver
W. Barnes, Frank M. Clute, Alfred P. Boller, Charles MacDonald, and Thomas S. King. Boller
and MacDonald were successful bridge engineers and they needed financiers and people

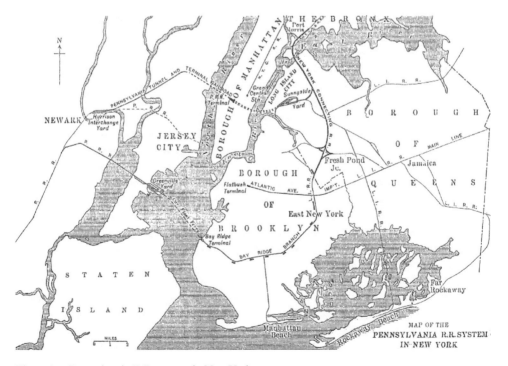

Figure 1.   Pennsylvania R.R. system in New York.

295

connected with the railroad industry to develop the project. The Hell Gate Bridge would allow the Pennsylvania Rail Company (PennRR) to travel to New England states which was not possible up until then.

In 1900, the PennRR acquired control of the Long Island Railroad (LIRR) and a connection with the LIRR across the East River and with the New England roads via the Connecting RR on the line originally mapped out, became vital along with the plan to enter Manhattan using two tunnels under the Hudson River.

In April 1902, the PennRR completed the purchase of the entire outstanding stock of the Connecting RR and a short time thereafter, as per a prior understanding with the New York, New Haven and Hartford Railroad Company (NY, NH & HRR), sold one-half of the stock to that company.

A map of the PennRR system in New York is shown in Figure 1 along with the route of the acquired Connecting RR.

2   VARIOUS DESIGNS OF THE HELL GATE BRIDGE (AMMANN 1918)

The first design prepared by Alfred P. Boller in 1900 (Figure 2) was a cantilever design with a central span of 840 ft supported on braced steel towers and carrying two tracks with open tie flooring and subjected to Cooper's E-40 loading.

In 1904, Gustav Lindenthal was selected by the PennRR to prepare different designs considering the ever-increasing weight of locomotives. He developed multiple designs until 1912 when permission was obtained to build the Hell Gate Bridge. Figures 3 through 7 show the evolution of the designs. The spandrel braced arch design developed by Lindenthal with the help of Consulting Architect Henry Hornbostel is shown in Figure 8. Lindenthal visualized

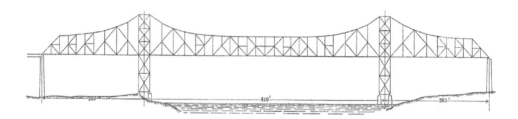

Figure 2.   Cantilever design (1900).

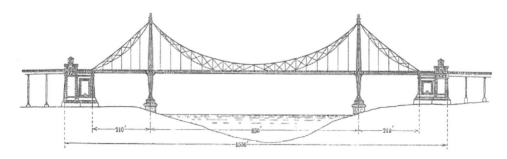

Figure 3.   Suspension bridge design (1904).

the bridge as a monumental portal for steamers entering New York Harbor from the Long Island Sound. The arch, flanked by massive masonry towers, was selected for construction by the PennRR and the NY, NH & HRR.

This arch span had four railroad tracks between its trusses and two highway tracks on brackets outside. The span distance between centers of skewbacks was 977.5 ft. The distance between near sides of the tower piers at the coping was 1015 ft. The distance between the centers of the trusses was 60 ft and the distance between outside railings was 93 ft. The bottom of the floor system would clear the river at high tide by 135 ft and the center of the top chord by 307.5 ft. The span was divided into 23 panels of 42.5 ft each and the arch was of the two-hinged type.

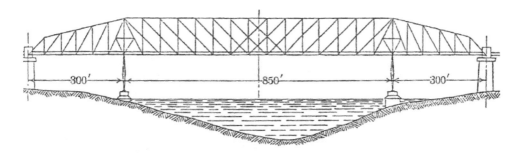

Figure 4.   Continuous truss design (1904).

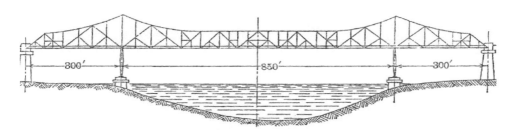

Figure 5.   Cantilever design (1904).

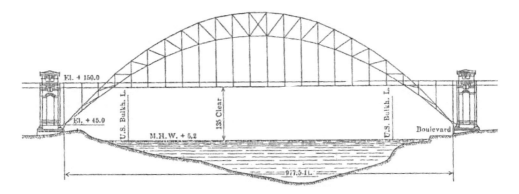

Figure 6.   Crescent arch design (1905).

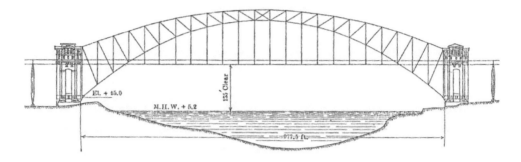

Figure 7.　Spandrel braced arch design (1905).

Figure 8.　Perspective view of final design (1912), Hell Gate Bridge.

3　MATERIAL (AMMANN 1918)

All the material in the steel superstructure of the Hell Gate Bridge and its approaches was rolled, forged, or cast steel made by the open-hearth process. The following four grades of steel were used:

1. Hard steel for all rolled parts and pins of the Hell Gate Arch bridge,
2. Structural steel for all rolled parts and pins of the approaches,
3. Rivet steel for all rivets, and
4. Cast steel for all castings for bearings

The different grades had to conform to the chemical and physical requirements in Table 1.

Table 1. Chemical and physical requirements for steel.

	Hard steel	Structural steel	Rivet steel	Cast steel
Phosphorus, max. (basic)	0.04	0.04	0.04	0.05
(acid)	0.06	0.06	0.04	0.08
Sulphur, max.	0.05	0.05	0.04	0.05
Ultimate tensile strength [lb/in^2] (max.)	76,000	70,000	58,000	
(desired)	71,000	66,000		
(min.)	66,000	62,000	50,000	33,000
Yield point, min.	38,000	35,000	28,000	33,000
Elongation, min. (2 in for cast steel)	*	22%	28%	20%
(8 in for other steel)	*	22%	28%	20%
Character of fracture	Silky	Silky	Fine, silky	Silky/fine granular
Cold bend without fracture	**	180° around pin of thickness of test piece	180° flat	90° around pin of thickness of test piece

*Minimum elongation for "hard steel": 1,400,000 divided by ultimate strength for thickness up to and including 0.75 in; 1% less for each additional 0.125 in in thickness, with a limit of 16% for thickness up to and including 2 in and 15% for thickness greater than 2 in.
**Cold bend test for "hard steel": 180° around a pin of double the thickness of the test piece for material up to and including 0.75 in, 180° around a pin 3 times that thickness for material greater than 0.75 in thick.

4 LOADS AND UNIT STRESSES (AMMANN 1918)

4.1 *Dead load*

The actual dead load varied from 45,000 lb/ft at the center to 62,000 lb/ft at the ends. The average dead load used for the final calculation was 51,000 lb/ft.

4.2 *Live load*

The following live loads were considered:

1. Cooper's E-60 loading was assumed for each of the four tracks, or an alternative three-axle load of 70,000 lb on each axle wherever this caused greater stress.
2. The arch trusses for a uniform load of 6000 lb/ft of track, or 24,000 lb/ft of bridge, placed in the most unfavorable position in either a single stretch or in two separate stretches, when the latter condition gave a greater stress.

4.3 *Impact*

The impact stresses or vertical dynamic effects of the locomotives and cars were determined according to Lindenthal's formula. This formula is provided in the paper by Ammann (1918).

4.4 *Lateral forces from live load*

The lateral force of lateral impact due to the swaying motion of fast-moving trains on tangents or the centrifugal force on curves up to 2 degrees was assumed at 600 lb/ft of single track, Figure 9. For the four-track bridge on tangent, a total lateral force of 1500 lb/ft was used.

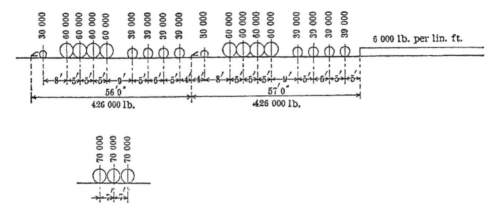

Figure 9.   Diagram of assumed live load.

4.5   *Wind pressure*

A wind load of 4600 lb/ft of bridge was considered for the design of the Hell Gate Bridge.

4.6   *Longitudinal force from traction and braking*

The longitudinal force acting along the rail from traction and braking was utilized either at 5000 lb for each of the eight driving axles of the two locomotives (125% of the load on each driving axle), or at 1000 lb/ft of train (approximately 15% of the average weight of the train), whichever gave the greater results.

4.7   *Temperature stresses*

The stresses from temperature were determined for a variation of ±72°F from the normal temperature of 60°F.

4.8   *Total stresses*

Various combinations of stresses were selected to determine the total stress.

4.9   *Permissible unit stresses*

The permissible unit stresses used for the Hell Gate Bridge are given in Table 2.

4.10   *Secondary stresses*

Care was taken in designing and detailing all the bridges to avoid large secondary stresses, and where this was not possible, cross-sectional areas of some of the members were increased. It was expected that most of the secondary stresses would be covered by the factor of safety.

4.11   *Erection stresses*

The erection stresses were well within the safe limits allowed for the total stresses in the completed bridge.

Table 2. Permissible unit stresses assumed.

		For trusses & Bracing of Hell Gate Bridge (Hard steel) in lb/in²	For approach spans & Floor system & Suspenders of Hell Gate Bridge (Structural steel) in lb/in²
Axial tension, net section		24,000	20,000
Bending on extreme fiber of beams, girders, and steel castings, net section		–	20,000
Axial compression, net section:			
a) Closed section, or section with two diaphragms, or one diaphragm and two planes of latticing	l/r = 20	24,000	20,000
	40	23,000	19,000
	60	22,000	17,000
	80	20,000	15,000
	100	18,000	14,000
	120	15,000	12,000
b) Half-open section with one cover and one latticing, or with one diaphragm without latticing	l/r = 20	23,000	20,000
	40	22,000	18,000
	60	20,000	16,000
	80	18,000	14,000
	100	16,000	13,000
	120	14,000	11,000
c) Open section, with two or more planes of latticing	l/r = 20	22,000	19,000
	40	20,000	17,000
	60	18,000	15,000
	80	16,000	13,000
	100	14,000	12,000
	120	12,000	10,000
Shearing stress:			
On plate girders, net section		–	15,000
Shop rivets and pins		15,000	–
Field rivets and turned bolts		12,000	–
Bearing stress:			
Pins		24,000	20,000
Shop rivets		30,000	–
Field rivets and turned bolts		24,000	–
Pressure on:			
Expansion rollers per linear inch		Diameter of roller, in inches, X 1000	
Granite masonry		800	800
Concrete masonry		600	800

5   FABRICATION OF THE HELL GATE BRIDGE

The fabrication of the 977.5 ft steel arch started at the Ambridge, PA plant of the American Bridge Company early in 1914. The fabricating process on this steelwork was unusual because of two specific requirements (Engineering News, 1914):

1. Practically all the holes needed to be drilled because punching was not possible for thicker plates, and
2. The structure had to be assembled in the yard in a horizontal position, with the field connections drilled to match before the parts were shipped to the site.

There were 840,000 shop rivets and 334,000 field rivets, of which 400,000 were 1.25 inches in diameter. Most of the 1 and 1.25-inch shop rivets were driven with hydraulic riveting machines having a pressure capacity of 100 tons (100 psi). The points of the long rivets were dipped in water after heating to ensure a more complete upsetting of the shank before the head was formed.

The main arch trusses were completely assembled on skids in the fabrication shop yard. Due to limited space, for the full length of the bridge the trusses were assembled in four panel sections. The last panel of each section stayed in place for the first panel of the next section, thereby assuring total continuity of construction (Engineering Record, 1914).

All field joints of the trusses were reamed in the yard by four portable electric reamers handled by gantry cranes. The heaviest finished pieces were shipped from the site on special cars of 150,000 lb capacity specifically built for this project.

Figure 10 shows one of the end shoes of the arch. Each shoe built up of steel castings with 4-inch metal thickness weighed 50,000 lb. In as much as it sits on a masonry seat inclined about 45° to the horizontal, provision was made for anchoring it in place until the arch was so far erected that the thrust of the structure would hold the shoe securely to its seat.

The end casting of the arch had a slightly convex cylindrical face to give hinge action. A dowel was connected to the hinge castings to prevent displacement transversely.

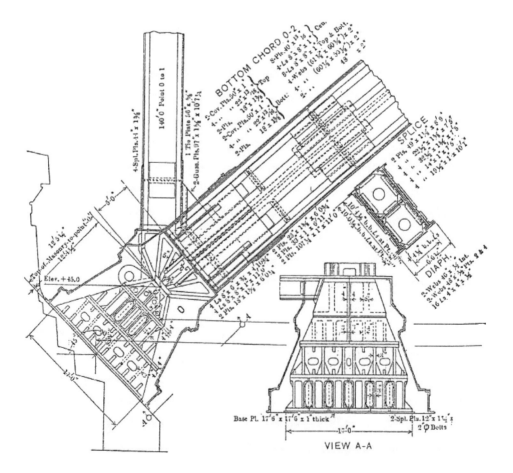

Figure 10.   Details of panel point O and bearing.

6   ERECTION OF HELL GATE BRIDGE

The erection plan was clearly thought out by Lindenthal. The river conditions ruled out the use of falsework except for a very short distance from each abutment. Therefore, erection on the cantilever principal with the use of temporary backstays (acting as counterweights) was adopted by the American Bridge Company. The bridge members planned for the subsequent construction were used as backstays to reduce the material handling twice. The total weight of steel in the backstays for both sides came to 15,500 tons of which about 2300 tons were not utilized in the permanent structure. Figure 11 shows the progress of the erection of the Hell Gate Bridge (Ammann 1918).

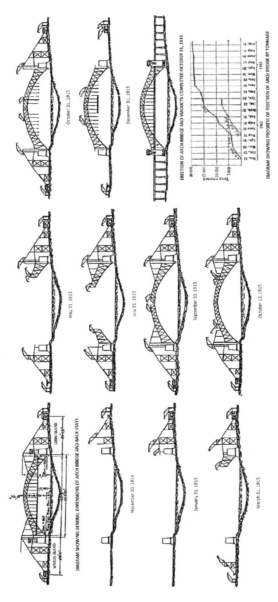

Figure 11   Progress diagram of erection of Hell Gate Bridge (Ammann 1918).

Adjustment of the arch trusses in height was required at various erection stages. For this purpose, four 3000-ton capacity hydraulic jacks were placed one on top of each of the four erection posts. There were multiple 500-ton capacity local jacks used to assist during the erection process.

Erection of the Hell Gate Arch is covered by Whitney (1915), Parsons (1915), Railway Review (1915), Skinner (1919), and Engineer (1915). The closing of the trusses at the center by jacking transformed the two individual cantilevers into a three-hinged arch. The transformation of the trusses from three-hinged into two-hinged arches required the connection of the top chord and one of the diagonals of the center panel of each truss at 60°F. These members were erected immediately after the closing of the arch, but had been left bolted at one end and free to move at the other.

The connection at the free end required drilling the rivet holes from the solid sections and riveting. This took several days. The bottom chords were fully riveted at this third hinge point and from there on the bridge had two hinges, one at each skewback.

The 1907 Quebec bridge disaster was in the minds of each of the 150 workers who erected the Hell Gate Bridge. The total construction cost of the project was about $30 million. The rule of thumb during those days was the loss of one life for every $1 million of construction cost. Although not perfect, the safety record of the Hell Gate Bridge was very good because only five lives were lost during construction.

7   STRESS MEASUREMENTS ON THE HELL GATE BRIDGE

Considering the scale and magnitude of the Hell Gate Bridge, Lindenthal decided to use his own money to measure stresses in the bottom chord at various stages of erection until the bridge was completed. He did this as scientific research. The details of his work are summarized here based on the paper by Steinman (1918).

The instrument used for the stress measurement was a 20-inch strain gage designed by James E. Howard. This instrument was essentially a micrometer caliper with an accuracy of 0.0001 inch.

The bottom chords of the arch had a double rectangular section (Figure 12) consisting of two compartments separated by a horizontal diaphragm. It was decided to take six readings at each cross-section, four at four corners and two at mid-height in the vertical webs. The measurements were taken during the first 12 stages of construction as shown in Figure 13.

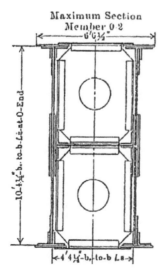

Figure 12.   Bottom chord member (Ammann 1918).

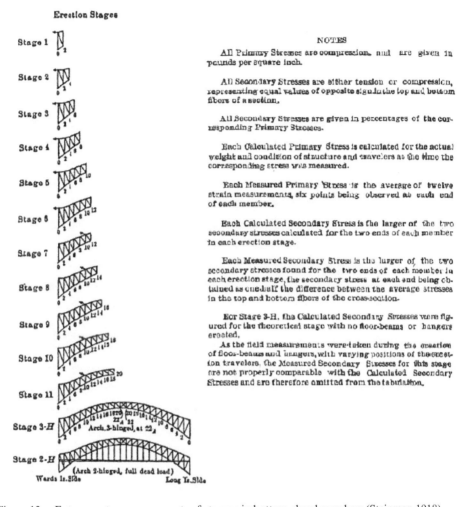

Figure 13.   Extensometer measurements of stresses in bottom chord members (Steinman 1918).

The measurements were limited to dead load only, and were stopped by Lindenthal due to lack of funding. At the end of his paper, Steinman provided a 10-point summary which compared the calculated and measured stresses. He considered the stress measurement a success because in most cases the calculated secondary stresses were lower than the measured ones, and where they exceeded, they were covered by the safety factor. This was the first major bridge where the stress measurements were experimentally carried out and, in most cases, the calculated and the measured stresses differed by less than ten percent.

8   GUSTAV LINDENTHAL, 1850–1936 (FRANKLAND, 1940)

Gustav Lindenthal (Figure 14) was born in Brunn, Austria on May 21, 1850. He was educated at Politechnicum College in Dresden, Germany and received practical training from 1866 to 1870. He came to the U.S. in 1874. Lindenthal worked at the 1876 World's Fair in Philadelphia as a laborer and then as a designer. In 1878 he joined the Atlantic Great Western Railroad as a bridge engineer. In 1881, Lindenthal started his own engineering practice in

Figure 14.   Gustav Lindenthal (Wildman 1921).

Pittsburgh and found assignments in the design and construction of important bridges for railroads and bridge companies.

He received recognition and very good publicity for replacing the Smithfield Street suspension bridge, originally built by John A. Roebling, with a 350 ft (106.7 m) span double-elliptical steel truss in 1882. This, and several other projects in Pittsburgh, brought him to the attention of Samuel Rae of the Pennsylvania Railroad who later supported Lindenthal in the proposal for the Hudson River Bridge. It was Lindenthal's paper in 1888 presented at the American Society of Civil Engineers convention, "The North River Bridge Problem, with a Discussion on Long Span Bridges," that excited not only the engineering community, but the general public as well.

Lindenthal spent his entire career in private practice except for a controversial two-year period (1902–1903) when he was appointed the Bridge Commissioner of New York City by Mayor Seth Low (Gandhi 2013). The most notable project of his professional practice was the Hell Gate Bridge in New York. On the Hell Gate arch project, Lindenthal hired Othmar H. Ammann, David. B. Steinman, and Charles S. Whitney, who would later become world-renowned engineers in their own right.

He combined his love of beauty in engineering works with his search for a structural solution. In his search for aesthetic design, he did not hesitate to consult architects whenever he had to deal with an important bridge project, such as the Hell Gate Bridge. Lindenthal died on July 31, 1935 at age 86.

9   RELATIONSHIP BETWEEN AMMANN AND STEINMAN

Ammann and his wife Lilly met Mr. and Mrs. Lindenthal at a social event in Philadelphia in 1910 (Rastorfer 2000). The Ammanns and the Lindenthals become friends over the next two years, even though there was an age difference of about 30 years between the two men. During that period, Lindenthal was designing the Hell Gate Bridge for the PennRR.

Lindenthal was authorized with construction of the Hell Gate Bridge in early 1912 and Ammann joined him as his assistant in June 1912. As the second in command, Ammann was responsible for about 2.5 miles of bridges, viaducts, trestles, embankmank structures, and the Hell Gate Bridge which carried four tracks and connected Queens to Wards Island.

Ammann was reliable, meticulous, and took his responsibilities seriously. He also kept the project on track and within budget. Lindenthal depended on Ammann, with whom he had a good working relationship due to their similar backgrounds and natural chemistry, for the success of the Hell Gate Bridge construction.

In August of 1914, Ammann was still a Swiss citizen and a reserve officer of the Swiss army. The news that the German army had taken position on the banks of the Rhine River across from Basel worried Ammann as his oldest son Werner, parents, and a brother were living in Basel at that time. Ammann left for Switzerland on August 6, 1914.

Ammann's abrupt departure created a major problem for Lindenthal who needed to fill his position with a suitable candidate who would be familiar with the project and equally capable as Ammann. Lindenthal promoted David B. Steinman to the position of First Assistant. Steinman was a brilliant engineer and mathematician and had earned his Ph.D. in Civil Engineering from Columbia University at the age of 24.

As it turned out, the war did not start on the Swiss border, and Ammann was released from the Swiss army in three months. Ammann returned to the U.S. with his son Werner in December 1914. Lindenthal immediately reinstalled Ammann as his First Assistant, and demoted Steinman to his former title. This did not sit well with Steinman and a bitter rivalry between the two brilliant engineers continued for the rest of their professional careers. According to Rastorfer, "the mention of Steinman's name was strictly forbidden in the Ammann household."

Figure 15 shows Lindenthal with his staff during the construction of the Hell Gate Bridge. Ammann is standing to right of Lindenthal wearing a felt hat and sporting a mustache. Steinman is fourth from the left in this photo.

Figure 15.   Gustav Lindenthal with staff (Rastorfer, 2000).

After the completion of the Hell Gate Bridge, Ammann wrote a paper on the design and construction of the Hell Gate Bridge and Steinman wrote a complementary paper on the stress measurements carried out by him on the project. Both papers were published in 1918 in the Transactions of the ASCE Volume 82.

Steinman acknowledges in his paper the help he received from Ammann "for suggestions during the prosecution of the investigation." It seems odd that Ammann, who was the supervisor of Steinman, would comment on Steinman's work in public by saying that in his opinion, the conclusion reached by Steinman "relative to secondary stresses were somewhat too far reaching," or in other words, "speculative."

Steinman, in his response to Ammann's criticism, politely asked Mr. Ammann "to point out anything in the summary of conclusions which can possibly be regarded as too far reaching from the results of the investigations."

10 RELATIONSHIP BETWEEN AMMANN AND WHITNEY

On the Hell Gate Bridge project, Ammann was Assistant Chief Engineer under Lindenthal and was in direct charge of the office, field, and inspection work. He had a staff of 95. Charles S. Whitney was one of the engineers who had received a Master's degree in Civil Engineering from Cornell University in 1914 and had joined the Hell Gate Bridge project as an inspector.

In 1942, Whitney wrote a paper on "Plastic Theory in Reinforced Concrete Design," which was published in Vol. 107 of the Transactions of the ASCE. His paper was well received as it simplified the ultimate strength design of reinforced concrete. To most engineers, he is well known for his "Whitney Stress block" which was adopted by the American Concrete Institute in 1963.

In 1946, Ammann formed a partnership with Whitney and the firm became known as Ammann and Whitney with a worldwide reputation in the design of suspension bridges, concrete shell structures, and special structural engineering projects.

11 CONCLUSIONS

From an engineering point of view, the Hell Gate Bridge project was a great success because it was the longest arch bridge in the world when completed. This was also the first time stresses were measured and compared with the calculated values for a major bridge. The agreement between the two within the permissible limits of experiments ($\pm 10\%$) gave confidence in the design approach adopted by Lindenthal and his assistants, Ammann and Steinman.

From a financial point of view, the New York Connecting Railroad, with its Hell Gate Bridge and few miles of track in Queens, was a disaster. It had cost $100 million to build. With the decline of railroad traffic and the increase in roadway traffic, it was certain even in 1917 that the net earnings from the passenger and freight business over the Hell Gate Bridge would never pay even a fraction of the interest charges on its cost (Railway Age Gazette 1917).

The author believes that with the experience and confidence gained, the rivalry developed between Ammann and Steinman on this project brought out the creativity in both of them during their professional careers and expanded the limits of long span bridges beyond most of the engineers' imagination.

ACKNOWLEDGEMENTS

The author thanks Kunal Kothawade, Livia Bennett, and Annie Sidou of Gandhi Engineering for their assistance during the preparation of this paper.

REFERENCES

Ammann, O.H. 1918. The Hell Gate Arch Bridge and Approaches of the New York Connecting Railroad over the East River in New York City. *Transactions, American Society of Civil Engineers.* V82: 852–1039.

Engineer 1915. Hell Gate Bridge, New York. 20(22): 495–497.

Engineering News 1898. Another East River Bridge. 39(10): 168.

Engineering News 1914. The Hell Gate Bridge in the shop. 72(23): 1116–1118.

Engineering Record, 1914. Fabricating Steelwork for the Hell Gate Arch. 70(26): 684–686.

Frankland, F.H. & Schmitt, F.E. 1940. Memoir of Gustav Lindenthal. *Transactions of the American Society of Civil Engineers.* 105: 1790–1794.

Gandhi, K. 2013. Lindenthal and the Manhattan Bridge eyebar chain controversy. In Khaled M. Mahmoud (ed.), *Durability of Bridge Structures.* CRC Press 2013: 285–300.

Parsons, Walter J. 1915. Methods and Equipment used in Erection of Hell Gate Arch over East River, New York City. *Wisconsin Engineer*, 20(3): 97–105.

Railway Age Gazette 1917. New York a Way Station. 62(14): 727–728.

Railway Review 1915. Hell Gate Arch Bridge and the New York Connecting R.R.: 453–461.

Rastorfer, Darl 2000. Six Bridges, The Legacy of Othmar H. Ammann, New Haven: Yale University Press: 7–11.

Skinner, Frank W. 1919. Hell Gate Bridge, New York. *Engineering* 108(16): 499–504.

Steinman, D.B. 1918. Stress Measurements on the Hell Gate Arch Bridge. *Transactions, American Society of Civil Engineers.* V82: 1040–1137.

Whitney, C.S. 1915. The Erection of the Hell Gate Arch, *Cornell Civil Engineer.* 24(3): 84–92.

Wildman, Edwin 1921. *Famous Leaders of Industry 2nd Series.* The Page Company (Boston): 339.

Chapter 26

The failure and reconstruction of the Quebec Bridge

K. Gandhi

Gandhi Engineering, Inc., New York, USA

ABSTRACT: At the time of its construction, the 1800 ft span cantilever bridge across the St. Lawrence River in Quebec, Canada was going to be the longest cantilever bridge in the world. However, on August 29, 1907, during its erection, the bridge collapsed killing 75 workers. A commission of prominent international engineers was formed by the Canadian Government to investigate the collapse of the Quebec Bridge. It was decided to build a new, but much heavier and stronger, cantilever bridge adjacent to the old failed bridge. On September 11, 1916, after the center span was raised successfully 12 to 15 feet, it suddenly fell into the St. Lawrence River killing eleven workers and injuring six. The St. Lawrence Bridge Company, which was erecting it, took full responsibility for the collapse of the second bridge and placed orders for the new steel. The new center span was successfully hoisted for the third time and put in place on September 18, 1917 using the same lifting procedure that was used in 1916. The new bridge was opened to traffic 100 years ago, on December 3, 1917. This paper provides the details of the old and new bridges, and the people connected with them.

1 INTRODUCTION

When a survey among bridge engineers in Europe, Canada, and the U.S. was made by Major Charles Bebe Stewart in 1846, only four engineers believed that it was possible to build a bridge across the Niagara Gorge, and they were John Roebling, Charles Ellett, Jr., Samuel Keefer, and Edward Serrell, (Gandhi 2006). Serrell was retained by the Corporation of the City of Quebec in Canada in 1851 to ascertain the feasibility of "throwing" a bridge over the St. Lawrence River. He made an examination and reported that a suspension bridge across that river was perfectly practical from a "scientific" point of view at about 6 miles above Quebec. The span would have been 1600 ft and the roadway would have been 100 ft above the water (Scientific American 1851).

Figure 1 from Middleton (2001) shows seven possible bridge locations investigated in the 19th century. The location selected by Serrell was at the St. Lawrence River's confluence with the Claudiere River (Location 1), the same location decided about 50 years later to build a bridge.

In 1887, the Quebec Bridge Company was incorporated by an Act of Parliament to build and operate a railway and highway bridge across the St. Lawrence River. Further Acts of Parliament extended the time for the construction of the bridge in 1891, 1897, and 1900. In 1903, the name was changed to the Quebec Bridge and Railway Co. and the government undertook to guarantee the bonds of the company up to $6,678,000 against conveyance of the property. The time for completion was fixed to July 1910.

In 1887, the government engineers of the Province of Quebec produced a design for a bridge where the river was about 2400 ft wide but was too deep for piers in the center. The design consisted of two granite piers to be built at a distance of 500 ft and 240 ft from the shores in about 40 ft of water with the cantilever ironwork to be built on them. The dimensions of the bridge were as follows (Railroad Gazette 1887):

Length of center (cantilever) span:	1442 ft
Length of shore spans:	487 ft
Total length of bridge and approaches:	3460 ft
Height from high water mark to bottom of bridge:	150 ft
Extreme height of top of cantilever above high water:	408 ft

In this design, the center span was 270 ft shorter than the 1710 ft single cantilever span of the Forth Bridge near Edinburgh, Scotland, the world's longest at the time.

To avoid founding the two piers supporting the cantilever span in deep water, numerous borings were made which indicated that while solid rock was beyond a depth feasible to reach, suitable material for foundation could be found well within the practicable limits of pneumatic work by locating the main piers near the shores. In 1900, Theodore Cooper, the Consulting Engineer to the Quebec Bridge Co., increased the length of the cantilever span to 1800 ft which would make the Quebec Bridge the longest cantilever span in the world when completed. He approved the stress sheets for the suspended and anchor spans in 1904. One-half elevation of the Quebec Bridge symmetrical about the centerline is shown in Figure 2 after the revisions made by Cooper.

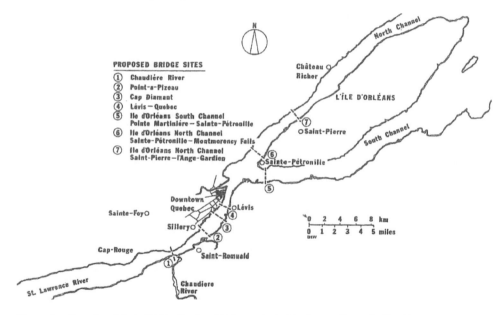

Figure 1.   Location 1 to build the Quebec Bridge, oriented in the north-south direction.

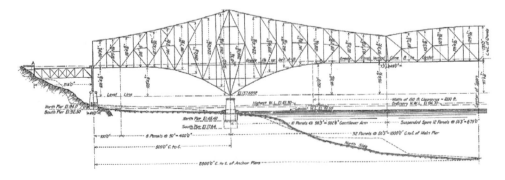

Figure 2.   Half diagram of the Quebec cantilever bridge.

The contract for the superstructure of the bridge was awarded to Phoenix Bridge Co. of Phoenixville, PA. and had an estimated weight of about 40,000 tons. The bridge was designed to carry two railroad tracks, two trolley tracks, and two roadways between the trusses on the single deck, and two sidewalks cantilevered outside the trusses. The trusses were placed 67 ft apart center to center. Mr. E.A. Hoare was Chief Engineer of the Quebec Bridge Co.

2   SUBSTRUCTURE WORK

Work on the Quebec Bridge was formally inaugurated on October 2, 1900 when Sir Wilfred Lawrence laid the cornerstone of the first abutment pier (Engineering News 1900b). The contract for the substructure was awarded to Mr. M.P. Davis of Ottawa. Mr. Davis estimated the quantity of masonry at 50,000 cubic yards (CY).

Besides the two abutments, one on the Quebec (north) and the other on the Levis (south) side, there were two anchor piers and two river piers, the latter supporting a span of 1800 ft. From each abutment to its neighboring pier the distance was 214 ft and the anchor spans were 500 ft each. The total length of the bridge from abutment to abutment was 3228 ft.

Mr. Davis planned to use 5000 CY of masonry before closing his operation for winter on November 15, 1900. He had until October 1902 in which to get the substructure built. In May of 1901 he planned to start work on one of the inner piers. During the winter of 1900 to 1901, he planned to build pneumatic caissons 168 ft (L) × 50 ft (W) × 50 ft (H) to be launched when required.

The abutments were U-shaped in plan and measured 80 ft at right angles to the bridge. Each of the two wing walls were 40 ft long. The wing walls were founded on solid rock, and together contained about 4,000 CY of masonry. Details of the south anchor pier and wooden caissons supporting the south main pier are excellently covered in Engineering News (1903). Figure 3 shows the plan and elevation of the Contractor's plant for south shore abutment and piers (Engineering News 1903).

The north caisson was launched on June 20, 1901 and the south caisson on May 26, 1902. The sinking of the south caisson was begun on June 7, 1902 and finished on October 17, 1902

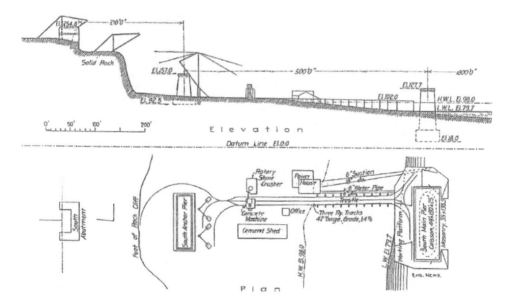

Figure 3.   Plan and elevation of contractor's plant for south shore abutment and piers.

thus requiring 131 days to sink 59 ft, or at an average sinking of 5.4 inches per day, varying from a minimum of 2 inches to a maximum of 10 inches. The number of men employed at the bridge site varied from 500 to 600. The construction plant for the north shore abutment and piers was in its general features a duplicate of that for the south shore work. In fact, the machinery and materials of the north shore plant were largely used in constructing the south shore plant.

3 SUPERSTRUCTURE

The superstructure consisted of pin connected members. In general, the eyebars were 15 and 16 inches in width and, for a few special members, 18 inches in width. The pins were 12 inches in diameter and the main lower pin at shoe was 24 inches in diameter. The main chords were 54 inches in depth and 68 inches in width by 4 ft deep. The main intermediate posts were from 40 to 48 inches in width and the main plate floor beams were 10 ft deep. The suspended span was 675 ft long and 130 ft deep at the center. Figure 4 shows a view of pedestals and shoes with main post and diagonals connected (Engineering News 1905).

   Figure 5 shows the south arm of the cantilever bridge on August 28, 1907, the day before its fall (Engineering News 1907).

Figure 4.   View of pedestals and shoes with main post and diagonals connected.

Figure 5.   South cantilever arm of the St. Lawrence Bridge at Quebec.

4   FALL OF THE PARTIALLY COMPLETED BRIDGE

On Thursday, August 29, 1907 at about 5:30 PM, 15 minutes before quitting time, there was a loud noise and without warning the river end of the south side of the cantilever slowly began to sink. When it was nearly down to the water's edge, the main traveler tower became unstable, broke its anchorage, and revolved towards the north shore. The 315 ft vertical posts over the main pier collapsed and the whole superstructure went down along with 86 men on the structures, 75 of whom were killed (Skinner 1907).

The trusses fell almost vertically and were terribly wrecked and mangled. The anchor span moved, longitudinally about 100 ft towards the river with its top chord almost intact, and except one eyebar which was sheared off, few of the eyebars suffered bending and twisting. The principal failures of the compression members were at the bottom chord splices and where latticing had yielded.

The substructure remained absolutely intact and uninjured except for small scars on the coping and where it had been chipped by the sharp corners of the steel members.

The collapse of the superstructure was due to the movement of the anchor arm trusses in their own plane about 100 ft towards the river, the vertical way they fell, and the fact that the buckled condition of the two lower chord panels, which had been reported buckled three days earlier, established the possibility of their failure.

The coroner's jury examined some witnesses and rendered a verdict which failed to fix the responsibility or cause of the disaster and left it to be investigated by a government commission.

5   REPORT OF THE ROYAL COMMISSION ON THE QUEBEC BRIDGE FAILURE

The Royal Commission was formed in 1907 and was composed of:

1. Mr. Henry Holgate, Civil Engineer, Montreal, Canada,
2. Mr. John G.G. Kerry, Civil Engineer, Toronto, Canada, and
3. Professor John Galbraith, Toronto, Canada.

The Royal Commission submitted its report to the Canadian Parliament on March 9, 1908. The single most important finding of the Commission was that "the bridge fell because the latticing of the lower floors near the main pier was too weak to carry the stresses to which it was subjected." The Commission further concluded that "although the lower chords 9-L and 9-R anchor arm, which in our judgement were the first to fail, failed from weakness of latticing, the stresses that caused the failure were to some extent due to the weak end details of the chords and to the looseness, or absence of, the splice plates arising partly from the necessities of the method of erection adopted and partly from a failure to appreciate the delicacy of the joints and the care with which they should be handled and watched during erection." The Royal Commission reported fifteen other conclusions that are given in Engineering—Contracting (1908).

The appendices to the report included information of interest to engineers. Appendix 13 contained a "Summary of Tests of Large Columns" (Engineering News, 1908a). The results of 176 tests were plotted and 10 conclusions were drawn. The last conclusion stated that "no tests have been made on columns of the form of the Quebec lower chords nor on any having more than about 1/25th of the cross-section of these chords."

Appendix 15 of the Quebec Bridge Commission's report described the tests of two compression chord models. One of the two test members was an exact model of chord 9 of the anchor arm which failed and the other was similar, differing only in having been strengthened at those points where the main weakness of the Quebec Bridge chord was thought to have been located. While the first test chord failed at a load just above that which caused the failure of the Quebec Bridge chords, the second test member failed at a load 38% higher than the first member (Engineering News, 1908b).

Commenting on Theodore Cooper who was the Consulting Engineer for the Quebec Bridge, Middleton (2000, p. 102) writes, "It was a role he was, in many ways, ill-equipped to fill. Poorly paid for his consulting duties, he could not afford the staff he needed to assist him. In ill health, he never once visited the bridge site after erection of the structure had commenced, and he was forced to rely upon letters, sketches, and photographs to understand the problems and questions that were presented to him for decision. Separated from the site of the work by almost 600 miles and subject to the vagaries of wire and postal communication of the time, Cooper was ill situated to provide the prompt and decisive action that the crisis of late August demanded." Cooper's illustrious career came to an abrupt and sad end because of the Quebec Bridge disaster.

6   REPORT OF SCHNEIDER ON THE DESIGN FOR THE QUEBEC BRIDGE

While the Royal Commission was conducting its investigation, the Government retained the services of Charles Conrad Schneider, a past president of the American Society of Civil Engineers (ASCE) and a well-known bridge engineer, to determine the sufficiency of the design for the original Quebec Bridge. The findings by Schneider are summarized below (Engineering News 1908):

1. The floor system and bracing are of sufficient strength to safely carry the traffic for which they were intended.
2. The trusses, as shown in the design submitted to the writer, do not conform to the requirements of the approved specifications, and are inadequate to carry the traffic or loads specified.
3. The latticing of many of the compression members is not in proportion to the sections of the members which they connect.
4. The trusses of the bridge, even if they had been designed in accordance with the approved specifications, would not be sufficient strength in all their parts to safely sustain the loads provided for in the specifications.

5. It is impracticable to use the fabricated material now on hand in the reconstruction of the bridge.
6. The present design is not well adapted to a structure of the magnitude of the Quebec Bridge and should, therefore, be discarded and a different design adopted for the new bridge, retaining only the length of the spans in order to use the present piers.
7. The writer considers the present piers strong enough to carry a heavier structure, assuming that the bearing capacity of the foundations is sufficient to sustain the increased pressure.

7 THE NEW BEGINNING

Exercising its rights under the 1903 legislation, the Canadian government took over the Quebec Bridge and Railway Co. and in August 1908 appointed a Board of Engineers to oversee the construction of the new bridge. The three international board members were:

1. Henri Etienne Vautelet, Chairman and Chief Engineer, Canada
2. Ralph Modjeski, Member, USA
3. Maurice Fitzmaurice, Member, England

The board checked the feasibility of a suspension bridge to occupy the place of the canti-lever bridge that had collapsed. The board members had differences of opinion as to the form, shape, and location of the new bridge. It was agreed to replace the old bridge with a new canti-lever bridge. As the new bridge was going to be 21 ft wider and weigh almost twice as much as the old bridge, it was evident that little or none of the existing substructure could be utilized.

In June 1910, Fitzmaurice resigned from the Board. He was replaced by Charles MacDon-ald, a well-established and well-known bridge engineer and former President of the ASCE. He was born in Canada, but moved to the United States at an early age.

In order to reduce the dead loads of the main members, the Board initiated tests on nickel-steel which permitted higher stresses compared to the mild steel (Engineering News, 1911b). The first series of tests covered a large number of riveted splices, both lap and butt, with dif-ferent thicknesses of plate packing, for comparison with prior tests of mild steel rivets. These tests were conducted at the University of Illinois at Urbana. The second series of teste were made of models of the principal compression members designed for the bridge. These models were tested at Phoenix Iron Co. in Phoenixville, PA.

The results of the first series of tests indicated that the ultimate strength of the rivets proved to be remarkably constant and the values ranged from 55,200 to 60,250 lbs per square inch (psi) in shear area. The results of a very large number of repetitions or reversals of stress proved conclusively that stress reversals produced very marked movements in the joints pos-sibly due to altered stress distribution. The results of the second series of tests on models showed that the elastic limits of the columns ranged from 37,000 to 45,600 psi. The ultimate strength ranged from 48,800 to 64,000 psi. The columns failed at loads slightly below the elastic limit of the tensile tests.

The removal of the superstructure weighing about 17,000 tons was undertaken to clear the channel for the construction of the new bridge. Oxy-acetylene torches were used to cut the large members into smaller ones for ease in handling and the rivets were cut out to separate different members. The scrap was sold in Montreal for about $12 per ton. The Quebec Bridge loss was estimated at $6,854,987 taking into account the value of unused steel at $300,000 (Railway Age Gazette 1911).

8 THE RECONSTRUCTION OF THE QUEBEC BRIDGE

The Board of Engineers prepared one design which is shown in Figure 6. The Board's design was for a cantilever bridge with anchor arms and cantilever arms of the same length and suspended spans to be erected by cantilevering out. Five modifications of this design were also prepared by the Board. A tender on any of the six propositions was going to be considered a tender on the Board's design. Two of these schemes were based on erecting the suspended span by cantilevering out while the remaining four were based on erecting the suspended span at an adjacent site and floating it into position. The design had a cantilever span of 1758 ft versus the 1800 ft span of the old bridge. Bids on this design and its modifications and alternate designs were invited from contractors in the USA, Canada, and Europe (Engineering Record 1910).

Figure 7 shows locations of the old and new main piers based on the design proposed by the Board of Engineers. On April 8, 1911, the contract for the Quebec Bridge was awarded to the St. Lawrence Bridge Company of Montreal for about $8,650,000. It was a combination of two Canadian firms, Dominion Bridge Co. of Lachine, Quebec and Canadian Bridge Co. of Walkerville, Ontario. In addition to the St. Lawrence Bridge Co., three other firms submitted bids:

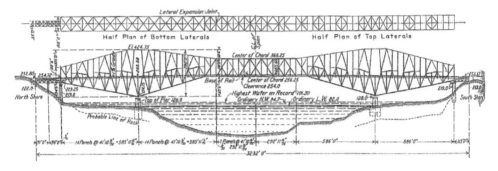

Figure 6.   Board of Engineers' design for suspended span bridge erected by cantilever method.

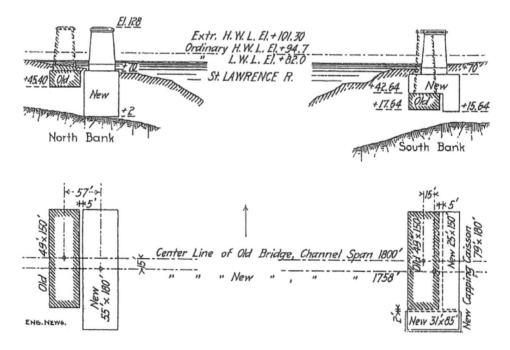

Figure 7.   Location sketch of old and new main piers.

1. Maschinen Augsburg-Nürnberg AG of Gustavburg, Germany,
2. British Empire Bridge Company of Montreal, and
3. Pennsylvania Steel Co. of Steelton, PA.

Three of the four firms submitted their own designs besides bidding on the Board's design. The St. Lawrence Bridge Co. submitted seven different designs with their respective bids. The German firm submitted bids on three of the designs of the Board and also bid on its own design. The British Empire Bridge Co. submitted six bids on the Board's six designs only. The Pennsylvania Steel Co. submitted ten bids on the Board's six designs using different erection schemes and a bid on an eyebar suspension span designed by Gustav Lindenthal.

Six of the designs submitted by the three competing firms along with the Board's design are shown in Figure 8 (Modjeski 1913). The design that was approved by a majority of the Board and additional experts appointed by the Government due to its ease in erection is

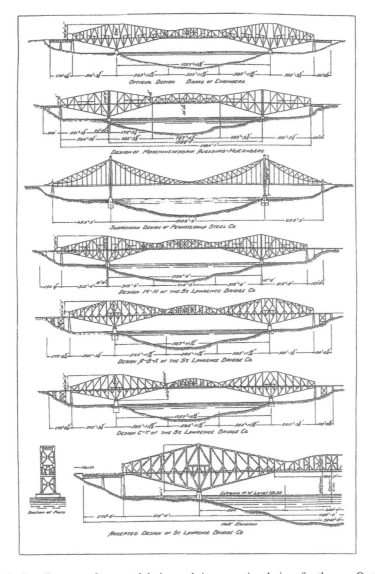

Figure 8.   Outline diagrams of accepted design and six competing designs for the new Quebec Bridge.

shown at the bottom. The old main piers were to be demolished to the mud-line and the granite blocks to be reused in the new piers.

The new bridge would be designed to accommodate two railway tracks and sidewalks for foot passengers. The government had decided not to support highway traffic on the bridge.

9   CHANGES IN THE BOARD OF ENGINEERS

On October 1, 1910 bids or tenders for the reconstruction of the Quebec Bridge were received and opened by the government. Vautelet had a difference of opinion with his original fellow board members Modjeski and MacDonald regarding the evaluation of multiple bids submitted by the four firms and selection of the winner. To resolve this matter, on January 20, 1911 the Government appointed two additional engineers, H.W. Hodge, a New York bridge engineer, and M.J. Butler, Vice President and General Manager of the Dominion Steel Corporation. On February 8, 1911, when the votes were taken, Vautelet selected the designs and bids submitted by the British Empire Co., whereas the other four members agreed that the design submitted by the St. Lawrence Co. was the best considering its constructability. The British Empire Co. had submitted six bids on the original design and five of its variations were prepared by the Board under Vautelet.

Vautelet submitted his resignation on February 28, 1911 which was accepted by the government (Engineering News, 1911). The remaining members changed the cantilever span of 1,758 ft selected by Vautelet back to 1800 ft. Both Hodge and Butler resigned from the Board as they confirmed the superiority of the design submitted by the St. Lawrence Bridge Co. over the Board's design and on April 4, 1911 the construction contract was signed by the government with the St. Lawrence Bridge Co.

In Vautelet's absence, MacDonald worked as Acting Chairman and Chief Engineer. When MacDonald joined the Board, it was with the understanding that soon after the award of the construction contract, he would retire. With his resignation, there were two vacancies in the Board. On May 6, 1911, Lt. Colonel Charles N. Monsarrat was appointed as the Chairman and Chief Engineer of the board and on May 15, 1911, Charles Schneider was appointed as a Board member. Following his death, his position was filled on January 8, 1916 by H.P. Borden who, up to then, had been working as Assistant to the Chief Engineer. The new and the final Board was as follows:

1. Charles Monserrat, Chairman and Chief Engineer, Canada
2. Ralph Modjeski, Member, USA
3. H.P. Borden, Member, Canada

10   CONSTRUCTION OF THE SUBSTRUCTURE

Work on the substructure continued over a four-year period from 1909 to 1913. At the end of the 1913 season, all that remained was finishing the bridge seats and minor details. The amount of masonry used in the four piers was 106,090 CY.

11   DESIGN OF THE SUPERSTRUCTURE

The bridge was designed for 5000 lb per lineal foot covering both entire tracks with two E60 engines. The engine and train loads were placed to give the maximum loading condition.

Wind load was assumed at 30 lb/ft^2 of exposed surface of the two trusses and 1.5 times the elevation of the floor and 300 lb per lineal ft as a moving load on the exposed surface of the train. A wind load of 30 lb/ft^2 parallel with the bridge was also assumed acting on one half of the area assumed for normal wind pressure.

12   SPECIAL SHOPWORK ON THE HEAVY MEMBERS OF THE NEW BRIDGE

Unprecedented steel fabrication was involved in the construction of the new Quebec Bridge. Among the factors were metal thickness of over 9 inches, 1-1/8-inch diameter rivets, mixing of nickel and carbon steels, the great weight of parts and completed members, drilling from solid metal for nearly all rivet holes, planed faces up to 10 ft × 20 ft, 45-inch pin-holes, and assembling in the shops. Engineering News (1914b) provides details and photos of the fabricated members and the fabrication shop.

All plates riveted together in the shop were given one coat of iron oxide and were allowed to dry before they were assembled. The shop coat was pure red lead to which 4 oz of lampblack was added for every 30 lb and mixed with pure linseed oil to the proper consistency. Each member was weighed individually and the weight was painted on in plain figures before being stored. The following conditions were assumed for temperature stresses (Engineering News 1914):

1. Variation of 150°F on the uniform temperature of the entire structure
2. A difference of 50°F between the temperature of steel and masonry
3. A difference of 25°F between the temperature of a shaded chord and the average temperature of a chord exposed to the sun
4. A difference of 25°F between the outer webs exposed to the sun and the inner webs of compression members

The superstructure of the bridge was partly constructed of carbon steel and partly of nickel steel. The floor throughout was made of carbon steel. The truss members of the suspended span were all nickel steel to reduce the dead load. Practically all members of the anchor arms were made of carbon steel except for a few members where it was necessary to use nickel steel in order to keep the grip of rivets down to practical limits.

The sizes of the rivets varied from 7/8-inch to 1-1/8 inch. All rivets were carbon steel. When the grip of the rivet exceeded four diameters, the allowable unit stress of the rivet was reduced by 1% for each 1/16 inch of additional grip. This did not apply to compression members having butt joints. All rivets over 5 inches long had a taper of 1/32 inch in 12 inches. The size under the head was 1/32 inch smaller than the diameter of the hole.

No material less than 1/2 inches in thickness was allowed in main members. Material 3/8-inch in thickness was allowed in details such as lattice bars and the tie-plates of the lateral and sway bracing, provided the requirements of the specification as to unsupported length were fulfilled.

13   HOISTING AND COLLAPSE OF THE SUSPENDED SPAN

At the end of July 1916, both the north and south cantilever arms of the new Quebec Bridge were completed, and the details for hoisting of the suspended span were finalized. Figure 9 shows the general scheme for hoisting the suspended span (Engineering News 1916). The suspended span was 640 ft long and 88 ft wide. The depth at the end was 70 ft and at the middle 110 ft.

The vertical distance through which the span was to be hoisted depended upon the water level in the St. Lawrence River, but was estimated to be about 145 ft. Each operation of the jacks hoisted the span 2 ft, and each jacking cycle took about 15 minutes to complete. There were approximately 73 separate lifting operations and the time it took from the moment of coupling up to the hanger lifting chains to the driving of the last pins connecting the two portions of the permanent eyebar suspenders was estimated at 20 hours barring unforeseen delays.

The center span fell into the St. Lawrence River at 10:50 AM on Monday, September 11, 1916 after it had been hung to the cantilever arms by the erection hanger chains and had been hoisted successfully 12 to 15 ft. Eleven lives were lost in the disaster and six men were injured.

By the process of elimination, the place of initial failure was narrowed down to where the truss rested on the girders which hung from the bottom of the lifting chains. A steel casting by which the weight of the southwest corner of the suspended span was transferred to the lifting girder broke in such a manner that the girder kicked back from under it. This corner of the span dropped into the water starting a chain reaction causing the entire span to fall in the St. Lawrence within a few seconds. Figures 10 through 12 show the sequence of events that took place in those few seconds (Engineering News 1916b).

Figure 9.   General scheme for hoisting the Quebec Bridge suspend span.

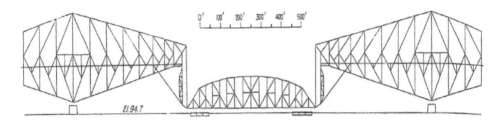

Figure 10.   The span linked to the chains and raised clear of the scows, attempting the last hazardous operation in the construction of the bridge.

Figure 11.   The falling span snapped by a lucky newspaper man a moment before the southeast corner of the span tore loose from its seat.

Figure 12.   A few seconds after the fall.

14 SUCCESSFUL HOISTING OF THE SUSPENDED SPAN

Immediately after the collapse of the suspended span, the St. Lawrence Bridge Co. took full responsibility, announcing they would bear the cost of replacing the fallen span and had placed orders for the new steel. Raising and reusing the fallen span, which was in water about 200 ft deep and possibly broken and twisted, was ruled out. Even though there was a scarcity of steel due to World War I, the carbon and nickel-alloy steel were made available to rebuild the Quebec Bridge.

Unlike the first attempt to raise the center span in September 1916, which the St. Lawrence Bridge Co. had planned to do in one day, in September 1917, they scheduled the hoisting of the suspended span over three to four days and there was no night work planned. The same lifting procedure that was used in 1916 was also used in 1917.

On Monday September 17, 1917, the four chairs under the ends of the spans were attached to the eyebar lifting-chains, and at 9:30 AM the jacking began, thereby raising the span 2 ft at each stroke. By 4:40 PM, twelve 2-ft lifts were made. The span was anchored against wind, and the work was stopped for the night.

On Tuesday September 18, 1917 22 more lifts were made. The work was not hurried, and it was interrupted to remove the freed links of the lifting chains. On the following day, 26 lifts were made, making 60 lifts over a three-day period. By noon on Thursday, September 20, 1917, seven lifts were already made and only eight more lifts were needed to raise the suspended span to its final position.

The 74th lift was taken very slowly (from 2:10 to 3:10 PM), as some of the wooden working platforms were taken down, the clearances inspected, and the eyebars guided into proper position. The 75th lift followed quickly and locomotive cranes were run out to all four corners with pin-driving cages and pins. At the end of the stroke, at 3:25 PM, the first of the eight pins were driven. The clearances were perfect and each long pin slipped through its eyebars with a few taps from a short rail swung by about ten men. At 4:10 PM the final pin was driven, and all restraint among the workers and onlookers was lost.

The river boats passed the signal to the City of Quebec and by the Mayor's proclamation, every whistle and bell and automobile horn was turned loose and flags and buntings were thrown to the breeze as Quebec's dream of thirty years had come true.

On Saturday September 22, 1917, the dismantling of the hoisting equipment started. The span floor system had yet to be erected, the footwalks laid, and some lateral bracing connections had to be riveted. It was expected trains would operate on the bridge in six to eight weeks. Although the official bridge opening ceremony was performed by the Prince of Wales (the future Edward VIII) on Aug. 22, 1919, trains started using the bridge almost 100 years ago on December 3, 1917.

15 PERSEVERANCE AND TRIUMPH OF CANADIAN ENGINEERING

It was not easy to live under the shadow of a highly-developed neighbor like the United States and claim credit for building the biggest cantilever bridge in the world because of the two previous failed attempts. The Government and engineers connected with this bridge in Canada were determined to make the third attempt the final and successful one. During the 1916 collapse, one end of the span was 2 ft higher than the other. Care was taken this time to lift the two ends of the span at the same time and at the same speed.

Every man was trained to perform duties of two positons which were of a different nature, and one not involving such nervous strain as the other. For example, the north central control operator and the north "end engineer" were interchangeable, both knowing each other's duties. Similarly, the assistant end engineer in charge of, say, the northwest corner was interchangeable with the assistant valve operator in charge of the northwest corner, and those two men were made to know each other's duties. It was made obligatory that the men in more serious positions were relieved at least every two hours.

There were consulting engineers and bridge company officials on both cantilevers carefully watching and closely monitoring every stage of the operation. Also, expert electricians, hydraulic engineers, and skilled mechanics were on hand ready to deal with any emergency.

The individuals deserving credit for the successful completion of this mammoth project were:

1. Board of Engineers

Henri E. Vautelet	Chairman and Chief Engineer (1908–1911)
Charles N. Monsarrat	Chairman and Chief Engineer (1911–1917)
Ralph Modjeski	Board Member (1908–1917)
Charles C. Schneider	Board Member (1911–1916)
Charles MacDonald	Board Member (1910–1911)
Maurice Fitzmaurice	Board Member (1908–1910)
H.P. Borden	Board Member (1916–1917)
Joseph Mayer	Principal Assistant Engineer
Archibald J. Meyers	Chief Draftsman

2. St. Lawrence Bridge Company

Phelphs Johnson	President and General Manager
George H. Duggan	Chief Engineer
George F. Porter	Engineer of Construction
J.D. Wilkens	Resident Engineer (1909–1915)
John Rankin	Resident Engineer (1915–1917)
E.H. Pacy	Assistant Engineer
H.E. Bates	Assistant Engineer for Shop & Field work
W.P. Copp	Chief Inspector of Erection
S.P. Mitchell	Consulting Engineer of Erection
Francois C. McMath	Consulting Engineer
Herbert W. McMillan	Chief Shop Inspector
C.J. Yarrell	Chief Mill Inspector
Walter P. Ladd	Superintendent of Manufacture
W.B. Fortune	Superintendent of Erection

16 CONCLUSIONS

The Ultimate success of this gigantic bridge was built on the ruins of the prior two collapses and painstaking analysis of what went wrong validated by laboratory tests on large compression members and tension bars. The effects of distortion in trusses were explored further than before, and means were devised for dealing with such effects. Much knowledge was added on the assembly of heavy members, and new standards were set as to the degree of precision and finish in shopwork. There was a quantum jump in our knowledge about planning and training for the erection of very large bridges.

According to the Engineering News Record (1917), "the great value of the achievement lies in the inspiration emanating from the courage of the men who have erected on the failure of 1907 and the loss of 1916 this greatest of bridges and in so doing not only have erected a monument to themselves and their courage and ability, but have vindicated the profession before a doubting world."

ACKNOWLEDGEMENTS

The author thanks Kunal Kothawade, Livia Bennett, and Annie Sidou of Gandhi Engineering for their assistance during the preparation, typing, and editing of this paper.

REFERENCES

Canadian Engineer 1917. Editorial: 33, 266.

Engineering-Contracting 1908. The Report of the Canadian Government Commission on the Quebec Bridge Failure: 29(12), 169–170.

Engineering News 1900a. The Quebec Bridge over the St. Lawrence River: 44(12), 189.

Engineering News 1900b. Work on the Quebec Bridge: 44(15), 241.

Engineering News 1903. The Substructure for the 1,800 ft Cantilever Bridge at Canada: 49(5), 92–97.

Engineering News 1905. The 1,800 ft span Cantilever Bridge across the St. Lawrence River at Quebec: 54(11), 272–274.

Engineering News 1907. The Fall of the Quebec Bridge: 58(11), 287–289.

Engineering News 1908a. A Summary of Tests of Large Columns: 59(15), 404–405.

Engineering News 1908b. Tests of Two Compression Chord Models: The Largest Column Tests Ever Made: 59(17), 455–459.

Engineering News 1908c. Report of C.C. Schneider on the Design for the Quebec Bridge: 60(5), 135.

Engineering News 1910. Caissons for the Main Piers of the New Quebec Bridge; Launch of the North Pier Caisson 64(10), 262–263.

Engineering News 1911a. Resignation of Mr. H.E. Vautelet from the Board of Engineers for the Quebec Bridge: 65(9), 271–272.

Engineering News 1911b. Tests of Nickel-Steel Details for the Board of Engineers, Quebec Bridge: 65(18), 526–531.

Engineering News 1914a. Design of the Superstructure of the New Quebec Bridge: 71(18), 942–945.

Engineering News 1916. Fig. 3. General Scheme for Hoisting the Quebec Bridge Suspended Span: 76(9), 422.

Engineering News-Record 1917. Quebec—The Final Chapter: 79(13), 579.

Engineering Record 1910. The Reconstruction of the Quebec Bridge: 62(11), 286–287.

Gandhi, K. 2006. Roebling's Railway Suspension Bridge over Niagara Gorge. *5th International Cable-Supported Bridge Operator's Conference*, New York, NY.

Middleton, William D. 2001. *The Bridge at Quebec*. Indiana University Press: 203pp.

Modjeski, Ralph 1913. Design of Large Bridges with Special Reference to the Quebec Bridge. *Journal of the Franklin Institute*: 176(3), 239–282.

Railway Age Gazette 1911. Quebec Bridge Loss $7,154,987: 50(5), 249.

Railroad Gazette 1887. Proposed Bridge at Quebec: 19(20), 341.

Scientific American 1851. Suspension Bridge over the St. Lawrence: 7(14), 106.

Skinner, Frank W (1907). Summary of the Quebec Bridge. *Cornell Civil Engineer*: 67(3), 67–75.

Author index

T - #0496 - 071024 - C340 - 246/174/15 - PB - 9780367735920 - Gloss Lamination